Cooperative and Noncooperative Multi-Level Programming

Masatoshi Sakawa • Ichiro Nishizaki

Cooperative and Noncooperative Multi-Level Programming

Springer

Masatoshi Sakawa
Department of Artificial Complex
Systems Engineering
Hiroshima University
1-4-1 Kagamiyama
Higashi-Hiroshima
739-8527 Japan
sakawa@hiroshima-u.ac.jp

Ichiro Nishizaki
Department of Artificial Complex
Systems Engineering
Hiroshima University
1-4-1 Kagamiyama
Higashi-Hiroshima
739-8527 Japan
nisizaki@hiroshima-u.ac.jp

ISSN 1387-666X
ISBN 978-1-4419-0675-5 e-ISBN 978-1-4419-0676-2
DOI 10.1007/978-1-4419-0676-2
Springer Dordrecht Heidelberg London New York

Library of Congress Control Number: 2009926796

Printed on acid-free paper

Springer is part of Springer Science+Business Media (www.springer.com)

To Our Parents and Families

Preface

To derive rational and convincible solutions to practical decision making problems in complex and hierarchical human organizations, the decision making problems are formulated as relevant mathematical programming problems which are solved by developing optimization techniques so as to exploit characteristics or structural features of the formulated problems. In particular, for resolving conflict in decision making in hierarchical managerial or public organizations, the multi-level formulation of the mathematical programming problems has been often employed together with the solution concept of Stackelberg equilibrium.

However, we conceive that a pair of the conventional formulation and the solution concept is not always sufficient to cope with a large variety of decision making situations in actual hierarchical organizations. The following issues should be taken into consideration in expression and formulation of decision making problems.

In formulation of mathematical programming problems, it is tacitly supposed that decisions are made by a single person while game theory deals with economic behavior of multiple decision makers with fully rational judgment. Because two-level mathematical programming problems are interpreted as static Stackelberg games, multi-level mathematical programming is relevant to noncooperative game theory; in conventional multi-level mathematical programming models employing the solution concept of Stackelberg equilibrium, it is assumed that there is no communication among decision makers, or they do not make any binding agreement even if there exists such communication. However, for decision making problems in such as decentralized large firms with divisional independence, it is quite natural to suppose that there exists communication and some cooperative relationship among the decision makers.

Moreover, in the real-world problems, diversity of evaluation has taken on a growing importance, and it is natural to suppose that decision makers desire to attain several simultaneous goals. Namely, they have multiple objectives and evaluate alternatives, considering trade-offs among the objectives. For such situations, formulations of mathematical models including multiple objectives are appropriate, and the multi-level programming methods under multiobjective environments should be developed.

When we model actual situations of decision making in mathematical programming problems, there are some difficulties that decision makers may face and need to handle; we should take into account imprecise data gathered to formulate problems, inaccuracies of human judgments in decision making, bounded rationality of decision makers, and uncertainties of events related to the decision making. Moreover, the experts do not always understand the nature of parameters of the problems precisely in the problem-formulation process, and their understanding may be somewhat fuzzy. From vagueness of judgments of the decision makers, we suppose that the decision makers have a fuzzy goal with respect to each of the multiple objectives. In some sort of circumstances, formulations of mathematical models including stochastic events are required.

Real-world decision problems in hierarchical organizations can be often formulated as difficult classes of optimization problems such as combinatorial problems and nonconvex nonlinear problems. For such problems, it is difficult to obtain exact optimal solutions, and thus it is quite natural for decision makers to require approximate optimal solutions instead. To meet this demand, recently several metaheuristics have been developed and their effectiveness is demonstrated. Among them, genetic algorithms are known to be one of the most practical and proven methods. Genetic algorithms initially proposed by Holland in early 1970's have made a wide variety of contributions to optimization, adaptation, and learning. Especially, applications to optimization continue to extend across difficult classes of optimization problems.

In this book, after presenting basic concepts in multi-level mathematical programming problems, the authors intend to introduce the latest advances in the new field of multi-level mathematical programming problems under fuzzy, multiobjective, and/or uncertain environments. Because the relation among decision makers in real-world hierarchical organizations can be expressed by either a cooperative or a noncooperative framework, we provide cooperative and noncooperative formulations of multi-level programming problems. For computational aspects, exact solution methods based on linear programming techniques are given if possible, but for complex problems difficult to search optimal solutions, computational techniques using genetic algorithms are utilized.

Hiroshima,
March 2009

Masatoshi Sakawa
Ichiro Nishizaki

Contents

Chapter 1
Introduction

1.1 Background

Decision making problems in decentralized organizations are often modeled as Stackelberg games, and they are formulated as two-level mathematical programming problems. In the Stackelberg game model, there are two players; the first player chooses a strategy at the start, and thereafter the second player with knowledge of the first player's strategy determines a strategy of the second player self (Simaan and Cruz, 1973a). It is assumed that each of the two players completely knows the objective functions and the constraints of them. In the context of two-level programming, the decision maker at the upper level first specifies a strategy, and then the decision maker at the lower level specifies a strategy so as to optimize the objective with full knowledge of the action of the decision maker at the upper level. For brevity, we abbreviate the decision makers at the upper level and at the lower level as the upper level DM and the lower level DM, respectively. Assuming that the lower level DM behaves rationally, that is, optimally responds to the decision of the upper level DM, the upper level DM also specifies the strategy so as to optimize the objective of self. Although a situation described as the above is called a Stackelberg equilibrium in the field of game theory or economics, in this book dealing with mathematical programming, we will refer to it as a Stackelberg solution.

Even if the objective functions of both decision makers and the common constraint functions are linear, such a two-level mathematical programming problem, i.e., a two-level linear programming problem is a non-convex programming problem with a special structure, and it is shown to be NP-hard (Jeroslow, 1985; Bard, 1991). In general, Stackelberg solutions do not satisfy Pareto optimality because of their noncooperative nature.

Computational methods for Stackelberg solutions are classified roughly into three categories: the vertex enumeration approach based on a characteristic that an extreme point of a set of rational responses of the lower level DM is also an extreme point of the feasible region, the Kuhn-Tucker approach in which the upper

M. Sakawa and I. Nishizaki (eds.), *Cooperative and Noncooperative Multi-Level Programming*, Operations Research/Computer Science Interfaces Series 48, DOI: 10.1007/978-1-4419-0676-2_1,

level problem with constraints including the optimality conditions of the lower level problem is solved, and the penalty function approach which adds a penalty term to the upper level objective function so as to satisfy optimality of the lower level problem.

The *k*th best method proposed by Bialas and Karwan (1984) can be classified as the vertex enumeration approach. The search procedure of the method starts from the first best point. Namely, an optimal solution to the upper level problem, which is the first best solution, is computed, and then it is verified whether the first best solution is also an optimal solution to the lower level problem. If the first best point is not a Stackelberg solution, the procedure continues to examine the second best solution to the upper level problem, and so forth. The parametric complementary pivot algorithm by Bialas and Karwan (1984) is considered to be one of computational methods in the Kuhn-Tucker approach, and this method is improved by Júdice and Faustino (1992) afterward. Bard and Falk (1982) transform the complementary slackness condition into a separable expression so as to apply a general branch-and-bound algorithm, and Bard and Moore (1990a) develop more efficient method in this line. Bard (1983a) formulates a two-level programming problem as an equivalent semi-infinite problem, and develops a grid search algorithm by utilizing a parametric linear program technique. Using a duality gap penalty function format, White and Anandalingam (1993) develop an algorithm for obtaining a Stackelberg solution to a two-level linear programming problem. The algorithm developed by Hansen, Jaumard and Savard (1992) is based on the branch-and-bound technique; it eliminates variables by finding binding constraints.

Extending the two-level linear programming, Bard (1984) and Wen and Bialas (1986) propose algorithms for obtaining Stackelberg solutions to three-level linear programming problems. Bard (1984) formulates the first level problem with constraints including the Kuhn-Tucker conditions for the third and the second level problems, and proposes a cutting plane algorithm employing a vertex search procedure for solving the formulated nonlinear problem. Wen and Bialas (1986) develop a hybrid algorithm to solve the three-level linear programming problem. The algorithm involves two existing solution methods for two-level linear programming problems: the *k*th best algorithm is employed to generate the *k*th best extreme point, and the complementary pivot algorithm is utilized to check feasibility of the point.

A two-level problem with a single decision maker at the upper level and two or more decision makers at the lower level is referred to as a decentralized two-level programming problem. In particular, for the case where the objective functions of all the decision makers and the constraint functions are linear, Simaan and Cruz (1973b) and Anandalingam (1988) suppose that the lower level DMs make decisions so as to equilibrate their objective function values with respect to a decision of the upper level DM on the assumption that all of the lower level DMs do not have motivation to cooperate mutually. Namely, assuming that a set of responses by the lower level DMs is a Nash equilibrium solution, the upper level DM selects a decision optimizing the objective function of self.

Concerning two-level programming problems with discrete decision variables, algorithms based on the branch-and-bound techniques are developed for obtaining

Stackelberg solutions to two-level 0-1 programming problems and two-level mixed integer programming problems (Bard and Moore, 1990b; Moore and Bard, 1990; Bard and Moore, 1992). Wen and Yang (1990) deal with a two-level mixed 0-1 programming problem with 0-1 decision variables of the upper level DM and continuous decision variables of the lower level DM, and propose an exact method and a heuristic method for obtaining Stackelberg solutions. The branch-and-bound techniques are also employed in their exact method where the 0-1 decision variables of the upper level DM are used as branching variables. Using a penalty function, Vicente, Savard and Judice (1996) show the equivalence between a multi-level programming problem with discrete decision variables and a corresponding two-level linear programming problem.

For solving large scale problems, there is fear that the computational time increases extremely, and therefore developing highly efficient approximate computational methods is important. The framework of genetic algorithms has been applied in order to obtain approximate Stackelberg solutions to the large scale two-level mathematical programming problems efficiently. As for researches on computational methods using genetic algorithms for two-level mathematical programming problems, Anandalingam *et al.* (1988) develop a computational method for obtaining Stackelberg solutions to two-level linear programming problems by employing vectors of floating-point numbers for the chromosomal representation in their artificial genetic systems. Niwa, Nishizaki and Sakawa (2001) employ the chromosomal representation by a binary string which corresponds to 0-1 variables in the mixed 0-1 programming problem equivalent to the two-level linear programming problem; considering the property of the two-level linear programming problem, they develop a method of generating an initial population and some procedures of the genetic operations. Using the framework of genetic algorithms, Nishizaki and Sakawa also develop computational methods for obtaining Stackelberg solutions to a two-level mixed 0-1 programming problem with 0-1 upper level decision variables and continuous lower level decision variables (Nishizaki and Sakawa, 2000) and a two-level integer programming problem (Nishizaki and Sakawa, 2005).

In noncooperative environments, it is difficult for the upper level DM to estimate or assess a single objective function or a utility function of the lower level DM. With this observation, Nishizaki and Sakawa (1999) formulate multiobjective two-level linear programming problems, and develop a computational method for obtaining Stackelberg solutions. Moreover, considering uncertain events, they formulate a two-level programming problem with random variable coefficients and develop algorithms for deriving Stackelberg solutions by employing some stochastic criteria such as the expectation maximization and the variance minimization (Nishizaki, Sakawa and Katagiri, 2003; Katagiri *et al.*, 2007).

Real-world applications under noncooperative situations are formulated by two-level mathematical programming problems, and their effectiveness is demonstrated (Marcotte, 1986; Ben-Ayed, Boyce and Blair, 1988; Bard and Moore, 1990c; Bard, 1998).

So far, we have considered situations where decision makers do not have motivation to cooperate mutually. As an illustration of such situations, the Stackelberg duopoly is often taken. The Stackelberg duopoly is summarized as follows. Suppose that Firm 1 and Firm 2 supply homogeneous goods to a market, and Firm 1 dominates Firm 2 in the market. Consequently, Firm 1 first determines a level of supply, and then Firm 2 decides its level of supply after realizing Firm 1's level of supply. Therefore, each of the two firms does not have a motivation to cooperate each other in this model.

However, when we intend to formulate a decision making problem in a decentralized firm as a two-level mathematical programming problem, different perspectives are required. In the decentralized firm, top management, an executive board, or headquarters is interested in overall management policy such as the long-term corporate growth or market share. In contrast, an operation division of the firm is concerned with coordination of daily activities. After the headquarters makes a decision in accordance with the overall management policy, the operation division determines a goal to be achieved and tries to attain the goal, fully understanding the decision by the headquarters. To formulate decision making problems in such decentralized firms, we have employed formulation of single-objective large scale mathematical programming problems with a special structure such as the brock angular structure or multiobjective programming problems with objective functions of all the levels. These formulations are interesting and important as Bialas and Karwan (1984) assert that the two-level mathematical programming formulation is intend to supplement the decomposition approach to the large scale problems, not supplant it. However, it is worth investigating to formulate a mathematical model explicitly including a hierarchical structure of the decision making problem and some coordination perspective.

From the viewpoint of taking into account a possibility of coordination or bargaining between decision makers, some attempts to derive Pareto optimal solutions to two-level linear programming problems are made (Wen and Hsu, 1991; Wen and Lin, 1996). Moreover, assuming communication and a cooperative relationship between decision makers, Lai (1996) and Shih, Lai and Lee (1996) develop a solution method for obtaining a Pareto optimal solution to a multi-level linear programming problem, in which a coordination procedure is included; therefore, an obtained solution is not always a Stackelberg solution. Their method is based on an idea that the lower level DM optimizes the objective function, taking a goal or preference of the upper level DM into consideration. Both of the decision makers identify membership functions of fuzzy goals for their objective functions, and particularly, the upper level DM also specifies those of fuzzy goals for the decision variables. The lower level DM solves a fuzzy programming problem with a constraint on a satisfactory degree of the upper level DM. Unfortunately, there is a possibility that their method leads a final solution to an undesirable one because of inconsistency between the fuzzy goals of the objective function and those of the decision variables.

By eliminating the fuzzy goals for the decision variables to avoid such problems in the methods of Lai (1996) and Shih, Lai and Lee (1996), Sakawa, Nishizaki and Uemura (1998) develop interactive fuzzy programming for two-level linear programming problems. Moreover, from the viewpoint of experts' imprecise or fuzzy understanding of the nature of parameters in a problem-formulation process, they extend it to interactive fuzzy programming for two-level linear programming problems with fuzzy parameters (Sakawa, Nishizaki and Uemura, 2000a). These results are also extended to deal with two-level linear fractional programming problems (Sakawa and Nishizaki, 1999, 2001a; Sakawa, Nishizaki and Uemura, 2000b) and two-level linear and linear fractional programming problems in multiobjective environments (Sakawa, Nishizaki and Oka, 2000; Sakawa, Nishizaki and Shen, 2001), considering diversity of evaluation by the decision makers. Taking into account uncertain events, interactive fuzzy programming is also extended into problems with random variable coefficients (Sakawa *et al.*, 2003; Perkgoz *et al.*, 2003; Kato *et al.*, 2004).

Interactive fuzzy programming can be also extended so as to manage decentralized two-level linear programming problems by taking into consideration individual satisfactory balance between the upper level DM and each of the lower level DMs as well as overall satisfactory balance between the two levels (Sakawa and Nishizaki, 2002a). Moreover, by using some decomposition methods which take advantage of the structural features of the decentralized two-level problems, efficient methods for computing satisfactory solutions are also developed (Kato, Sakawa and Nishizaki, 2002; Sakawa, Kato and Nishizaki, 2001).

Consider computational aspects of interactive fuzzy programming. Because interactive fuzzy programming for two-level linear programming problems and the derivative and related problems is based on linear programming techniques, it is relatively easy to obtain satisfactory solutions to such problems. However, concerning two-level programming problems with discrete decision variables or two-level nonlinear programming problems, this argument is not true. Therefore, it becomes important to develop highly efficient approximate computational methods.

As an efficient meta-heuristics, genetic algorithms initiated by Holland (1975) have been attracted attention of many researchers with applicability in optimization, as well as in search and learning. Furthermore, publications of books by Goldberg (1989) and Michalewicz (1996) bring heightened and increasing interest in applications of genetic algorithms to complex function optimization. Focusing on multiobjective mathematical problems, Sakawa and his colleagues have been advancing genetic algorithms to derive satisficing solutions to multiobjective problems (Sakawa, 2001).

By using their techniques in the genetic algorithms which are proven to be efficient approximate computational methods for multiobjective problems, Sakawa, Nishizaki and Hitaka (1999, 2001) develop interactive fuzzy programming for multi-level programming problems with discrete decision variables. Moreover, they also tackle difficult multi-level problems such as two-level nonlinear nonconvex programming problems (Sakawa and Nishizaki, 2002b) and multi-level nonlinear integer programming problems (Azad *et al.*, 2005, 2006).

Their diversified development of interactive fuzzy programming approach produces real-world applications (Sakawa, Nishizaki and Uemura, 2001, 2002), where a transportation problem and a production problem including work force assignment in a housing material manufacturer are dealt with.

1.2 Description of contents

Although researches on multi-level mathematical programming problems have been accumulated and a few books on multi-level mathematical programming problems have been published recent years (e.g. Shimizu, Ishizuka and Bard (1997); Bard (1998); Dempe (2002)), there seems to be no book which concerns fuzziness of judgements, uncertainty of environments, and diversity of evaluations, as well as dealing with noncooperative and cooperative behaviors of decision makers in decision problems in decentralized organizations.

In this book, to utilize multi-level programming for resolution of conflict in decision making problems in real-world decentralized organizations, we introduce several representations of multi-level problems by taking into consideration the following aspects: fuzziness and ambiguity of human judgements, uncertainty of events characterizing decision making problems, multiplicity of evaluation criteria, and noncooperative or cooperative behaviors of decision makers. We present algorithms for deriving solutions in various situations of decision making, and also provide some applications on the basis of the continuing research works by the authors.

For hierarchical decision making in noncooperative situations, after reviewing basic concepts in two-level linear programming problems, we present computational methods for obtaining Stackelberg solutions to various types of multi-level problems. As for cooperative hierarchical decision making, interactive fuzzy programming methods are introduced in order to derive satisfactory solutions to each of the decision makers.

The main aim of this book is to provide computational methods for obtaining Stackelberg solutions and algorithms for deriving mutual cooperative solutions (satisfactory solutions) to multi-level programming problems in fuzzy, stochastic, and/or multiobjective environments. Selecting and applying an appropriate method so as to meets a decision situation from the various methods introduced in this book, decision makers can expect to resolve conflicts in decision making problems in real-world decentralized organizations. Organization of each chapter is briefly summarized as follows.

In chapter 2, we review three optimization concepts incorporating fuzziness and ambiguity of human judgements, multiplicity of evaluation criteria, and uncertainty of events characterizing decision making problems; these optimization concepts underlie the formulations presented in subsequent chapters.

Real-world decision problems are often formulated as difficult classes of optimization problems such as combinatorial problems and nonconvex nonlinear problems. When formulated problems are such difficult classes of optimization problems and consequently it is difficult to obtain exact optimal solutions to the problems, decision makers may require an approximate optimal solution. To cope with this request, recently abilities of several meta-heuristics have been tested and their effectiveness is evaluated. In this book, genetic algorithms, which are considered to be a most practical and proven method, are employed to search approximate optimal solutions; for convenience of readers, we provide the fundamentals of genetic algorithms in the final section of chapter 2.

Chapter 3 deals with noncooperative decision making in hierarchical organizations. First, to review the historical background of multi-level programming, we show a couple of conventional computational methods for obtaining Stackelberg solutions to two-level linear programming problems. Because there is fear that, in proportion as scale of the problem, computational time of searching Stackelberg solutions exceedingly increases in exact methods such as the branch-and-bound based methods, developing highly efficient approximate computational methods is important for large scale problems. To efficiently search approximate Stackelberg solutions to the large scale two-level programming problems, computational methods based on genetic algorithms are developed. Such methods can be also applicable to difficult classes of optimization problems such as combinatorial problems. From this viewpoint, the subsequent three sections are concerned with computational methods based on genetic algorithms for obtaining Stackelberg solutions to two-level linear, 0-1 and integer programming problems.

In the rest of chapter 3, we deal with multiobjective two-level linear programming and two-level linear stochastic programming. In multiobjective two-level linear programming problems, it is assumed that each of the decision makers has multiple objective functions conflicting with each other. In such multiobjective environments, the upper level DM must take account of multiple or infinite rational responses of the lower level DM. We examine three types of situations based on anticipation of the upper level DM: an optimistic anticipation, a pessimistic anticipation, and an anticipation arising from the past behaviors of the lower level DM. Mathematical programming problems for obtaining Stackelberg solutions based on the three types of anticipation are formulated, and algorithms for solving the problems are presented. Concerning two-level linear stochastic programming, we reduce a two-level linear programming problem with random variable coefficients into two types of deterministic problems by employing the expectation and the variance models, and computational methods for solving the reduced deterministic problems are given.

If the decision makers bargain with each other and as a consequence they can make decisions cooperatively, it is not always appropriate to use the concept of Stackelberg solutions. In the first five sections of chapter 4, we intend to give solution methods, assuming cooperative behavior of decision makers in decision problems with hierarchical structure. First, we present interactive fuzzy programming for two-level linear programming problems. In the interactive method, after determining the fuzzy goals of the decision makers at both levels, a satisfactory solution is derived efficiently by updating the minimal satisfactory level of the upper level DM with considerations of overall satisfactory balance between both levels. Second, from the viewpoint of experts' imprecise or fuzzy understanding of the nature of parameters in a problem-formulation process, the solution method is extended to interactive fuzzy programming for two-level linear programming problems with fuzzy parameters. Furthermore, these results are extended to deal with multi-level linear programming problems, two-level linear fractional programming problems and decentralized two-level linear programming problems in which there are a single decision maker at the upper level and two or more decision makers at the lower level.

In the subsequent two sections of chapter 4, we deal with two-level 0-1 or nonlinear nonconvex programming problems. While, in the interactive fuzzy programming shown in the previous sections, the linear programming techniques are utilized to solve the formulated mathematical programming problems, in these sections, we apply genetic algorithms to solving 0-1 and nonlinear nonconvex programming problems. For the two-level 0-1 programming problems, we apply the genetic algorithms with the double string representation of individuals proposed by Sakawa et al. (Sakawa and Shibano, 1996; Sakawa *et al.*, 1997; Sakawa, 2001) to deriving satisfactory solutions to the two-level 0-1 programming problems. In contrast, for the two-level nonlinear nonconvex programming problems, we utilize a floating point genetic algorithm, called the revised GENOCOP III which employs a vector of floating point numbers for the chromosomal representation (Sakawa and Yauchi, 1998, 1999, 2000; Sakawa, 2001).

In the last two sections of chapter 4, we examine multiobjective two-level linear programming and two-level linear stochastic programming. For multiobjective two-level linear programming, assuming that each of the decision makers at both levels has partial information of preference, tentative solutions are derived and evaluated by using the partial information on preferences of the decision makers. For two-level linear stochastic programming, to cope with the problems with random variable coefficients, the expectation and the variance optimization problems with the chance constraint are formulated, and interactive fuzzy programming for deriving a satisfactory solution is presented. We also outline formulations of the maximum probability model and the fractile criterion model as alternative solution methods.

Finally, chapter 5 is devoted to applying cooperative and noncooperative decision making methods to real-world decision making problems in decentralized organizations. First, we deal with decision making problems on production and work force assignment in a housing material manufacturer and a subcontracting company. We formulate two types of two-level programming problems: one is a profit maximization problem of both the housing material manufacturer and the subcontracting company, and the other is a profitability maximization problem of them. Applying the interactive fuzzy programming for two-level linear and linear fractional programming problems, we derive satisfactory solutions to the problems. Second, we deal with a transportation problem in the housing material manufacturer and derive a satisfactory solution to the problem by taking into account not only the degree of satisfaction with respect to objectives of the housing material manufacturer but also those of two forwarding agents to which the housing material manufacturer entrusts transportation of products. Third, as a noncooperative decision making problem, we consider a purchase problem for food retailing, and formulate a two-level linear programming problem with a food retailer and a middle trader. The food retailer deals with vegetables and fruits which are purchased from the middle trader; the middle trader buys vegetables and fruits ordered from the food retailer at the central wholesale markets in several cities, and transports them from each of the central wholesaler markets to a storehouse of the food retailer by truck. We compute Stackelberg solutions to the two-level linear programming problem in which the profits of the food retailer and the middle trader are maximized.

Chapter 2
Optimization Concepts and Computational Methods

This chapter is devoted to reviewing optimization concepts and the related computational methods that will be used in the remaining chapters. In particular, we deal with three optimization concepts incorporating fuzziness and ambiguity of human judgements, multiplicity of evaluation criteria, and uncertainty of events characterizing decision making problems. Moreover, we provide the fundamentals of genetic algorithms which are considered to be one of the most practical and proven meta heuristics for difficult classes of optimization problems.

2.1 Fuzzy programming

Zimmermann (1976) introduces the concept of fuzzy set theory into linear programming. Assuming that the membership functions for fuzzy sets are linear, he shows that, by employing the principle of the fuzzy decision by Bellman and Zadeh (1970), a linear programming problem with a fuzzy goal and fuzzy constraints can be solved by using standard linear programming techniques.

A linear programming problem is represented as

$$\text{minimize} \quad z(x_1,\ldots,x_n) = c_1x_1 + \cdots + c_nx_n \tag{2.1a}$$

$$\text{subject to} \quad a_{11}x_1 + \cdots + a_{1n}x_n \leqq b_1 \tag{2.1b}$$

$$\cdots\cdots\cdots$$

$$a_{m1}x_1 + \cdots + a_{mn}x_n \leqq b_m \tag{2.1c}$$

$$x_j \geqq 0, \quad j = 1,\ldots,n, \tag{2.1d}$$

where x_j is a decision variable, and c_j, a_{ij} and b_i are given coefficients of the objective function and the constraints. Let $\boldsymbol{x} = (x_1,\ldots,x_n)^T$ denote a column vector of the decision variables, and let $\boldsymbol{c} = (c_1,\ldots,c_n)$, $A = [a_{ij}]$, and $\boldsymbol{b} = (b_1,\ldots,b_m)^T$ denote an n-dimensional row vector of the coefficients of the objective function, an $m \times n$ matrix of the coefficients of the left-hand side of the constraints, and an m-dimensional

M. Sakawa and I. Nishizaki (eds.), *Cooperative and Noncooperative Multi-Level Programming*, 11
Operations Research/Computer Science Interfaces Series 48, DOI: 10.1007/978-1-4419-0676-2_2,

column vector of the coefficients of the right-hand side of the constraints, respectively; the superscript T means transposition of a vector or a matrix. Then, problem (2.1) is simply rewritten in the following vector and matrix representation:

$$\text{minimize } z(\boldsymbol{x}) = \boldsymbol{c}\boldsymbol{x} \tag{2.2a}$$
$$\text{subject to } A\boldsymbol{x} \leqq \boldsymbol{b} \tag{2.2b}$$
$$\boldsymbol{x} \geqq \boldsymbol{0}. \tag{2.2c}$$

To a standard linear programming problem (2.2), taking into account the imprecision or fuzziness of a decision maker's judgment, Zimmermann considers the following linear programming problem with a fuzzy goal and fuzzy constraints.

$$\boldsymbol{c}\boldsymbol{x} \lesssim z_0 \tag{2.3a}$$
$$A\boldsymbol{x} \lesssim \boldsymbol{b} \tag{2.3b}$$
$$\boldsymbol{x} \geqq \boldsymbol{0}, \tag{2.3c}$$

where the symbol $\lesssim$ denotes a relaxed or fuzzy version of the ordinary inequality $\leqq$. From the decision maker's preference, the fuzzy goal (2.3a) and the fuzzy constraints (2.3b) mean that the objective function $\boldsymbol{c}\boldsymbol{x}$ should be essentially smaller than or equal to a certain level z_0, and that the values of the constraints $A\boldsymbol{x}$ should be essentially smaller than or equal to $\boldsymbol{b}$, respectively. Assuming that the fuzzy goal and the fuzzy constraints are equally important, he employs the following unified formulation.

$$B\boldsymbol{x} \lesssim \boldsymbol{b}' \tag{2.4a}$$
$$\boldsymbol{x} \geqq \boldsymbol{0}, \tag{2.4b}$$

where

$$B = \begin{bmatrix} \boldsymbol{c} \\ A \end{bmatrix}, \quad \boldsymbol{b}' = \begin{bmatrix} z_0 \\ \boldsymbol{b} \end{bmatrix}. \tag{2.5}$$

To express the imprecision or fuzziness of the decision maker's judgment, the ith fuzzy constraint $(B\boldsymbol{x})_i \lesssim b'_i$ is interpreted as the following linear membership function:

$$\mu_i((B\boldsymbol{x})_i) = \begin{cases} 1, & \text{if } (B\boldsymbol{x})_i \leqq b'_i \\ 1 - \dfrac{(B\boldsymbol{x})_i - b'_i}{d_i}, & \text{if } b'_i < (B\boldsymbol{x})_i \leqq b'_i + d_i \\ 0, & \text{if } (B\boldsymbol{x})_i > b'_i + d_i, \end{cases} \tag{2.6}$$

where d_i is a subjectively specified constant expressing the limit of the admissible violation of the ith constraint, and it is depicted in Figure 2.1.

On the basis of the principle of the fuzzy decision by Bellman and Zadeh (1970), the problem of finding the maximum decision is represented as

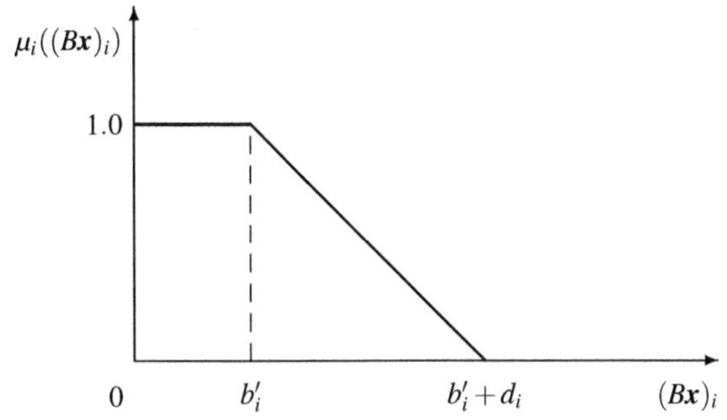

Fig. 2.1 Linear membership function for the ith fuzzy constraint.

$$\text{maximize} \quad \min_{0 \leqq i \leqq m+1} \mu_i((\boldsymbol{Bx})_i) \tag{2.7a}$$

$$\text{subject to } \boldsymbol{x} \geqq \boldsymbol{0}. \tag{2.7b}$$

With the variable transformation $b''_i = b'_i/d_i$ and $(\boldsymbol{B}'\boldsymbol{x})_i = (\boldsymbol{Bx})_i/d_i$, and an auxiliary variable λ, problem (2.7) can be transformed into the following conventional linear programming problem:

$$\text{maximize } \lambda \tag{2.8a}$$

$$\text{subject to } \lambda \leqq 1 + b''_i - (\boldsymbol{B}'\boldsymbol{x})_i, \ i = 0, \ldots, m+1 \tag{2.8b}$$

$$\boldsymbol{x} \geqq \boldsymbol{0}. \tag{2.8c}$$

Because the fuzzy decision is represented as $\min_{0 \leqq i \leqq m+1} \mu_i((\boldsymbol{Bx})_i)$, it is often called the minimum operator.

2.2 Multiobjective programming

2.2.1 Multiobjective programming problem

A problem to optimize multiple conflicting linear objective functions simultaneously under a given linear constraints is called a multiobjective linear programming problem. Let $\boldsymbol{c}_i = (c_{i1}, \ldots, c_{in})$, $i = 1, \ldots, k$ denote a vector of coefficients of the i objective function. Then, the multiobjective linear programming problem is represented as

$$\text{minimize } z_1(\boldsymbol{x}) = c_{11}x_1 + \cdots + c_{1n}x_n \tag{2.9a}$$
$$\cdots\cdots\cdots$$
$$\text{minimize } z_k(\boldsymbol{x}) = c_{k1}x_1 + \cdots + c_{kn}x_n \tag{2.9b}$$
$$\text{subject to } a_{11}x_1 + \cdots + a_{1n}x_n \leqq b_1 \tag{2.9c}$$
$$\cdots\cdots\cdots$$
$$a_{m1}x_1 + \cdots + a_{mn}x_n \leqq b_m \tag{2.9d}$$
$$x_j \geqq 0, \quad j = 1, \ldots, n. \tag{2.9e}$$

Alternatively, it is expressed by

$$\text{minimize } \boldsymbol{z}(\boldsymbol{x}) = \boldsymbol{C}\boldsymbol{x} \tag{2.10a}$$
$$\text{subject to } \boldsymbol{A}\boldsymbol{x} \leqq \boldsymbol{b} \tag{2.10b}$$
$$\boldsymbol{x} \geqq \boldsymbol{0}. \tag{2.10c}$$

Following the convention in the literature of multiobjective optimization, we introduce the following binary relations:

$$\boldsymbol{x} = \boldsymbol{y} \Leftrightarrow x_i = y_i, \; i = 1, \ldots, k \tag{2.11a}$$
$$\boldsymbol{x} \geqq \boldsymbol{y} \Leftrightarrow x_i \geqq y_i, \; i = 1, \ldots, k \tag{2.11b}$$
$$\boldsymbol{x} \geq \boldsymbol{y} \Leftrightarrow x_i \geqq y_i, \; i = 1, \ldots, k, \text{ and } \boldsymbol{x} \neq \boldsymbol{y} \tag{2.11c}$$
$$\boldsymbol{x} > \boldsymbol{y} \Leftrightarrow x_i > y_i, \; i = 1, \ldots, k. \tag{2.11d}$$

First, we give the notion of optimality in a multiobjective linear programming problem. Because there does not always exist a solution minimizing all of the objective functions simultaneously, a solution concept of Pareto optimality plays an important role in multiobjective optimization, and it is defined as follows. Let S denote the nonempty set of all feasible solutions of problem (2.10), i.e., $S \triangleq \{\boldsymbol{x} \in \mathbb{R}^n \mid \boldsymbol{A}\boldsymbol{x} \leqq \boldsymbol{b}, \, \boldsymbol{x} \geqq \boldsymbol{0}\}$.

Definition 2.1. A point $\boldsymbol{x}^*$ is said to be a *Pareto optimal solution* if and only if there does not exist another $\boldsymbol{x} \in S$ such that $\boldsymbol{z}(\boldsymbol{x}) \leq \boldsymbol{z}(\boldsymbol{x}^*)$.

By substituting the strict inequality $<$ for the inequality $\leq$ in Definition 2.1, weak Pareto optimality is defined as a slightly weaker solution concept.

2.2.2 Interactive multiobjective programming

As seen from Definition 2.1, in general there exist a infinite number of Pareto optimal solutions if the feasible region S is not empty. In real-world decision making problems, to make a reasonable decision or implement a desirable scheme, the decision maker should select one point among the set of Pareto optimal solutions. For this end, several interactive multiobjective programming methods were developed from the 1970s to the 1980s, and it is known that the reference point method developed by Wierzbicki (1979) is relatively practical.

For each of the objective functions $z(\boldsymbol{x}) = (z_1(\boldsymbol{x}), \ldots, z_k(\boldsymbol{x}))^T$ in problem (2.10), the decision maker specifies the reference point $\bar{z} = (\bar{z}_1, \ldots, \bar{z}_k)^T$ which reflects the desired values of the objective functions, and it is thought that by changing the reference points in the interactive solution procedure, the decision maker can perceive, understand and learn the decision maker's own preference.

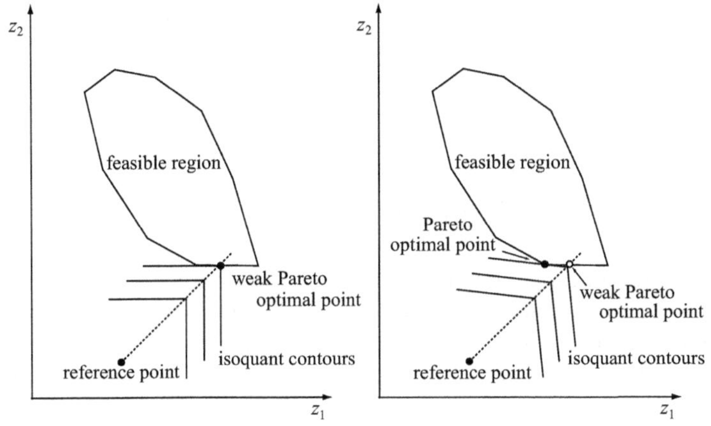

Fig. 2.2 Reference point method.

After the reference point $\bar{z}$ is specified, the following minimax problem is solved:

$$\text{minimize} \quad \max_{i=1,\ldots,k} \{z_i(\boldsymbol{x}) - \bar{z}_i\} \tag{2.12a}$$

$$\text{subject to} \quad A\boldsymbol{x} \leqq \boldsymbol{b},\ \boldsymbol{x} \geqq \boldsymbol{0}. \tag{2.12b}$$

An optimal solution to problem (2.12) is a Pareto optimal solution closest to the reference point in the L_∞ norm; the L_∞ norm is also called the Tchebyshev norm or the Manhattan distance. Introducing an auxiliary variable v, problem (2.12) is equivalently expressed as follows:

$$\text{minimize} \quad v \tag{2.13a}$$

$$\text{subject to} \quad z_i(\boldsymbol{x}) - \bar{z}_i \leqq v,\ i = 1, \ldots, k \tag{2.13b}$$

$$A\boldsymbol{x} \leqq \boldsymbol{b},\ \boldsymbol{x} \geqq \boldsymbol{0}. \tag{2.13c}$$

Because the isoquant contour of the objective function $\max_{i=1,\ldots,k}\{z_i(\boldsymbol{x}) - \bar{z}_i\}$ is orthogonal, there is a possibility that an optimal solution to problem (2.13) is not a Pareto optimal solution but a weak a Pareto optimal solution due to a location of the reference point and the shape of the feasible region as seen in the left-hand graph of Figure (2.2). Let ρ be a given small positive number. By adding the augmented term $\rho\sum_{i=1}^{k}(z_i(\boldsymbol{x})-\bar{z}_i)$ to the objective function, the isoquant contour of the revised objective function has an obtuse angle as seen in the right-hand graph of Figure (2.2). From this fact, a Pareto optimal solution to multiobjective linear programming problem (2.10), which is closest to the reference point $\bar{z}$, can be obtained by solving the following revised problem:

$$\text{minimize} \quad v+\rho\sum_{i=1}^{k}(z_i(\boldsymbol{x})-\bar{z}_i) \tag{2.14a}$$

$$\text{subject to} \quad z_i(\boldsymbol{x})-\bar{z}_i \leqq v,\ i=1,\ldots,k \tag{2.14b}$$

$$A\boldsymbol{x} \leqq \boldsymbol{b},\ \boldsymbol{x} \geqq \boldsymbol{0}. \tag{2.14c}$$

2.2.3 Fuzzy multiobjective programming

To multiobjective linear programming problem (2.10), Zimmermann (1978) extends the fuzzy programming shown in the previous section by introducing fuzzy goals for all the objective functions. Assuming that the decision maker has a fuzzy goal for each of the objective functions, the corresponding linear membership function is defined as

$$\mu_i(z_i(\boldsymbol{x})) = \begin{cases} 1, & \text{if } z_i(\boldsymbol{x}) \leqq z_i^1 \\ \dfrac{z_i(\boldsymbol{x})-z_i^0}{z_i^1-z_i^0}, & \text{if } z_i^1 < z_i(\boldsymbol{x}) \leqq z_i^0 \\ 0, & \text{if } z_i(\boldsymbol{x}) > z_i^0, \end{cases} \tag{2.15}$$

where z_i^0 and z_i^1 denote the values of the ith objective function $z_i(\boldsymbol{x})$ such that the degrees of membership function are 0 and 1, respectively, and it is depicted in Figure 2.3.

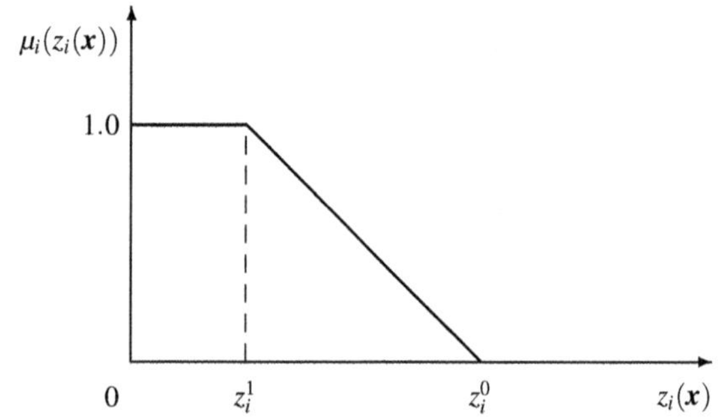

Fig. 2.3 Linear membership function for the ith fuzzy goal.

Following the principle of the fuzzy decision by Bellman and Zadeh (1970), the multiobjective linear programming problem (2.10) can be interpreted as the following maximin problem:

$$\text{maximize} \quad \min_{i=1,\ldots,k} \{\mu_i(z_i(\boldsymbol{x}))\} \tag{2.16a}$$

$$\text{subject to} \quad A\boldsymbol{x} \leqq \boldsymbol{b},\ \boldsymbol{x} \geqq \boldsymbol{0}. \tag{2.16b}$$

Problem (2.16) is equivalently rewritten as a conventional linear programming problem:

$$\text{maximize} \quad \lambda \tag{2.17a}$$

$$\text{subject to} \quad \mu_i(z_i(\boldsymbol{x})) \geqq \lambda,\ i = 1,\ldots,k \tag{2.17b}$$

$$A\boldsymbol{x} \leqq \boldsymbol{b},\ \boldsymbol{x} \geqq \boldsymbol{0}, \tag{2.17c}$$

where λ is an auxiliary variable.

Sakawa, Yano and Yumine (1987) propose an interactive method for a multiobjective linear programming problem with fuzzy goals. After identifying the membership functions $\mu_i(z_i(\boldsymbol{x}))$, $i = 1,\ldots,k$ for the fuzzy goals of the objective functions $z_i(\boldsymbol{x})$, $i = 1,\ldots,k$, the decision maker is asked to specify the reference membership values which are the aspiration levels of achievement for the values of the membership functions. It follows that the reference membership value is a natural extension of the reference point in the reference point method by Wierzbicki (1979).

Let $\bar{\boldsymbol{\mu}} = (\bar{\mu}_1,\ldots,\bar{\mu}_k)^T$ denote the reference membership values for the membership functions $\boldsymbol{\mu}(\boldsymbol{z}(\boldsymbol{x})) = (\mu_1(z_1(\boldsymbol{x})),\ldots,\mu_k(z_k(\boldsymbol{x})))^T$. Then, by solving the minimax problem

$$\text{minimize} \quad \max_{i=1,\ldots,k} \{\bar{\mu}_i - \mu_i(z_i(\boldsymbol{x}))\} \tag{2.18a}$$

$$\text{subject to} \quad A\boldsymbol{x} \leqq \boldsymbol{b},\ \boldsymbol{x} \geqq \boldsymbol{0}, \tag{2.18b}$$

a Pareto optimal solution closest to the vector of the reference membership values in the L_∞ norm can be obtain.

Problem (2.18) is equivalently expressed as follows:

$$\text{minimize} \quad v \tag{2.19a}$$

$$\text{subject to} \quad \bar{\mu}_i - \mu_i(z_i(\boldsymbol{x})) \leqq v,\ i = 1,\ldots,k \tag{2.19b}$$

$$A\boldsymbol{x} \leqq \boldsymbol{b},\ \boldsymbol{x} \geqq \boldsymbol{0}. \tag{2.19c}$$

2.3 Stochastic programming

When some elements governed by stochastic events are included in the constraints of a linear programming problem, such constraints are difficult to completely hold. To formulate deterministic problems from the problems with random variables,

Charnes and Cooper (1963) propose chance constraint programming which admits random data variations and permits constraint violations up to specified probability limits.

Let $\tilde{A}$ and $\tilde{\boldsymbol{b}}$ denote an $m \times n$ matrix and an m-dimensional column vector of the coefficients of the left-hand side and the right-hand side of the constraints, respectively; and suppose that some or all of the elements of $\tilde{A}$ and $\tilde{\boldsymbol{b}}$ are random variables. Then, a chance constraint formulation for the constraint $\tilde{A}\boldsymbol{x} \leqq \tilde{\boldsymbol{b}}$ of a linear programming problem is represented as

$$P[\tilde{A}\boldsymbol{x} \leqq \tilde{\boldsymbol{b}}] \geqq \boldsymbol{\alpha}, \tag{2.20}$$

where P means a probability measure. The vector $\boldsymbol{\alpha}$ are probabilities of the extents to which constraint violations are admitted. Then, the element α_i is associated with the ith constraint $\sum_{j=1}^{n} a_{ij}x_j \leqq b_i$, and the ith constraint is interpreted as follows:

$$P\left[\sum_{j=1}^{n} \tilde{a}_{ij}x_j \leqq \tilde{b}_i\right] \geqq \alpha_i. \tag{2.21}$$

Inequality (2.21) means that the ith constraint may be violated, but at most $\beta_i = 1 - \alpha_i$ proportion of the time.

First, assume that only $\tilde{b}_i$ in the right-hand side of the chance constraint condition (2.21) is a random variable and $\tilde{a}_{ij}$ is a constant, i.e, $\tilde{a}_{ij} = a_{ij}$. Let $F_i(\tau)$ denote its probability distribution function. From the fact that

$$P\left(\sum_{j=1}^{n} a_{ij}x_j \leqq \tilde{b}_i\right) = 1 - F_i\left(\sum_{j=1}^{n} a_{ij}x_j\right),$$

the chance constraint condition (2.21) can be rewritten as

$$F_i\left(\sum_{j=1}^{n} a_{ij}x_j\right) \leqq 1 - \alpha_i. \tag{2.22}$$

Let $K_{1-\alpha_i}$ denote the maximum of τ such that $\tau = F_i^{-1}(1-\alpha_i)$, and then inequality (2.22) can be simply expressed as

$$\sum_{j=1}^{n} a_{ij}x_j \leqq K_{1-\alpha_i}. \tag{2.23}$$

Second, consider a more general case where not only $\tilde{b}_i$ but also $\tilde{a}_{ij}$ in the left-hand side of (2.21) are random variables, and particularly we assume that $\tilde{b}_i$ and $\tilde{a}_{ij}$ are normal random variables. Let $\mu_{\tilde{b}_i}$ and $\sigma_{\tilde{b}_i}$ be the mean and the variance of $\tilde{b}_i$, and let $\mu_{\tilde{a}_{ij}}$ and V_{ij} be the mean and the variance-covariance matrix of $\tilde{a}_{ij}$. Moreover, assume that $\tilde{b}_i$ and $\tilde{a}_{ij}$ are independent. Then, the chance constraint condition (2.21) can be transformed into

$$\sum_{j=1}^{n} \mu_{\tilde{a}_{ij}} x_j - \bar{K}_{1-\alpha_i} \sqrt{\sigma^2_{\tilde{b}_i} + \boldsymbol{x}^T V \boldsymbol{x}} \leqq \mu_{\tilde{b}_i}, \tag{2.24}$$

where $\bar{K}_{1-\alpha_i} = F_N^{-1}(1-\alpha_i)$ and $F_N(\cdot)$ is the standardized normal distribution function with parameter $(0,1)$.

Charnes and Cooper (1963) also consider three kinds of decision rules for optimizing objective functions with random variables: (i) the minimum or maximum expected value model, (ii) the minimum variance model, and (iii) the maximum probability model, which are referred to as the E-model, the V-model, and the P-model, respectively. Moreover, Kataoka (1963) and Geoffrion (1967) individually propose the fractile criterion model.

Let $\tilde{\boldsymbol{c}} = (\tilde{c}_1, \ldots, \tilde{c}_n)$ denote an n-dimensional row vector of the coefficients of the objective function, and suppose that some or all of coefficients $\tilde{c}_j$, $j = 1, \ldots, n$ are random variables. Then, the objective function in the E-model is represented as

$$E[\tilde{\boldsymbol{c}}\boldsymbol{x}] = E\left[\sum_{j=1}^{n} \tilde{c}_j x_j\right], \tag{2.25}$$

where E means the function of expectation. Let m_i denote the mean value of $\tilde{c}_j$. Then, the objective function of the E-model can be transformed to

$$E\left[\sum_{j=1}^{n} \tilde{c}_j x_j\right] = \sum_{j=1}^{n} m_j x_j. \tag{2.26}$$

The realization value of the objective function may vary quite widely even if the expected value of the objective function is minimized. In such a case, it may be suspicious if a plan based on the solution of the E-model would work well because uncertainty is large. Some decision makers would prefer to plans with lower uncertainty. The objective function in the V-model is represented as

$$Var[\tilde{\boldsymbol{c}}\boldsymbol{x}] = Var\left[\sum_{j=1}^{n} \tilde{c}_j x_j\right], \tag{2.27}$$

where Var means the function of variance. Let V denote an $n \times n$ variance-covariance matrix for the vector of the random variables $\tilde{\boldsymbol{c}}$, then the objective function of the V-model can be transformed into

$$Var\left[\sum_{j=1}^{n} \tilde{c}_j x_j\right] = \boldsymbol{x}^T V \boldsymbol{x}. \tag{2.28}$$

In the P-model, the probability that the objective function value is smaller than a ceratin target value is maximized, and then the objective function of the P-model is represented as

$$P\left[\tilde{\boldsymbol{c}}\boldsymbol{x} \leqq f_0\right], \tag{2.29}$$

where f_0 is a given target value for the objective function.

The fractile criterion model is considered as complementary to the P-model; a target variable to the objective function is minimized after the probability that the objective function value is smaller than the target variable is guaranteed to be larger than a given assured level. Then, the objective function of the fractile criterion model is represented as

$$f \text{ subject to } P\left[\tilde{\boldsymbol{c}}\boldsymbol{x} \leqq f\right] \geqq \alpha, \tag{2.30}$$

where f and α are the target variable to the objective function and the given assured level for the probability that the objective function value is smaller than the target variable.

2.4 Genetic algorithms

It is hard to obtain exact optimal solutions of difficult classes of optimization problems such as combinatorial problems and nonconvex nonlinear problems, and thus it is quite natural for decision makers to require approximate optimal solutions instead. To meet this demand, recently several meta-heuristics have been developed and their effectiveness is demonstrated. Among them, genetic algorithms are known to be one of the most practical and proven methods. A computational framework of genetic algorithms initiated by Holland (1975) has been attracted attention of many researchers with applicability in optimization, as well as in search and learning. Furthermore, publications of books by Goldberg (1989) and Michalewicz (1996) bring heightened and increasing interests in applications of genetic algorithms to complex function optimization.

The fundamental procedure of genetic algorithms is shown as a flowchart in Figure 2.4, and it is summarized as follows:

Step 0: Initialization Generate a given number of individuals randomly to form the initial population.

Step 1: Evaluation Calculate the fitness value of each individual in the population.

Step 2: Reproduction According to the fitness values and a reproduction rule specified in advance, select individuals from the current population to form the next population.

Step 3: Crossover Select two individuals randomly from the population, and exchange some part of the string of one individual for the corresponding part of the other individual with a given probability for crossover.

Step 4: Mutation Alter one or more genes in the string of an individual with a given probability of mutation.

Step 5: Termination Stop the procedure if the condition of termination is satisfied, and an individual with the maximum fitness value is determined as an approximate optimal solution. Otherwise, go to Step 1.

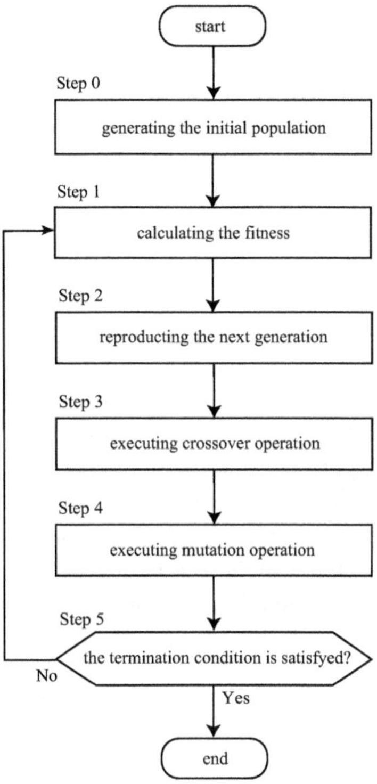

Fig. 2.4 Flowchart of genetic algorithms.

Representation of individuals

When genetic algorithms are applied to optimization problems, a vector of decision variables corresponds to an individual in the population, which is represented by a string as in Figure 2.5.

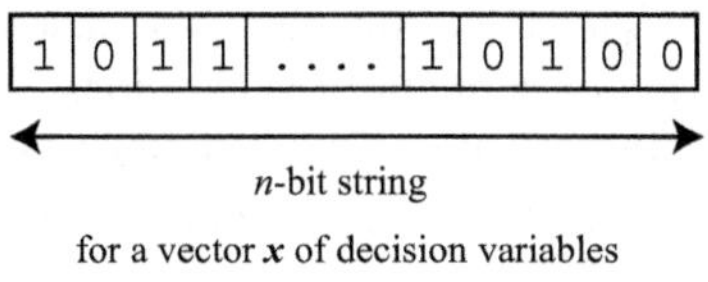

Fig. 2.5 Individual represented by a string.

As seen in Figure 2.5, each element of the string is either 1 or 0 usually, but real numbers, integers, alphabets, or some other symbols can also be used to represent individuals.

Let $\boldsymbol{s}$ and $\boldsymbol{x}$ denote an individual represented by a string and a vector of the decision variables, respectively. The string $\boldsymbol{s}$ which means a chromosome in the context of biology is called the genotype of an individual, and the decision variables $\boldsymbol{x}$ is called the phenotype. The mapping from phenotypes to genotypes is called coding, and the reverse mapping is called decoding.

Fitness function and scaling

In an optimization problem where the objective function is minimized or maximized, a solution with the lowest objective function value or the highest objective function value is searched. When genetic algorithms are applied to solving an optimization problem, a solution to the optimization problem is associated with an individual in the genetic algorithm, and the objective function value of the solution corresponds to the fitness of the individual. Thus, an individual with a higher fitness value has a higher probability of surviving in the next generation.

Let $z(\boldsymbol{x})$ denote an objective function to be minimized in an optimization problem. The corresponding fitness function in genetic algorithms is commonly defined as (Goldberg, 1989)

$$f(\boldsymbol{s}_i) = \begin{cases} C_{\max} - z(\boldsymbol{x}) & \text{if } z(\boldsymbol{x}) < C_{\max} \\ 0 & \text{otherwise,} \end{cases} \tag{2.31}$$

where $\boldsymbol{s}_i$ denotes the ith individual in the population, and $C_{\max}$ is a given constant. For example, the value of $C_{\max}$ is determined as the largest objective function value $z(\boldsymbol{x})$ observed thus far, the largest value $z(\boldsymbol{x})$ in the current population, or the largest value $z(\boldsymbol{x})$ in the last t generations. Similarly, in maximization problems, to prevent the fitness value from being negative, the constant $C_{\min}$ is introduced, and the following fitness function is often used:

$$f(\boldsymbol{s}_i) = \begin{cases} z(\boldsymbol{x}) + C_{\min} & \text{if } z(\boldsymbol{x}) + C_{\min} > 0 \\ 0 & \text{otherwise.} \end{cases} \tag{2.32}$$

For example, the value of $C_{\min}$ is determined as the absolute value of the smallest $z(\boldsymbol{x})$ in the current population or in the last t generations.

To properly distribute fitness values in the population, fitness scaling is employed. The linear scaling, which is a simple and useful procedure, is represented by

$$f'(\boldsymbol{s}_i) = af(\boldsymbol{s}_i) + b, \tag{2.33}$$

where f and f' are the raw fitness value and the scaled fitness value, respectively; a and b are coefficients. To perform the operation of reproduction appropriately, the coefficients a and b may be chosen in such a way that the average scaled fitness value f'_{ave} is equal to the average raw fitness value f_{ave}, and the maximum scaled fitness value is determined as $f'_{\max} = C_{\text{mult}} f_{\text{ave}}$, where C_{mult} is a given constant.

Genetic operators

The three genetic operators, reproduction, crossover, and mutation, are outlined below. Individuals are copied into the next generation according to their fitness values by a reproduction operator. The roulette wheel selection is one of the most popular reproduction operators, and in this method, each individual in the current population has a roulette wheel slot sized in proportion to its fitness value. Let *pop_size* be the number of individuals in the population. The percentage of the roulette wheel given to an individual $\boldsymbol{s}_i$ is $100f(\boldsymbol{s}_i)/\sum_{l=1}^{pop\_size} f(\boldsymbol{s}_l)\%$. Namely, the individual $\boldsymbol{s}_i$ is reproduced with the probability $p(\boldsymbol{s}_i) = f(\boldsymbol{s}_i)/\sum_{l=1}^{pop\_size} f(\boldsymbol{s}_l)$ each spin of the roulette wheel. An example of the roulette wheel is given in Figure 2.6; the numbers in the wheel are fitness values of individuals, and the decimal numbers outside of the wheel are the corresponding probabilities.

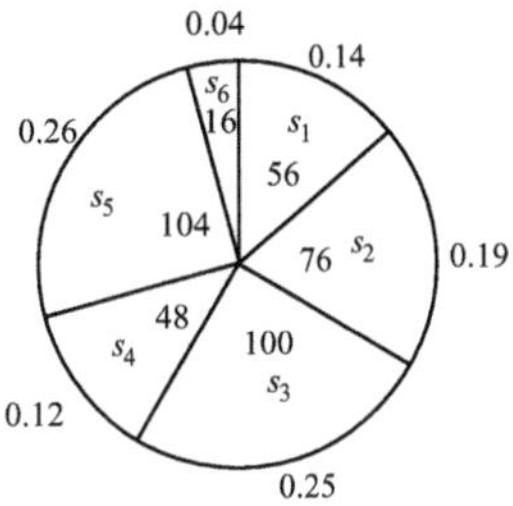

	s_1	s_2	s_3	s_4	s_5	s_6
$p(s_i)$	0.14	0.19	0.25	0.12	0.26	0.04
$f(s_i)$	56	76	100	48	104	16

Fig. 2.6 Biased roulette wheel.

Crossover creates offsprings into the next population by combining the genetic material of two parents. The variation caused by the crossover process may bring offsprings better fitness values, and thus it is thought that crossover plays an important roll in genetic algorithms. Although there are many different types of crossover, we provide a simple example here: a single-point crossover operator is the most simple operator of crossover. In this operation, two parent strings $\boldsymbol{s}_1$ and $\boldsymbol{s}_2$ are randomly chosen from a mating pool in which newly reproduced individuals are entered temporarily, and then one crossover point in the strings is chosen at random. Two offsprings are made by exchanging the substrings which are parts of the left side of the parent strings $\boldsymbol{s}_1$ and $\boldsymbol{s}_2$ from the crossover point. The crossover operation is illustrated in Figure 2.7.

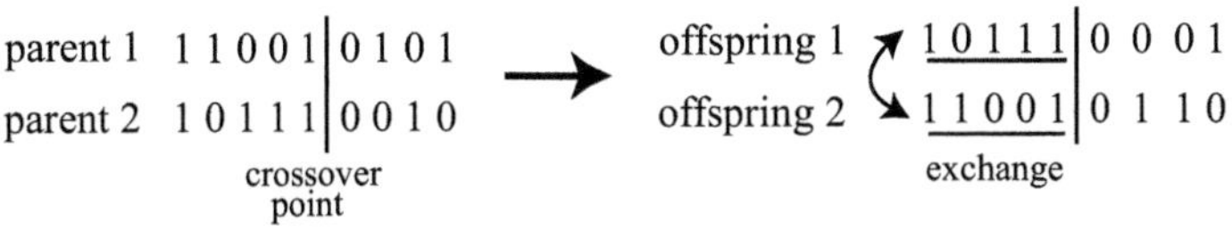

Fig. 2.7 Crossover operation.

With a small probability, an operation of mutation provides the string of an individual with a randomly tiny alteration, and it is recognized that mutation serves as local search. In the representation of the 0-1 bit strings, mutation means changing a 1 to a 0 and vice versa. A simple version of mutation operator is illustrated in Figure 2.8.

parent 1 0 1 1 1 0 1 0 0 → offspring 1 0 0 1 1 0 1 0 0

Fig. 2.8 Mutation operation.

Chapter 3
Noncooperative Decision Making in Hierarchical Organizations

This chapter deals with situations of noncooperative decision making in hierarchical organizations. First, to review the historical background of multi-level programming, we show a couple of conventional computational methods for obtaining Stackelberg solutions to two-level linear programming problems. The subsequent three sections are concerned with computational methods based on genetic algorithms for obtaining Stackelberg solutions to two-level linear, 0-1 and integer programming problems. In the rest of this chapter, we consider formulations and computational methods for multiobjective two-level linear programming and two-level linear stochastic programming.

3.1 Historical background

In the real world, we often encounter situations where there are two or more decision makers in an organization with a hierarchical structure, and they make decisions in turn or at the same time so as to optimize their objective functions. In particular, consider a case where there are two decision makers; one of the decision makers first makes a decision, and then the other who knows the decision of the opponent makes a decision. Such a situation is formulated as a two-level programming problem. We call the decision maker who first makes a decision the leader, and the other decision maker the follower in this chapter which is devoted to dealing with noncooperative decision making situations.

For two-level programming problems, the leader first specifies a decision and then the follower determines a decision so as to optimize the objective function of the follower with full knowledge of the decision of the leader. According to this rule, the leader also makes a decision so as to optimize the objective function of self. The solution defined as the above mentioned procedure is a Stackelberg equilibrium solution, and we call it a Stackelberg solution shortly.

A two-level linear programming problem for obtaining the Stackelberg solution is formulated as:

M. Sakawa and I. Nishizaki (eds.), *Cooperative and Noncooperative Multi-Level Programming*, 25
Operations Research/Computer Science Interfaces Series 48, DOI: 10.1007/978-1-4419-0676-2_3,

$$\underset{\boldsymbol{x}}{\text{minimize}} \;\; z_1(\boldsymbol{x},\boldsymbol{y}) = \boldsymbol{c}_1\boldsymbol{x} + \boldsymbol{d}_1\boldsymbol{y} \tag{3.1a}$$

$$\text{where } \boldsymbol{y} \text{ solves}$$

$$\underset{\boldsymbol{y}}{\text{minimize}} \;\; z_2(\boldsymbol{x},\boldsymbol{y}) = \boldsymbol{c}_2\boldsymbol{x} + \boldsymbol{d}_2\boldsymbol{y} \tag{3.1b}$$

$$\text{subject to } \; A\boldsymbol{x} + B\boldsymbol{y} \leqq \boldsymbol{b} \tag{3.1c}$$

$$\boldsymbol{x} \geqq \boldsymbol{0} \tag{3.1d}$$

$$\boldsymbol{y} \geqq \boldsymbol{0}, \tag{3.1e}$$

where $\boldsymbol{c}_i$, $i = 1,2$ are n_1-dimensional row coefficient vector, $\boldsymbol{d}_i$, $i = 1,2$ are n_2-dimensional row coefficient vector, A is an $m \times n_1$ coefficient matrix, B is an $m \times n_2$ coefficient matrix, $\boldsymbol{b}$ is an m-dimensional column constant vector. In the two-level linear programming problem (3.1), $z_1(\boldsymbol{x},\boldsymbol{y})$ and $z_2(\boldsymbol{x},\boldsymbol{y})$ represent the objective functions of the leader and the follower, respectively, and $\boldsymbol{x}$ and $\boldsymbol{y}$ represent the decision variables of the leader and the follower, respectively.

Each decision maker knows the objective function of the opponent as well as the objective function of self and the constraints. The leader first makes a decision, and then the follower makes a decision so as to minimize the objective function with full knowledge of the decision of the leader. Namely, after the leader chooses $\boldsymbol{x}$, the follower solves the following linear programming problem:

$$\underset{\boldsymbol{y}}{\text{minimize}} \;\; z_2(\boldsymbol{x},\boldsymbol{y}) = \boldsymbol{d}_2\boldsymbol{y} + \boldsymbol{c}_2\boldsymbol{x} \tag{3.2a}$$

$$\text{subject to } \; B\boldsymbol{y} \leqq \boldsymbol{b} - A\boldsymbol{x} \tag{3.2b}$$

$$\boldsymbol{y} \geqq \boldsymbol{0}, \tag{3.2c}$$

and chooses an optimal solution $\boldsymbol{y}(\boldsymbol{x})$ to problem (3.2) as a rational response. Assuming that the follower chooses the rational response, the leader also makes a decision such that the objective function $z_1(\boldsymbol{x},\boldsymbol{y}(\boldsymbol{x}))$ is minimized. Then, the solution defined as the above mentioned procedure is a Stackelberg solution.

We show the basic concepts of the two-level linear programming problem (3.1) as follows:

(i) The feasible region of the two-level linear programming problem:

$$S = \{(\boldsymbol{x},\boldsymbol{y}) \mid A\boldsymbol{x} + B\boldsymbol{y} \leqq \boldsymbol{b},\; \boldsymbol{x} \geqq \boldsymbol{0},\; \boldsymbol{y} \geqq \boldsymbol{0}\} \tag{3.3}$$

(ii) The decision space of the follower after $\boldsymbol{x}$ is specified by the leader:

$$S(\boldsymbol{x}) = \{\boldsymbol{y} \geqq \boldsymbol{0} \mid B\boldsymbol{y} \leqq \boldsymbol{b} - A\boldsymbol{x},\; \boldsymbol{x} \geqq \boldsymbol{0}\} \tag{3.4}$$

(iii) The decision space of the leader:

$$S_X = \{\boldsymbol{x} \geqq \boldsymbol{0} \mid \text{there is a } \boldsymbol{y} \text{ such that } A\boldsymbol{x} + B\boldsymbol{y} \leqq \boldsymbol{b},\; \boldsymbol{y} \geqq \boldsymbol{0}\} \tag{3.5}$$

(iv) The set of rational responses of the follower for $\boldsymbol{x}$ specified by the leader:

$$R(\boldsymbol{x}) = \left\{ \boldsymbol{y} \geqq \boldsymbol{0} \mid \boldsymbol{y} \in \arg\min_{\boldsymbol{y} \in S(\boldsymbol{x})} z_2(\boldsymbol{x}, \boldsymbol{y}) \right\} \tag{3.6}$$

(v) Inducible region:

$$IR = \{(\boldsymbol{x}, \boldsymbol{y}) \mid (\boldsymbol{x}, \boldsymbol{y}) \in S,\ \boldsymbol{y} \in R(\boldsymbol{x})\} \tag{3.7}$$

(vi) Stackelberg solution:

$$\left\{ (\boldsymbol{x}, \boldsymbol{y}) \mid (\boldsymbol{x}, \boldsymbol{y}) \in \arg\min_{(\boldsymbol{x}, \boldsymbol{y}) \in IR} z_1(\boldsymbol{x}, \boldsymbol{y}) \right\} \tag{3.8}$$

To understand the basic concepts of the two-level linear programming problem (3.1), consider the following numerical example:

$$\begin{aligned}
&\underset{x}{\text{minimize}} \quad z_1(x, y) = -x - 8y \\
&\underset{y}{\text{minimize}} \quad z_2(x, y) = -4x + y \\
&\text{subject to} \quad (x, y) \in S \triangleq \{-x + 2y \leqq 13,\ 2x + 3y \leqq 37,\ 2x - y \leqq 17, \\
&\qquad\qquad\qquad 2x - 3y \leqq 11,\ x + 4y \geqq 11,\ 5x + 2y \geqq 19\}.
\end{aligned}$$

The feasible region S of this numerical example of a two-level linear programming problem together with directions decreasing the objective function values of the two decision makers is depicted in Figure 3.1.

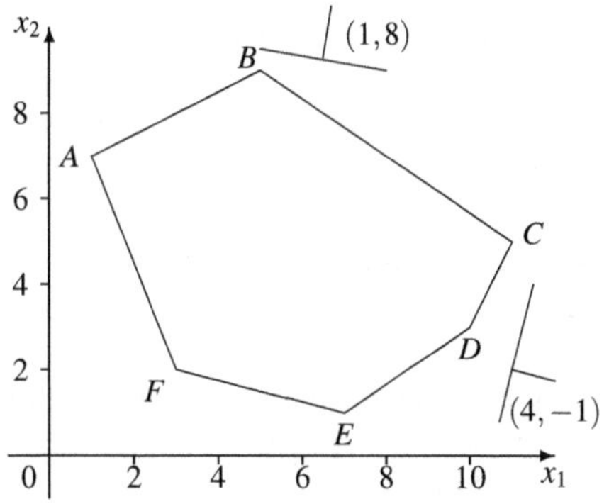

Fig. 3.1 Feasible region of the numerical example.

Let DM1 and DM2 denote the decision makers having the objective functions $z_1(x, y) = -x - 8y$ and $z_2(x, y) = -4x + y$, respectively. If DM1 can determines not only the value of x but also that of y, the optimal solution for DM1 is $(x, y) = (5, 9)$ at point B which yields the objective function values $(z_1, z_2) = (-77, -11)$. Similarly, the optimal solution for DM2 is $(x, y) = (11, 5)$ of point C, and the corresponding objective function values are $(z_1, z_2) = (-51, -39)$. Conversely, the solution yielding the maximal (worst) objective function value, $\max_{(x,y)\in S}(-x - 8y)$, for DM1 is $(x, y) = (7, 1)$ of point E, and the corresponding objective function values are $(z_1, z_2) = (-15, -27)$. For DM2, the worst solution is $(x, y) = (1, 7)$ of point A and the corresponding objective function values are $(z_1, z_2) = (-57, 3)$.

If DM1 and DM2 are the leader and the follower, respectively, the problem is formulated as

$$
\begin{aligned}
\underset{x}{\text{minimize}} \quad & z_1(x, y) = -x - 8y \\
& \text{where } y \text{ solves} \\
\underset{y}{\text{minimize}} \quad & z_2(x, y) = -4x + y \\
\text{subject to} \quad & (x, y) \in S,
\end{aligned}
$$

and the inducible region IR is the piecewise-linear boundary C-D-E-F-A. The solution minimizing the objective function of DM1 in the inducible region IR, that is, the Stackelberg solution is $(x, y) = (1, 7)$ of point A. Conversely, if DM2 and DM1 are the leader and the follower, respectively, the inducible region IR is the piecewise-linear boundary B-C-D-E. The solution minimizing the objective function of DM2 in the inducible region IR, that is, the Stackelberg solution is $(x, y) = (11, 5)$ of point C. It should be noted that Stackelberg solutions depend on which decision maker to become the leader.

Computational methods for obtaining Stackelberg solutions to two-level linear programming problems are classified roughly into three categories: the vertex enumeration approach (Bialas and Karwan, 1984), the Kuhn-Tucker approach (Bard and Falk, 1982; Bard and Moore, 1990a; Bialas and Karwan, 1984; Hansen, Jaumard and Savard, 1992), and the penalty function approach (White and Anandalingam, 1993). The vertex enumeration approach takes advantage of the property that there exists a Stackelberg solution in a set of extreme points of the feasible region. In the Kuhn-Tucker approach, the leader's problem with constraints involving the optimality conditions of the follower's problem is solved. In the penalty function approach, a penalty term is appended to the objective function of the leader so as to satisfy the optimality of the follower's problem. It is well-known that a two-level linear programming problem is an NP-hard problem (Shimizu, Ishizuka and Bard, 1997). In the following, we outline a couple of conventional computational methods for obtaining Stackelberg solutions.

The kth best method proposed by Bialas and Karwan (1984) can be thought of as the vertex enumeration approach, and it is based on a very simple idea. The solution search procedure of the method starts from a point which is an optimal solution to the problem of the leader and checks whether it is also an optimal solution to the problem of the follower or not. If the first point is not the Stackelberg solution, the procedure continues to examine the second best solution to the problem of the leader, and so forth.

At the beginning, the following linear programming problem is solved:

$$\underset{x}{\text{minimize}} \; z_1(\boldsymbol{x},\boldsymbol{y}) = \boldsymbol{c}_1\boldsymbol{x} + \boldsymbol{d}_1\boldsymbol{y} \tag{3.9a}$$

$$\text{subject to} \; A\boldsymbol{x} + B\boldsymbol{y} \leqq \boldsymbol{b} \tag{3.9b}$$

$$\boldsymbol{x} \geqq \boldsymbol{0}, \; \boldsymbol{y} \geqq \boldsymbol{0}, \tag{3.9c}$$

Assuming that the feasible region of problem (3.9) is not empty and there are N extreme points, i.e., N basic solutions of (3.9). Let $(\hat{\boldsymbol{x}}^{[1]},\hat{\boldsymbol{y}}^{[1]})$ denote an optimal solution to the problem (3.9), and $(\hat{\boldsymbol{x}}^{[2]},\hat{\boldsymbol{y}}^{[2]})$, ..., $(\hat{\boldsymbol{x}}^{[N]},\hat{\boldsymbol{y}}^{[N]})$ be the rest of $N-1$ basic feasible solutions such that $z_1(\boldsymbol{x}^{[j]},\boldsymbol{y}^{[j]}) \leqq z_1(\boldsymbol{x}^{[j+1]},\boldsymbol{y}^{[j+1]})$, $j = 1,\dots,N-1$. It is verified if the solution $(\hat{\boldsymbol{x}}^{[j]},\hat{\boldsymbol{y}}^{[j]})$ is optimal to problem (3.2) of the follower from $j = 1$ to $j = N$ in turn. Then, the first solution found to be optimal to problem (3.2) is the Stackelberg solution.

In the Kuhn-Tucker approach, the leader's problem with constraints involving the optimality conditions of the follower's problem (3.2) is solved. The Kuhn-Tucker conditions for problem (3.2) are shown as follows:

$$\boldsymbol{u}B - \boldsymbol{v} = -\boldsymbol{d}_2 \tag{3.10a}$$

$$\boldsymbol{u}(A\boldsymbol{x} + B\boldsymbol{y} - \boldsymbol{b}) - \boldsymbol{v}\boldsymbol{y} = 0 \tag{3.10b}$$

$$A\boldsymbol{x} + B\boldsymbol{y} \leqq \boldsymbol{b} \tag{3.10c}$$

$$\boldsymbol{y} \geqq \boldsymbol{0}, \; \boldsymbol{u}^T \geqq \boldsymbol{0}, \; \boldsymbol{v}^T \geqq \boldsymbol{0}, \tag{3.10d}$$

where $\boldsymbol{u}$ is an m-dimensional row vector and $\boldsymbol{v}$ is an n_2-dimensional row vector.

Then, the follower's problem (3.2) for a two-level linear programming problem can be replaced by the above conditions (3.10), and problem (3.1) is rewritten as the following equivalent single-level mathematical programming problem:

$$\text{minimize} \; z_1(\boldsymbol{x},\boldsymbol{y}) = \boldsymbol{c}_1\boldsymbol{x} + \boldsymbol{d}_1\boldsymbol{y} \tag{3.11a}$$

$$\text{subject to} \; \boldsymbol{u}B - \boldsymbol{v} = -\boldsymbol{d}_2 \tag{3.11b}$$

$$\boldsymbol{u}(A\boldsymbol{x} + B\boldsymbol{y} - \boldsymbol{b}) - \boldsymbol{v}\boldsymbol{y} = 0 \tag{3.11c}$$

$$A\boldsymbol{x} + B\boldsymbol{y} \leqq \boldsymbol{b} \tag{3.11d}$$

$$\boldsymbol{x} \geqq \boldsymbol{0}, \; \boldsymbol{y} \geqq \boldsymbol{0}, \boldsymbol{u}^T \geqq \boldsymbol{0}, \; \boldsymbol{v}^T \geqq \boldsymbol{0}. \tag{3.11e}$$

(3.11f)

From the equality constraint (3.11b) of the problem (3.11), $\boldsymbol{v}$ is eliminated and the equality constraint (3.11c) is transformed into

$$\boldsymbol{u}(\boldsymbol{b} - A\boldsymbol{x} - B\boldsymbol{y}) + (\boldsymbol{u}B + \boldsymbol{d}_2)\boldsymbol{y} = 0. \tag{3.12}$$

Moreover, the complementarity condition (3.12) implies that $\boldsymbol{b} - A\boldsymbol{x} - B\boldsymbol{y} \geqq \boldsymbol{0}$, $\boldsymbol{u}^T \geqq \boldsymbol{0}$, $(\boldsymbol{u}B + \boldsymbol{d}_2)^T \geqq \boldsymbol{0}, \boldsymbol{y} \geqq \boldsymbol{0}$. Let A_i and B_i be the ith row vector of the matrix A and the matrix B, respectively, and let B^j and d_{2j} be the jth column vector of the matrix B and the jth element of the vector $\boldsymbol{d}_2$, respectively. Then, the condition of either $u_i = 0$ or $b_i - A_i\boldsymbol{x} - B_i\boldsymbol{y} = 0$ for $i = 1,\dots,m$ and the condition of either $\boldsymbol{u}B^j + d_{2j} = 0$ or $y_j = 0$ for $j = 1,\dots,n_2$ must be satisfied simultaneously. By introducing zero-one vectors $\boldsymbol{w}_1 = (w_{11}, \ldots, w_{1m})$ and $\boldsymbol{w}_2 = (w_{21}, \ldots, w_{2n_2})$, the equality constraint (3.12) can be expressed as follows (Fortuny-Amat and McCarl, 1981):

$$\boldsymbol{u} \leqq M\boldsymbol{w}_1 \quad (3.13a)$$

$$\boldsymbol{b} - A\boldsymbol{x} - B\boldsymbol{y} \leqq M(\boldsymbol{e} - \boldsymbol{w}_1^T) \quad (3.13b)$$

$$\boldsymbol{u}B + \boldsymbol{d}_2 \leqq M\boldsymbol{w}_2 \quad (3.13c)$$

$$\boldsymbol{y} \leqq M(\boldsymbol{e} - \boldsymbol{w}_2^T), \quad (3.13d)$$

where $\boldsymbol{e}$ is an m-dimensional vector of ones, and M is a large positive constant.

Therefore, the mathematical programming problem (3.11) is equivalent to the following mixed zero-one programming problem , and it can be solved by a zero-one mixed integer solver.

$$\text{minimize} \quad z_1(\boldsymbol{x}, \boldsymbol{y}) = \boldsymbol{c}_1\boldsymbol{x} + \boldsymbol{d}_1\boldsymbol{y} \quad (3.14a)$$

$$\text{subject to} \quad \boldsymbol{0} \leqq \boldsymbol{u}^T \leqq M\boldsymbol{w}_1^T \quad (3.14b)$$

$$\boldsymbol{0} \leqq \boldsymbol{b} - A\boldsymbol{x} - B\boldsymbol{y} \leqq M(\boldsymbol{e} - \boldsymbol{w}_1^T) \quad (3.14c)$$

$$\boldsymbol{0} \leqq (\boldsymbol{u}B + \boldsymbol{d}_2)^T \leqq M\boldsymbol{w}_2^T \quad (3.14d)$$

$$\boldsymbol{0} \leqq \boldsymbol{y} \leqq M(\boldsymbol{e} - \boldsymbol{w}_2^T) \quad (3.14e)$$

$$\boldsymbol{x} \geqq \boldsymbol{0}. \quad (3.14f)$$

In the penalty function approach, the duality gap of the follower's problem (3.2) is appended to the objective function of the leader. The dual problem to problem (3.2) ignoring the constant term $\boldsymbol{c}_2\boldsymbol{x}$ is written as

$$\text{minimize} \quad \boldsymbol{u}(A\boldsymbol{x} - \boldsymbol{b}) \quad (3.15a)$$

$$\text{subject to} \quad -\boldsymbol{u}B \leqq \boldsymbol{d}_2 \quad (3.15b)$$

$$\boldsymbol{u}^T \geqq \boldsymbol{0}, \quad (3.15c)$$

where $\boldsymbol{u}$ is an m-dimensional row vector. Because the duality gap $\boldsymbol{d}_2\boldsymbol{y} - \boldsymbol{u}(A\boldsymbol{x} - \boldsymbol{b})$ is zero if $\boldsymbol{y}$ is a rational responses of the follower with respect to a choice $\boldsymbol{x}$ of the leader, i.e., $\boldsymbol{y} \in R(\boldsymbol{x})$, the following mathematical programming problem is formulated:

$$\text{minimize} \quad \boldsymbol{c}_1\boldsymbol{x} + \boldsymbol{d}_1\boldsymbol{y} + K\boldsymbol{u}(A\boldsymbol{x} - \boldsymbol{b}) \quad (3.16a)$$

$$\text{subject to} \quad A\boldsymbol{x} + B\boldsymbol{y} \leqq \boldsymbol{b} \quad (3.16b)$$

$$-\boldsymbol{u}B \leqq \boldsymbol{d}_2 \quad (3.16c)$$

$$\boldsymbol{x} \geqq \boldsymbol{0},\ \boldsymbol{y} \geqq \boldsymbol{0},\ \boldsymbol{u}^T \geqq \boldsymbol{0}, \quad (3.16d)$$

where K is a constant value. By repeatedly solving problem (3.16) for updated values of K and $\boldsymbol{u}$, problem (3.16) yields an optimal solution to problem (3.1), i.e., the Stackelberg solution.

3.2 Two-level linear programming

In this section, we provide a computational method based on genetic algorithms for obtaining Stackelberg solutions to two-level linear programming problems (Niwa, Nishizaki and Sakawa, 2001).

3.2.1 Mixed zero-one programming problem corresponding to two-level linear programming problem

First, we consider the characteristics of a two-level linear programming problem (3.14) which is given in the following again.

$$
\begin{aligned}
\text{minimize } & z_1(\boldsymbol{x},\boldsymbol{y}) = \boldsymbol{c}_1\boldsymbol{x} + \boldsymbol{d}_1\boldsymbol{y} && (3.14\text{a})\\
\text{subject to } & \boldsymbol{0} \leqq \boldsymbol{u}^T \leqq M\boldsymbol{w}_1^T && (3.14\text{b})\\
& \boldsymbol{0} \leqq \boldsymbol{b} - A\boldsymbol{x} - B\boldsymbol{y} \leqq M(\boldsymbol{e} - \boldsymbol{w}_1^T) && (3.14\text{c})\\
& \boldsymbol{0} \leqq (\boldsymbol{u}B + \boldsymbol{d}_2)^T \leqq M\boldsymbol{w}_2^T && (3.14\text{d})\\
& \boldsymbol{0} \leqq \boldsymbol{y} \leqq M(\boldsymbol{e} - \boldsymbol{w}_2^T) && (3.14\text{e})\\
& \boldsymbol{x} \geqq \boldsymbol{0}. && (3.14\text{f})
\end{aligned}
$$

In the constraints (3.14b) and (3.14d) including the variable $\boldsymbol{u}$, there do not exist both the decision variables $\boldsymbol{x}$ of the leader and the decision variables $\boldsymbol{y}$ of the follower. Moreover, the objective function $z_1(\boldsymbol{x},\boldsymbol{y})$ does not include the variables $\boldsymbol{u}$. Therefore, if the zero-one variables $\boldsymbol{w}_1$ and $\boldsymbol{w}_2$ are fixed at certain values, a pair of vectors of the decision variables $\boldsymbol{x}$ and $\boldsymbol{y}$ and the variables $\boldsymbol{u}$, which are the Lagrange multipliers in the Kuhn-Tucker optimality condition, are separable.

If the following set U is empty, any pair of $\boldsymbol{x}$ and $\boldsymbol{y}$ do not satisfy the Kuhn-Tucker conditions.

$$U = \{\boldsymbol{u} \mid \boldsymbol{u}^T \leqq M\boldsymbol{w}_1^T,\ (\boldsymbol{u}B + \boldsymbol{d}_2)^T \leqq M\boldsymbol{w}_2^T\}. \tag{3.17}$$

In other words, there does not exist any pair of $\boldsymbol{x}$ and $\boldsymbol{y}$ in the inducible region. If the set U is not empty, an optimal solution to the linear programming problem with some fixed zero-one variables $\boldsymbol{w}_1$ and $\boldsymbol{w}_2$ exists in the inducible region.

Next, we consider the rest of the constraints including the decision variables $\boldsymbol{y}$ of the follower in problem (3.14). The lower bound of each constraint is zero. The upper bound of each constraint is either zero or unbounded depending on the values of $\boldsymbol{w}_1$ and $\boldsymbol{w}_2$. This means that a constraint with $\boldsymbol{y}$ is either an equality constraint or an inequality constraint. We give the following example in order to understand how the feasible region and the rational responses change depending on the values of the zero-one variables $\boldsymbol{w}_1$ and $\boldsymbol{w}_2$:

$$\underset{x}{\text{minimize}} \; z_1(x,y) = 5x + y \tag{3.18a}$$

where y solves

$$\underset{y}{\text{minimize}} \; z_2(x,y) = -x + 7y \tag{3.18b}$$

$$\text{subject to} \; x - y \geqq 2 \tag{3.18c}$$

$$x + 3y \leqq 15 \tag{3.18d}$$

$$4x + y \leqq 26 \tag{3.18e}$$

$$x - 2y \leqq 4 \tag{3.18f}$$

$$x \geqq 0 \tag{3.18g}$$

$$y \geqq 0, \tag{3.18h}$$

where x is a decision variable of the leader, and y is a decision variable of the follower. The feasible region and the inducible region for problem (3.18) is shown in Figure 3.2.

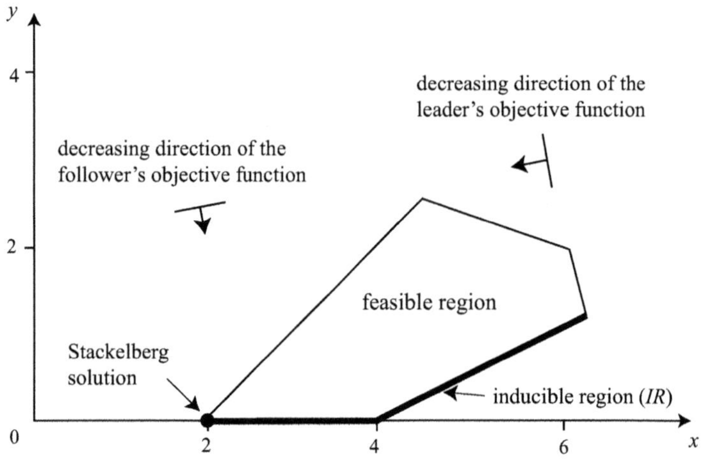

Fig. 3.2 Feasible region and inducible region.

For this numerical example, the constraints (3.18c), (3.18d), (3.18e) and (3.18f) become equality constraints when $\boldsymbol{w}_1 = \mathbf{1}$, and the constraint (3.18h) becomes an equality constraint when $w_2 = 1$. For given fixed values of $\boldsymbol{w}_1$ and w_2, the combination of inequality constraints and equality constraints is considered in the following cases.

(i) If there exist more than three equality constraints, the feasible region of the corresponding problem (3.14) with fixed $\boldsymbol{w}_1$ and w_2 is empty.

(ii) If there exist two equality constraints, the feasible region corresponds to an extreme point of the original feasible region S of the two-level linear programming problem or it is empty.
(iii) If there exists one equality constraint or does not exist any equality constraint, the feasible region is non-empty.

For the numerical example depicted in Figure 3.2, there exist sixteen combinations of $\boldsymbol{w}_1$ and w_2 corresponding to (i), ten combinations corresponding to (ii), and six combinations corresponding to (iii). The number of combinations yielding feasible solutions among these 32 combinations is 11, and only the six combinations yield feasible solutions in the inducible region.

In order to appropriately specify the values of zero-one variables $\boldsymbol{w}_1$ and $\boldsymbol{w}_2$ and find the Stackelberg solution efficiently, the above mentioned consideration gives the following suggestion. If most of the elements of $\boldsymbol{w}_1$ and $\boldsymbol{w}_2$ are set at zero, problem (3.14) is likely to be infeasible because the number of the equality constraints in the set U (3.17) is large. In contrast, if the number of the variables $\boldsymbol{w}_1$ and $\boldsymbol{w}_2$ whose values are one is large, the objective function value does not become small, and it is difficult to find the optimal solution because the number of active constraints, i.e., equality constraints in $A\boldsymbol{x}+B\boldsymbol{y} \leqq \boldsymbol{b}$ is also large.

The optimal solution to problem (3.14) is also an extreme point of the feasible region S satisfying the following constraints of the original two-level linear programming problem (3.1):

$$A\boldsymbol{x}+B\boldsymbol{y} \leqq \boldsymbol{b} \tag{3.19a}$$
$$\boldsymbol{x} \geqq \boldsymbol{0} \tag{3.19b}$$
$$\boldsymbol{y} \geqq \boldsymbol{0}. \tag{3.19c}$$

Under the assumption of the non-degeneration, the fact that a solution is an extreme point of S requires the n_1+n_2 equality constraints in (3.19). Because $\boldsymbol{w}_1$ and $\boldsymbol{w}_2$ correspond to (3.19a) and (3.19c), the minimum of the number of equality constraints in (3.19a) and (3.19c) is zero and the maximum of it is $m+n_2$. If all of (3.19b) with n_1 constraints are active, the constraints of (3.19a) and (3.19c) must include at least n_2 equality constraints. Therefore, we must set at least n_2 elements of $\boldsymbol{w}_1$ and $\boldsymbol{w}_2$ at one.

In the computational method, we set n_2 elements of $\boldsymbol{w}_1$ and $\boldsymbol{w}_2$ at one, and by using the simplex method, a linear programming problem with fixed values of $\boldsymbol{w}_1$ and $\boldsymbol{w}_2$ is solved.

3.2.2 Computational methods based on genetic algorithms

We describe a representation of individuals and genetic operators used in the genetic algorithm for solving two-level linear programming problems.

Representation of individuals and initial population

In the computational method based on the framework of genetic algorithms for obtaining Stackelberg solutions, the zero-one variables $\boldsymbol{w}_1$ and $\boldsymbol{w}_2$ of problem (3.14) are represented as binary strings, and then, the total length of the string is $m+n_2$ bits. An instance of the string in the method is shown in Figure 3.3.

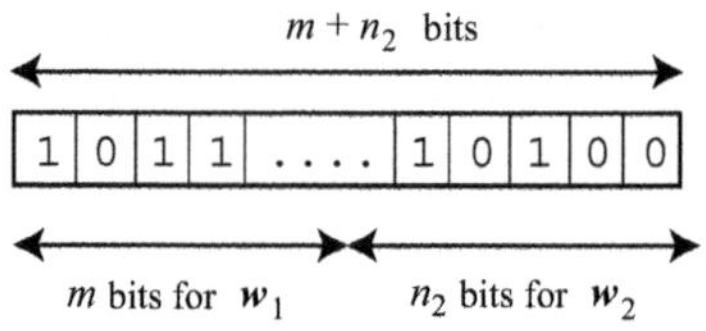

Fig. 3.3 Zero-one bit string.

In the initial population, individuals represented by the strings as in Figure 3.3 are generated. Let *pop_size* denote the size of the population. As we mentioned the previous subsection, we set n_2 bits in the string at 1s. Namely, we select n_2 bits at random and put 1s in those bits; and then, we also put 0s in the remaining m bits. By performing this procedure, we generate the initial population with *pop_size* individuals.

The procedure for generating the initial population is summarized as follows:

Step 1 Let $\boldsymbol{s}_l$ denote the l th individual in the initial population. Set $l := 1$.

Step 2 Select n_2 bits in a string $\boldsymbol{s}_l$ with $m+n_2$ bits at random, and put 1s in those bits. Put 0s in the remaining m bits.

Step 3 In order to check whether the individual $\boldsymbol{s}_l$ generated in Step 2 is feasible or not, execute the first phase of the simplex method to the set U (3.17) with fixed values of $\boldsymbol{w}_1$ and $\boldsymbol{w}_2$. If the set U is empty, return to Step 2.

Step 4 If $l > pop\_size$, stop the algorithm. Otherwise, set $l := l+1$ and return to Step 1.

Fitness and reproduction

When the variables $\boldsymbol{w}_1$ and $\boldsymbol{w}_2$ are fixed for a given individual $\boldsymbol{s}_l$, problem (3.14) is reduced into a linear programming problem. Let z_l be the optimal value of the objective function of problem (3.14) with the fixed values of $\boldsymbol{w}_1$ and $\boldsymbol{w}_2$. After computing the optimal values z_l, $l = 1, \ldots, pop\_size$ of all the individuals in the population, to avoid negative values of z_l, the linear scaling is performed. Let z_l' denote the scaled value. Then, the fitness $f(\boldsymbol{s}_l)$ for each individual $\boldsymbol{s}_l$ is defined as the reciprocal of the scaled value z_l', i.e., $f(\boldsymbol{s}_l) = 1/z_l'$.

As a reproduction operator, the elitist roulette wheel selection, in which the elitism and the roulette wheel selection are combined together, is adopted. The individual s_l is reproduced into the next generation using a roulette wheel with slots sized according to fitness values. Namely, s_l is reproduced with a probability of $f(s_l)/\sum_{i=1}^{pop\_size} f(s_i)$.

Genetic operations

We describe the crossover operator and the mutation operator in the computational method for obtaining Stackelberg solutions. In the crossover operation, the number of bits whose values are 1s is kept at n_2, and an example of the crossover operation is depicted in Figure 3.4.

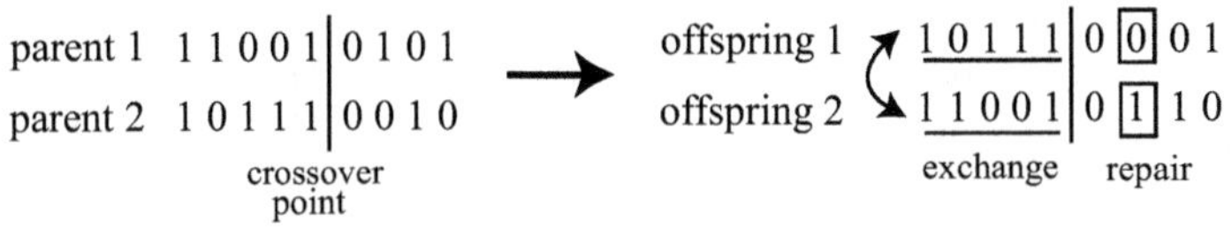

Fig. 3.4 Crossover operation.

This procedure is summarized as follows.

Step 1 For two strings s_1 and s_2 of parents, choose one crossover point in the strings at random. Let the crossover point be between the kth and the $k+1$th bits.

Step 2 Count the number of bits with 1s from the leftmost bit to the kth bit, and let q_1 and q_2 be the numbers of bits with 1s for the two parents. Calculate the difference between q_1 and q_2, i.e., $q' = q_1 - q_2$.

Step 3 Exchange the substrings which are parts of the left side of the whole strings s_1 and s_2 from the crossover point.

Step 4 If $q' > 0$, randomly select one bit such that the value of the bit of s_1 is 0 and that of s_2 is 1, and invert them. This operation is repeated q' times. If $q' < 0$, do the opposite.

Similarly, in the mutation operation, to keep the number of bits with 1s unchanged, a pair of one bit of 1 and one bit of 0 are selected, and they are exchanged. The mutation operation is illustrated in Figure 3.5.

Fig. 3.5 Mutation operation.

3.2.3 Computational Experiments

To demonstrate the efficiency of the computational method based on the framework of genetic algorithms for obtaining Stackelberg solutions, computational experiments are carried out. In the computational experiments, the number of zero-one variables $\boldsymbol{w}_1$ and $\boldsymbol{w}_2$ in problem (3.14) is regarded as a measure of the size of two-level linear programming problems. Two kinds of two-level linear programming problems are generated for each size of problems with zero-one variables: 10, 20, 30, 40, 50 and 60 variables.

Each entry of $\boldsymbol{c}_1$, $\boldsymbol{c}_2$, $\boldsymbol{d}_1$, $\boldsymbol{d}_2$, A, B and $\boldsymbol{b}$ is a random value in the interval $[-100, 100]$. The parameters in the genetic algorithm are specified as follows: the population size $pop\_size = 20$; the probability of crossover $p_c = 0.4$; the probability of mutation $p_m = 0.01$; and the maximum generation $max\_gen = 100$. The computational experiments are performed on a computer with MMX Pentium 200MHz. For comparison, the exact computational method proposed by Hansen, Jaumard and Savard (1992), which is reported to be an effective algorithm (Shimizu, Ishizuka and Bard, 1997), and the approximate computational method based on Tabu search proposed by Gendreau, Marcotte, and Savard (1996) are taken up. For the sake of simplicity, let GA, HJS , and TS denote the method based on genetic algorithms, the exact method by Hansen, Jaumard and Savard, and the method based on Tabu search by Gendreau, Marcotte and Savard.

For GA and TS, each problem is solved 10 times. In the result of computational time, the average of two problems of the same scale and the 10 runs is shown in Table 3.1 and Figure 3.6. Table 3.2 is given to show the accuracy of the solutions obtained by GA and TS because GA and TS are approximate computational methods; the accuracy is expressed by the ratio of the approximate method to the exact method, i.e., 100×(the objective function value obtained by HJS)/(the objective function value obtained by GA or TS).

Table 3.1 Computational time [sec.].

size	HJS	TS	GA
10	0.050	0.995	1.212
20	0.971	4.824	4.263
30	7.050	7.293	7.542
40	19.924	14.962	14.462
50	49.827	30.878	26.059
60	131.925	43.017	47.834

Table 3.1 and Figure 3.6 show that on computational time, GA (or TS) is more effective than HJS for problems with more than 40 variables, and the difference between GA and HJS (or TS and HJS) becomes large as the size of problems grows large. As for the accuracy of the solutions, as seen in Table 3.2, GA can find the exact Stackelberg solution in the 10 runs. Moreover, the mean and the worst case are also close to the exact Stackelberg solution, and GA seems to show a good performance.

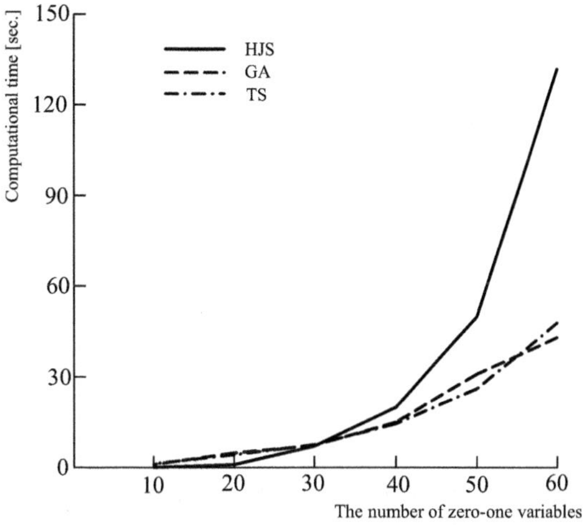

Fig. 3.6 Computational time.

Table 3.2 Accuracy of solutions.

	GA			TS		
size	best	worst	mean	best	worst	mean
10	100.00	100.00	100.00	100.00	100.00	100.00
20	100.00	99.36	99.59	100.00	100.00	100.00
30	100.00	99.10	99.63	100.00	99.99	99.99
40	100.00	99.67	99.91	99.78	99.29	99.54
50	100.00	98.93	99.66	100.00	98.69	99.34
60	100.00	98.50	99.41	100.00	95.42	97.71

As for comparison of GA with TS, GA and TS show similar performances, but in the size 60 of problem the accuracy of the solution of GA is slightly better than that of TS.

To see the effect of the specific representation of binary strings in order to efficiently find Stackelberg solutions, we perform an additional computational experiment with a straightforward GA in which the binary strings are randomly formed, that is, 0s and 1s are randomly put in the strings. The result of the additional computational experiment is shown in Table 3.3. As seen in Table 3.3, the straightforward GA cannot find the exact Stackelberg solutions to problems with more than 20 variables, and the obtained solutions to problem with 60 variables are obviously inferior in accuracy to that of GA shown in Table 3.2.

Table 3.3 Accuracy of solutions by the straightforward GA.

size	best	worst	mean
10	100.00	95.53	99.30
20	100.00	98.00	99.34
30	99.42	97.42	98.35
40	98.89	97.62	98.36
50	97.52	95.21	96.35
60	95.44	90.60	93.11

3.3 Two-level mixed zero-one programming

Concerning two-level programming problems with discrete decision variables, Bard and Moore (1990b, 1992) develop algorithms based on the branch-and-bound techniques for obtaining Stackelberg solutions to two-level zero-one programming problems and two-level mixed integer programming problems. Wen and Yang (1990) deal with a two-level mixed zero-one programming problem with zero-one decision variables of the leader and real-valued decision variables of the follower, and they propose an exact method and a heuristic method for obtaining Stackelberg solutions. Their exact method also utilizes the branch-and-bound techniques, and the leader's zero-one decision variables are used as branching variables. However, there is fear that the computational time increases extremely as the problem size becomes larger, and therefore developing highly efficient approximate computational methods is important for solving large scale problems. In this section, we present a computational method based on the framework of genetic algorithms for obtaining Stackelberg solutions to two-level mixed zero-one programming problems in which the leader has zero-one decision variables and the follower controls real-valued decision variables (Nishizaki and Sakawa, 2000).

After formulating a facility location and transportation problem in order to illustrate the two-level mixed zero-one programming form, we show two methods on the basis of genetic algorithms for obtaining Stackelberg solutions to two-level mixed zero-one programming problems. In the methods, zero-one bit strings are employed for the chromosomal representation. Namely, zero-one decision variables in the problem correspond to individuals expressed by zero-one bit strings in artificial genetic systems. Therefore, the decisions of the leader are represented directly by the individuals in the genetic algorithms. Conversely, the follower's decisions, i.e., rational responses for the decisions specified by the leader are determined by solving linear programming problems.

A two-level mixed zero-one programming problem, in which the leader and the follower control zero-one decision variables and real-valued decision variables, respectively, is represented as

$$\underset{x}{\text{minimize}} \quad z_1(\boldsymbol{x},\boldsymbol{y}) = \boldsymbol{c}_1\boldsymbol{x} + \boldsymbol{d}_1\boldsymbol{y} \tag{3.20a}$$

where $\boldsymbol{y}$ solves

$$\underset{y}{\text{minimize}} \quad z_2(\boldsymbol{x},\boldsymbol{y}) = \boldsymbol{c}_2\boldsymbol{x} + \boldsymbol{d}_2\boldsymbol{y} \tag{3.20b}$$

$$\text{subject to} \quad A\boldsymbol{x} + B\boldsymbol{y} \leqq \boldsymbol{b} \tag{3.20c}$$

$$\boldsymbol{x} \in \{0,1\}^{n_1} \tag{3.20d}$$

$$\boldsymbol{y} \geqq \boldsymbol{0}, \tag{3.20e}$$

where $\boldsymbol{x} = (x_1,\ldots,x_{n_1})^T$ is an n_1-dimensional variable column vector, $\boldsymbol{y} = (y_1,\ldots, y_{n_2})^T$ an n_2-dimensional variable column vector; $\boldsymbol{c}_i$, $i = 1,2$ is an n_1-dimensional coefficient row vector, $\boldsymbol{d}_i$, $i = 1,2$ is an n_2-dimensional coefficient row vector; A and B are $m \times n_1$ and $m \times n_2$ coefficient matrices, respectively, and $\boldsymbol{b}$ is an m-dimensional constant row vector; $\boldsymbol{x}$ is the leader's decision variable vector and $\boldsymbol{y}$ is the follower's decision variable vector; $z_1(\boldsymbol{x},\boldsymbol{y})$ is the leader's objective function, and $z_2(\boldsymbol{x},\boldsymbol{y})$ is the follower's objective function.

For two-level mixed zero-one programming problems (3.20), the basic concepts are the same as those for two-level linear programming problems except for the feasible region and the solution space of the leader which are rewritten as follows:

(i) The feasible region:

$$S = \{(\boldsymbol{x},\boldsymbol{y}) \mid A_1\boldsymbol{x} + A_2\boldsymbol{y} \leqq \boldsymbol{b},\ \boldsymbol{x} \in \{0,1\}^{n_1},\ \boldsymbol{y} \geqq \boldsymbol{0}\} \tag{3.21}$$

(iii) The decision space of the leader:

$$S_X = \{\boldsymbol{x} \in \{0,1\}^{n_1} \mid \text{there exists a } \boldsymbol{y} \text{ such that } A_1\boldsymbol{x} + A_2\boldsymbol{y} \leqq \boldsymbol{b},\ \boldsymbol{y} \geqq \boldsymbol{0}\} \tag{3.22}$$

The decision space of the follower $S(\boldsymbol{x})$, the set of rational responses $R(\boldsymbol{x})$, the inducible region IR, and Stackelberg solutions are defined as in the same expressions (3.4), (3.6), (3.7), and (3.8) for the two-level linear programming problems.

3.3.1 Facility location and transportation problem

Before describing the computational methods based on genetic algorithms, we formulate a facility location and transportation problem in order to illustrate the two-level mixed zero-one programming problems. In the facility location and transportation problem, the leader is a manufacturer which distributes products from production plants (factories) to distribution centers (warehouses or stores). The manufacturer subcontracts to a forwarding agent in order to transport the products from the factories to the warehouses, and therefore the forwarding agent is dealt with as the follower in the two-level mathematical model. Suppose that the manufacturer has several factories and warehouses, and it should operate them so as to minimize the total cost. Taking into account locations of the factories and the warehouses, the forwarding agent makes a transportation plan of products from the factories to the warehouses so as to minimize the transportation costs. The two-level programming problem for the facility location and transportation is characterized by the following decision variables, the objective function and the constraints.

Decision variables

Assume that there are n_1 factories and n_2 warehouses. Let $\boldsymbol{x} = (x_1, \ldots, x_{n_1})^T$ denote decision variables of the manufacturer, the leader, for the n_1 factories. If the factory i is open or built, $x_i = 1$; otherwise $x_i = 0$.

The forwarding agent, the follower, must carry the products from the n_1 factories to the n_2 warehouses. Let $\boldsymbol{y} = (y_{11}, \ldots, y_{1n_2}, \ldots, y_{n_1 1}, \ldots, y_{n_1 n_2})^T$ denote decision variables of the follower; y_{ij} is the volume of transportation from the factory i to the warehouse j.

Objective function

The leader wants to minimize the total cost which is the sum of the production cost and the running and maintenance costs from operating factories. Let c_{1i} be the running and maintenance costs of the factory i, and d_{1i} the cost of one unit of the product. Then, the objective function of the leader is represented by

$$z_1(\boldsymbol{x}, \boldsymbol{y}) = \sum_{i=1}^{n_1} \left(c_{1i} x_i + d_{1i} \sum_{j=1}^{n_2} y_{ij} \right). \tag{3.23}$$

Because the follower chooses the values of $\boldsymbol{y}$ after the values of $\boldsymbol{x}$ have been determined by the leader, the values of $\boldsymbol{y}$ are not determined yet when the leader chooses the values of $\boldsymbol{x}$. Thus, it follows that the leader determines the values of $\boldsymbol{x}$ on the assumption that the follower will choose the rational reaction, i.e., an optimal solution to the linear programming problem with the fixed parameters $\boldsymbol{x}$.

The follower also intends to minimize the sum of the transportation costs and the expenditure at the factories such as office works. Let c_{2i} be the expenditure at the factory i, and d_{2ij} the cost of one unit of transformation from the factory i to the warehouse j. The the objective function of the follower is represented by

$$z_2(\boldsymbol{x}, \boldsymbol{y}) = \sum_{i=1}^{n_1} c_{2i} x_i + \sum_{i=1}^{n_1} \sum_{j=1}^{n_2} d_{2ij} y_{ij}. \tag{3.24}$$

Constraints

Let f_i be the production capacity of the factory i, and g_j the demand of the region in which the warehouse j exists. A limitation on production at the factory i is expressed as

$$\sum_{j=1}^{n_2} y_{ij} \leqq f_i x_i, \tag{3.25}$$

and a condition for supplying the products to satisfy the local demand in the region of the warehouse j can be expressed as

$$\sum_{i=1}^{n_1} y_{ij} \leqq g_j. \tag{3.26}$$

There might be a limitation of the number of operating factories if the labor force is not enough to operate all the factories; such a limitation can be expressed as

$$\sum_{i=1}^{n_1} x_i \leqq M, \tag{3.27}$$

where M is the upper bound of the number of operating factories.

Formulation

Then, the problem can now be written in two-level mixed zero-one programming problem form (3.20) as follows:

$$\underset{\boldsymbol{x}}{\text{minimize}} \ z_1(\boldsymbol{x},\boldsymbol{y}) = \sum_{i=1}^{n_1} \left(c_{1i} x_i + d_{1i} \sum_{j=1}^{n_2} y_{ij} \right) \tag{3.28a}$$

where $\boldsymbol{y}$ solves

$$\underset{\boldsymbol{y}}{\text{minimize}} \ z_2(\boldsymbol{x},\boldsymbol{y}) = \sum_{i=1}^{n_1} c_{2i} x_i + \sum_{i=1}^{n_1} \sum_{j=1}^{n_2} d_{2ij} y_{ij} \tag{3.28b}$$

$$\text{subject to} \ \sum_{j=1}^{n_2} y_{ij} \leqq f_i x_i \tag{3.28c}$$

$$\sum_{i=1}^{n_1} y_{ij} \leqq g_j \tag{3.28d}$$

$$\sum_{i=1}^{n_1} x_i \leqq M \tag{3.28e}$$

$$x_i \in \{0,1\}, \ i = 1,\ldots,n_1 \tag{3.28f}$$

$$y_{ij} \geqq 0, \ i = 1,\ldots,n_1, \ j = 1,\ldots,n_2. \tag{3.28g}$$

3.3.2 Computational methods based on genetic algorithms

To develop computational methods for obtaining Stackelberg solutions to two-level mixed zero-one programming problems, we need some devices for treating constraints of the problems because simple genetic algorithms are not enough to meet mathematical programming problems with constraints. Michalewicz (1996) classifies algorithms handling constraints into three categories: algorithm based on penalty functions; algorithms based on repair methods; and algorithms based on decoders.

In this subsection, we present two algorithms for obtaining Stackelberg solutions to two-level mixed zero-one programming problems: the simple genetic algorithm with the penalty function and the genetic algorithm with double strings. Following the above classification by Michalewicz, the former is classified as one of algorithms based on penalty functions, and the latter would be classified as an algorithm based on some repair method rather than an decoder.

In the proposed methods, zero-one decision variables $\boldsymbol{x}$ of the leader correspond to individuals expressed as zero-one bit strings in artificial genetic systems. For each individual corresponding to a decision $\boldsymbol{x}$ of the leader, the follower's rational response $\boldsymbol{y} \in R(\boldsymbol{x})$ can be obtained by solving a linear programming problem. We assume that, for any $\boldsymbol{x} \in S(X)$, the set $R(\boldsymbol{x})$ of the follower's rational responses is a singleton. The basic structure of the computational methods is summarized as follows.

Step 1 For the leader's decision variable vector $\boldsymbol{x}$, generate *pop_size* individuals at random, and form the initial population.

Step 2 Decode each individual $\boldsymbol{s}$ into a decision variable vector $\hat{\boldsymbol{x}}$, and solve the follower's linear programming problem:

$$\underset{\boldsymbol{y}}{\text{minimize}} \quad z_2(\hat{\boldsymbol{x}},\boldsymbol{y}) = \boldsymbol{c}_{22}\boldsymbol{y} + \boldsymbol{c}_{21}\hat{\boldsymbol{x}} \tag{3.29a}$$

$$\text{subject to} \quad B\boldsymbol{y} \leqq \boldsymbol{b} - A\hat{\boldsymbol{x}} \tag{3.29b}$$

$$y_j \geqq 0, \; j = 1,\ldots,n_2. \tag{3.29c}$$

If there exists an optimal solution $\hat{\boldsymbol{y}}$, it should be in the set $R(\hat{\boldsymbol{x}})$ of follower's rational responses, i.e., $\hat{\boldsymbol{y}} \in R(\hat{\boldsymbol{x}})$.

Step 3 Evaluate each individual $\boldsymbol{s}$ by the fitness function:

$$f(\boldsymbol{s}) = \begin{cases} f_s(z_1(\hat{\boldsymbol{x}},\hat{\boldsymbol{y}})) & \text{if } \hat{\boldsymbol{x}} \in S(X) \\ p(\hat{\boldsymbol{x}}) & \text{otherwise,} \end{cases} \tag{3.30}$$

where $f_s(z_1(\hat{\boldsymbol{x}},\hat{\boldsymbol{y}}))$ is a scaled fitness value, and $p(\hat{\boldsymbol{x}})$ is a penalty value.

Step 4 If the termination condition is satisfied, the algorithm stops and an individual with the maximal fitness value is decoded as an approximate Stackelberg solution to problem (3.20).

Step 5 After the reproduction, operations of the crossover and the mutation are performed. Return to Step 2.

3.3.2.1 Simple genetic algorithm

The simple genetic algorithm (Goldberg, 1989) is composed of three genetic operators: reproduction, crossover and mutation. To implement the simple genetic algorithm with the penalty function for obtaining Stackelberg solutions, the following zero-one bit string representation of individuals and genetic operators are employed.

Representation of individuals

For a decision variable vector $\boldsymbol{x}$ of the leader, an individual $\boldsymbol{s}$ is directly represented as an n_1 zero-one bits long string. An instance of the string in this method is shown in Figure 3.7.

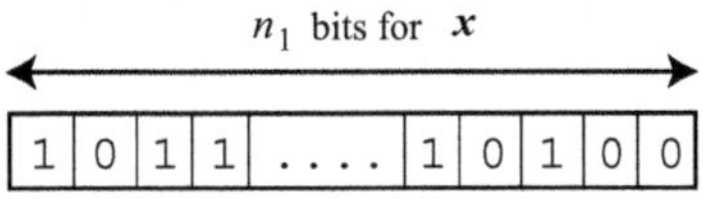

Fig. 3.7 Zero-one bit string.

Fitness and reproduction

For the decision of the leader $\hat{\boldsymbol{x}}$ decoded from an individual $\boldsymbol{s}$, the fitness function is defined as

$$f(\boldsymbol{s}) = \begin{cases} C_{\max} - z_1(\hat{\boldsymbol{x}}, \hat{\boldsymbol{y}}) & \text{if } \hat{\boldsymbol{x}} \in S(X) \\ -9999 & \text{otherwise,} \end{cases} \tag{3.31}$$

where $\hat{\boldsymbol{y}}$ is the optimal solution to problem (3.29), and $C_{\max}$ is a maximal value of the objective function $z_1(\hat{\boldsymbol{x}}, \hat{\boldsymbol{y}})$ in the population.

As a reproduction operator, the elitist roulette wheel selection is adopted. The elitist roulette wheel selection is a combination of the elitism and the roulette wheel selection. Let $\boldsymbol{x}^*(t)$ be the best individual, which yields the minimum of the leader's objective function, up to the generation t. If there does not exist the best individual $\boldsymbol{x}^*(t)$ in the population of the next generation $t+1$, $\boldsymbol{x}^*(t)$ is included in the next population additionally, and then the population size temporarily increases by one. The roulette wheel selection is the most popular way of selection. This reproduction generates individuals in the next generation using a roulette wheel with slots sized according to fitness values. The size of a slot is given by a probability of $f(\boldsymbol{s}_i)/\sum_{i=1}^{pop\_size} f(\boldsymbol{s}_i)$.

Genetic operations

In the crossover operation, two reproduced individuals are chosen at random. A point for crossing over is selected in the strings of the individuals, and then two new offsprings are created by swapping substrings which are parts of the right side from the crossover point on the original strings.

As for the mutation operation, with a given small probability, each zero-one bit in a string is randomly changed, i.e., a 1 is changed to a 0, and vice versa.

3.3.2.2 Genetic algorithm with double strings

In the simple genetic algorithm, there exist some individuals $\boldsymbol{s}$ decoded into infeasible solutions $\boldsymbol{x}$ which do not satisfy the constraints of problem (3.20), i.e., $\boldsymbol{x} \notin S(X)$, and the fitness values of such individuals are penalized for violating the constraints. Unfortunately, the exploration by such penalty methods in genetic algorithms does not work efficiently.

Assuming that all the coefficients of the constraint functions are positive, any individual $\boldsymbol{s}$ can be decoded into a feasible solution $\boldsymbol{x} \in S(X)$ by using genetic algorithms with the double string representation of individuals (Sakawa and Shibano, 1996; Sakawa *et al.*, 1997; Sakawa, 2001).

Encoding and decoding

To generate only feasible solutions, the double string representation shown in Figure 3.8 is adopted for the individual representation. In a double string, $i(j)$ and $s_{i(j)}$ denote the index of an element in a decision variable vector and the value of the element, respectively, i.e., for a given index $i(j) \in \{1,\ldots,n_1\}$ of the element of the decision variable vector $\boldsymbol{x}$, the value of the element $x_{i(j)}$ is $s_{i(j)} \in \{1,0\}$. A string $\boldsymbol{s}$ can be transformed into a solution $\boldsymbol{x} = (x_1,\ldots,x_{n_1})^T$ as $x_{i(j)} := s_{i(j)}$, $j = 1,\ldots,n_1$.

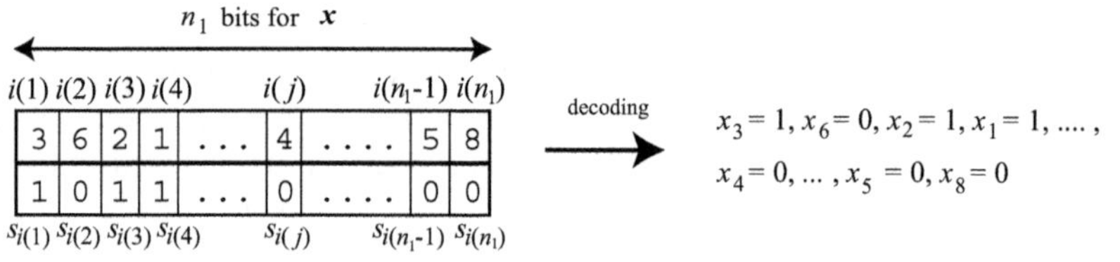

Fig. 3.8 Double string representation.

Unfortunately, because this mapping may generate solutions $\hat{\boldsymbol{x}}$ corresponding to problems (3.29) without any feasible solution, the following decoding algorithm for eliminating such solutions is developed. For a given string $\boldsymbol{s}$, in the algorithm, n_1, j, $i(j)$, $x_{i(j)}$ and $\boldsymbol{a}_{i(j)}$ denote, respectively, the length of the string, a position in the string, an index of a variable, a 0-1 value of the variable with the index $i(j)$ decoded from the string, and the $i(j)$th column vector of the coefficient matrix A. Let $\boldsymbol{sum}$ be an m-dimensional temporal vector.

Step 1 Set $j := 1$ and $\boldsymbol{sum} := \boldsymbol{0}$.
Step 2 If $s_{i(j)} = 1$, set $j := j+1$, and go to step 3. Otherwise, i.e., if $s_{i(j)} = 0$, set $j := j+1$, and go to step 4.
Step 3 If $\boldsymbol{sum} + \boldsymbol{a}_{i(j)} \leqq \boldsymbol{b}$, set $x_{i(j)} := 1$ and $\boldsymbol{sum} := \boldsymbol{sum} + \boldsymbol{a}_{i(j)}$, and go to step 4. Otherwise, set $x_{i(j)} := 0$ and go to step 4.
Step 4 If $j > n_1$, stop the algorithm, and a solution $\boldsymbol{x} = (x_1, \ldots, x_{n_1})^T$ is obtained from the individual represented by the double string. Otherwise, return to step 2.

In the decoding algorithm, when $s_{i(j)}$ is equal to 1, $x_{i(j)}$ is set at 1 as long as the constraints are satisfied; otherwise, $x_{i(j)}$ is set at 0. The solution $\hat{\boldsymbol{x}}$ derived by using this procedure is in the leader's decision space $S(X)$, i.e., $\hat{\boldsymbol{x}} \in S(X)$. By solving problem (3.29), a feasible solution $(\hat{\boldsymbol{x}}, \hat{\boldsymbol{y}})$ can be obtained.

Fitness and reproduction

For a decision $\hat{\boldsymbol{x}}$ of the leader derived by using the double string decoding algorithm, the same fitness function (3.31) as that of the simple genetic algorithm is employed. The elitist roulette wheel selection is also adopted in a way similar to in the simple genetic algorithm.

Genetic operations

When a single-point or multi-point crossover operator is applied to individuals represented by double strings, there is some possibility that multiple indices take the same number. Recall that the same violation occurs in solving traveling salesman problems or scheduling problems through genetic algorithms. As one of possible approaches to circumvent such violations, so-called "partially matched crossover (PMX)" is useful (Goldberg and Lingle, 1985). It enables us to generate desirable offsprings without changing the double string structure. However, in order to handle each element $s_{i(j)}$ in a double string efficiently, it is necessary to revise some parts of the procedure. The revised PMX for treating double strings is given as follows:

Step 1 For two individuals $\boldsymbol{s}^1$ and $\boldsymbol{s}^2$ represented by double strings, choose two crossover points; let h and k be the first bit and the last bit of the substring to be exchanged.
Step 2 According to the following procedure of the PMX, reorder the upper string of $\boldsymbol{s}^1$ together with the corresponding lower string.

Step 2-1 Set $j := h$.

Step 2-2 Find $\hat{j}$ such that $i^1(\hat{j}) = i^2(j)$. Then, exchange $\left(i^1(j), s^1_{i^1(j)}\right)^T$ with $\left(i^1(\hat{j}), s^1_{i^1(\hat{j})}\right)^T$, and set $j := j+1$.

Step 2-2 If $j > k$, stop. Otherwise, return to Step 2-2.

Similarly, reorder s^2. Let $s^{1'}$ and $s^{2'}$ be the reordered double strings.

Step 3 Offsprings $s^{1''}$ and $s^{2''}$ are obtained by exchanging the lower substrings between two crossover points of $s^{1'}$ and $s^{2'}$.

An example of the operation of the revised PMX for treating the double string structure is given in Figure 3.9.

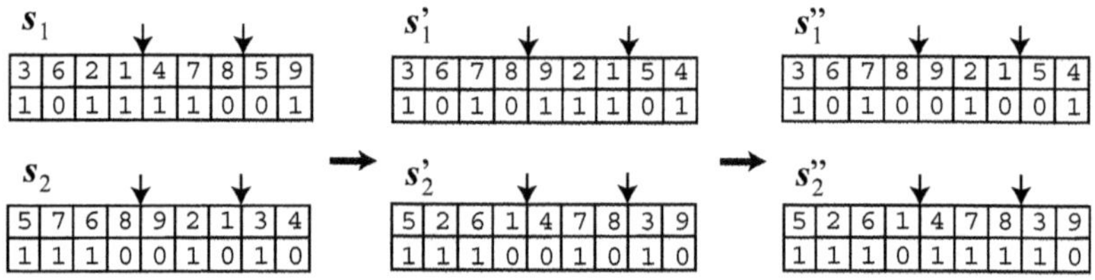

Fig. 3.9 Revised PMX for treating the double string structure.

It is well recognized that a mutation operator plays a role of local random search in genetic algorithms. In the genetic algorithm with double strings, for the lower string of a double string, the operation of mutation of bit-reverse type is adopted, and another genetic operator, inversion, can be introduced together with the PMX operator. The procedure of the inversion is shown as follows:

Step 1 For an individual s, choose two inversion points at random.

Step 2 Invert both the upper and lower substrings between two inversion points.

An example of the operation of the inversion for individuals in the double string representation is given in Figure 3.10.

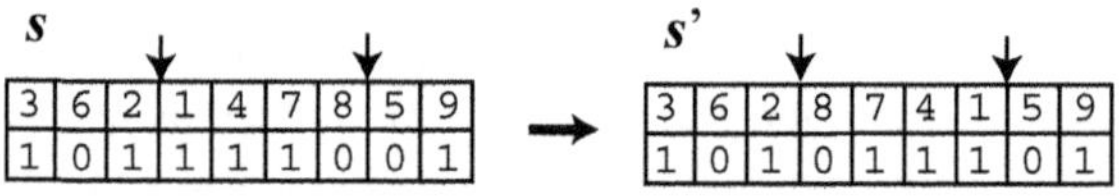

Fig. 3.10 Inversion for individuals of the double string structure.

3.3.3 Computational Experiments

To demonstrate the feasibility and efficiency of the aforementioned methods based on the simple genetic algorithm with the penalty function and the genetic algorithm with double strings for large scale problems, computational experiments are performed. These two methods using genetic algorithms are compared with the methods proposed by Wen and Yang (1990) which are an exact algorithm based on the branch-and-bound techniques and a heuristic algorithm.

The termination condition of the computational methods based on the genetic algorithms is shown as follows. Let $I_{\min}$, $I_{\max}$, ε, $F_{\max}$ and F_{mean} denote the minimum and the maximum numbers of generations, a tolerance of convergence, the maximum fitness value, and the mean fitness value, respectively. Let t denote the current number of generation. If $t > I_{\min}$ and $(F_{\max} - F_{\text{mean}})/F_{\max} < \varepsilon$, or $t > I_{\max}$, the procedure stops, and then an individual with the maximum fitness value in the population is regarded as an approximate Stackelberg solution.

In the computational experiments, parameters of the genetic algorithms are set as follows: the minimum and the maximum numbers $I_{\min}$ and $I_{\max}$ of generations are 20 and 200, respectively; the tolerance of convergence ε is 0.06; the size of the population is 200; and probabilities of the genetic operations are shown in Table 3.4. The parameters are determined and adjusted on the basis of the results of the related studies (Sakawa and Shibano, 1996; Sakawa *et al.*, 1997; Sakawa and Shibano, 1997).

Table 3.4 Probabilities of the genetic operations.

	simple GA	GA with double strings
crossover	0.800	0.800
mutation	0.030	0.015
inversion	——	0.015

Five different-sized problems are solved: dimensions of the decision variable vectors $\boldsymbol{x}$ and $\boldsymbol{y}$ of the leader and the follower are (10, 10), (15, 15), (20, 20), (25, 25), and (30, 30); five problems with different coefficients are prepared for each size; each problem is solved ten times; and the methods using genetic algorithms and Wen and Yang's methods are compared on computational time (CPU time) and the accuracy of obtained solutions.

Each entry of the coefficient matrices A and B of a two-level mixed zero-one programming problem is randomly chosen from the set of all integers in the interval $[0,100]$; that of the constant vector $\boldsymbol{b}$ is a sum of entries of the corresponding row vector of A and B multiplied by 0.6; and that of the coefficient vectors $\boldsymbol{c}_i$ and $\boldsymbol{d}_i$, $i = 1,2$ of the objective functions are also randomly chosen from the set of all integers in the interval $[-100,100]$. The computational experiments are performed on a personal computer Gateway 2000 P5-90 (CPU: Pentium; 90 MHz), and computer programs are written in C++.

Table 3.5 Computational time (CPU time): [sec.].

size (n_1, n_2)	method	mean	minimum	maximum
	branch-and-bound	0.2	0.0	1.0
20-dimension	heuristics	0.0	0.0	0.0
(10, 10)	simple GA	10.2	6.0	12.0
	GA with double strings	1.7	0.0	4.0
	branch-and-bound	3.0	0.0	8.0
30-dimension	heuristics	0.0	0.0	0.0
(15, 15)	simple GA	14.6	10.0	23.0
	GA with double strings	2.5	1.0	6.0
	branch-and-bound	337.2	19.0	517.0
40-dimension	heuristics	0.0	0.0	0.0
(20, 20)	simple GA	20.0	15.0	26.0
	GA with double strings	11.8	2.0	25.0
	branch-and-bound	15383.6	3.0	22439.0
50-dimension	heuristics	0.0	0.0	0.0
(25, 25)	simple GA	23.4	2.0	35.0
	GA with double strings	18.1	3.0	38.0
	branch-and-bound	—	—	—
60-dimension	heuristics	0.0	0.0	0.0
(30, 30)	simple GA	33.9	25.0	43.0
	GA with double strings	26.1	4.0	50.0

* Computational times smaller than 0.05 second are shown as 0.0.

In Table 3.5, the mean, the minimum and the maximum values of the computational time (CPU time) for each method are shown. Wen and Yang's methods are expressed as "branch-and-bound" and "heuristics" and the two methods using genetic algorithms "simple GA" and "GA with double strings."

For the smallest problem with 20-dimensional decision variables, because the computational time of the branch-and-bound method is smaller than those of the two genetic algorithms and the branch-and-bound method finds the exact Stackelberg solution, the branch-and-bound method is more effective than the two genetic algorithms. However, we cannot find a significant difference on the computational times of the branch-and-bound method and the two genetic algorithms for the 30-dimensional problem, and for the problems with more than 30-dimensional decision variables, the branch-and-bound method evidently needs computational time larger than those of the two genetic algorithms. Moreover, the computational time of the branch-and-bound method for the 60-dimensional problem finally exceeds 24 hours, which is specified as the limit of the computational time. In comparison between the two genetic algorithms, the computational time of the genetic algorithm with double strings is smaller than that of the simple genetic algorithm for all the problems. The heuristics method is the quickest method to get an approximate Stackelberg solution, and it does not take more than 0.05 second for all the problems.

On the accuracy of solutions obtained by the heuristic method and the two genetic algorithms, ratios of the approximate solutions to the exact solutions, i.e., 100×(the objective function value obtained by the branch-and-bound method)/(the objective function value obtained by the GAs or the heuristics method) are given in Table 3.6. We solve five problems for each size, and the mean, the minimum and the maximum values of the ratios for five problems are shown in Table 3.6. Because we cannot obtain the exact Stackelberg solutions to the 60-dimensional problems, we compare them with the minimal objective function value obtained in the computational experiments.

Table 3.6 Accuracy of obtained solutions [%].

size (n_1, n_2)	method	mean	minimum	maximum
20-dimension	heuristics	85.23	61.52	97.26
$(10,10)$	simple GA	100.00	100.00	100.00
	GA with double strings	100.00	100.00	100.00
30-dimension	heuristics	96.13	92.77	100.00
$(15,15)$	simple GA	100.00	100.00	100.00
	GA with double strings	100.00	100.00	100.00
40-dimension	heuristics	92.83	77.85	100.00
$(20,20)$	simple GA	100.00	100.00	100.00
	GA with double strings	100.00	100.00	100.00
50-dimension	heuristics	93.24	90.18	99.53
$(25,25)$	simple GA	98.90	91.94	100.00
	GA with double strings	99.98	99.41	100.00
60-dimension	heuristics	89.72	76.89	98.71
$(30,30)$	simple GA	98.21	95.68	100.00
	GA with double strings	99.09	92.33	100.00

Table 3.7 Numbers of finding the exact Stackelberg solutions.

size (n_1, n_2)	method	mean	minimum	maximum
20-dimension	simple GA	10.0	10	10
$(10,10)$	GA with double strings	10.0	10	10
30-dimension	simple GA	10.0	10	10
$(15,15)$	GA with double strings	10.0	10	10
40-dimension	simple GA	10.0	10	10
$(20,20)$	GA with double strings	10.0	10	10
50-dimension	simple GA	4.6	0	8
$(25,25)$	GA with double strings	9.6	9	10
60-dimension	simple GA	0.4	0	2
$(30,30)$	GA with double strings	5.4	1	9

As seen in Tables 3.5 and 3.6, although the computational times of the heuristic method are smaller than those of the two genetic algorithms, on the accuracy of computation, the two genetic algorithms are superior to the heuristic method. Because the data of Table 3.6 are not enough to judge which genetic algorithm is better than the other one, the number of finding the exact Stackelberg solutions to the five problems in the ten runs is shown in Table 3.7. The mean, the minimum and the maximum values of the numbers of finding the exact Stackelberg solutions are shown in Table 3.7, and, for the 60-dimensional problems, the comparison with the minimal objective function value obtained in the computational experiments is given. As seen in Table 3.7, for the 50-dimensional and 60-dimensional problems, it becomes clear that the solutions obtained by the genetic algorithm with double strings is more accurate than those by the simple genetic algorithm.

3.4 Two-level linear integer programming

In this section, we deal with two-level integer programming problems. Moore and Bard (1990) develop an algorithm based on the branch-and-bound techniques based on the depth first rule for obtaining Stackelberg solutions to two-level mixed integer programming problems. In their method, the two-level linear programming problems relaxed from the original integer problems must be solved repeatedly, and then there is fear that, in proportion as the size of the problem, the computational time exceedingly increases. Therefore, developing highly efficient approximate computational methods is important for large scale problems, and in this section, we give a computational method based on the framework of genetic algorithms for obtaining Stackelberg solutions to two-level integer programming problems (Nishizaki and Sakawa, 2005).

Assuming that there exist the upper and lower bounds for each of integer decision variables, the zero-one bit string representation is employed to express individuals corresponding to the integer decision variables in artificial genetic systems. Because it is required that each individual satisfies the constraints of a given problem and a response of the follower with respect to a decision of the leader is rational, individuals not satisfying these two conditions are penalized in the artificial genetic systems.

Let $\boldsymbol{x} = (x_1, \ldots, x_{n_1})^T \in \mathbb{Z}^{n_1}$ and $\boldsymbol{y} = (y_1, \ldots, y_{n_2})^T \in \mathbb{Z}^{n_2}$ denote the decision variables of the leader and the follower, respectively, where $\mathbb{Z}^{n_1}$ and $\mathbb{Z}^{n_2}$ are the sets of all the n_1- and n_2-dimensional integer vectors. Then, two-level integer programming problems can be represented by

$$\underset{x}{\text{minimize}}\ z_1(\boldsymbol{x},\boldsymbol{y}) = \boldsymbol{c}_1\boldsymbol{x} + \boldsymbol{d}_1\boldsymbol{y} \tag{3.32a}$$

where $\boldsymbol{y}$ solves

$$\underset{y}{\text{minimize}}\ z_2(\boldsymbol{x},\boldsymbol{y}) = \boldsymbol{c}_2\boldsymbol{x} + \boldsymbol{d}_2\boldsymbol{y} \tag{3.32b}$$

$$\text{subject to}\ A\boldsymbol{x} + B\boldsymbol{y} \leqq \boldsymbol{b} \tag{3.32c}$$

$$\boldsymbol{x} \geqq \boldsymbol{0},\ \boldsymbol{y} \geqq \boldsymbol{0} \tag{3.32d}$$

$$\boldsymbol{x} \in \mathbb{Z}^{n_1},\ \boldsymbol{y} \in \mathbb{Z}^{n_2}, \tag{3.32e}$$

where $z_1(\boldsymbol{x},\boldsymbol{y})$ and $z_2(\boldsymbol{x},\boldsymbol{y})$ are the objective functions of the leader and the follower, respectively; $\boldsymbol{c}_1 = (c_{11},\ldots,c_{1n_1})$ and $\boldsymbol{d}_1 = (d_{11},\ldots,d_{1n_2})$ are n_1- and n_2-dimensional coefficient row vectors of the objective function of the leader; $\boldsymbol{c}_2 = (c_{21},\ldots,c_{2n_1})$ and $\boldsymbol{d}_2 = (d_{21},\ldots,d_{2n_2})$ are n_1- and n_2-dimensional coefficient row vectors of the objective function of the follower; A and B are $m \times n_1$ and $m \times n_2$ coefficient matrices in the constraints; $\boldsymbol{b} = (b_1,\ldots,b_m)^T$ is an m-dimensional constant column vector in the constraints.

For two-level integer programming problems (3.32), the basic concepts are the same as those for two-level linear programming problems except for the feasible region, the decision space of the follower, and the decision space of the leader which are rewritten as follows:

(i) The feasible region:

$$S = \{(\boldsymbol{x},\boldsymbol{y}) \mid A\boldsymbol{x} + B\boldsymbol{y} \leqq \boldsymbol{b},\ \boldsymbol{x} \in \mathbb{Z}_+^{n_1},\ \boldsymbol{y} \in \mathbb{Z}_+^{n_2}\}, \tag{3.33}$$

where $\mathbb{Z}_+^{n_1}$ and $\mathbb{Z}_+^{n_2}$ are the sets of all of the n_1- and n_2-dimensional non-negative integer vectors.

(ii) The decision space of the follower:

$$S(\boldsymbol{x}) = \{\boldsymbol{y} \in \mathbb{Z}_+^{n_2} \mid A\boldsymbol{x} + B\boldsymbol{y} \leqq \boldsymbol{b},\ \boldsymbol{x} \in \mathbb{Z}_+^{n_1}\} \tag{3.34}$$

(iii) The decision space of the leader:

$$S_X = \{\boldsymbol{x} \in \mathbb{Z}_+^{n_1} \mid \text{there exists}\ \boldsymbol{y} \in \mathbb{Z}_+^{n_2}\ \text{such that}\ A\boldsymbol{x} + B\boldsymbol{y} \leqq \boldsymbol{b}\} \tag{3.35}$$

The set of rational responses $R(\boldsymbol{x})$, the inducible region IR, and Stackelberg solutions are defined as in the same expressions (3.6), (3.7), and (3.8) for the two-level linear programming problems.

To understand the concept of Stackelberg solutions to two-level integer programming problems geometrically, consider the following numerical example with only two decision variables x and y:

$$\underset{x}{\text{minimize}}\ z_1(x,y) = 3x + y \tag{3.36a}$$

where y solves

$$\underset{y}{\text{minimize}}\ z_2(x,y) = -x + 5y \tag{3.36b}$$

$$\text{subject to}\ x - 4y \leqq 1,\ 3x + 8y \geqq 24 \tag{3.36c}$$

$$3x + 2y \geqq 12,\ -3x + 5y \leqq 20 \tag{3.36d}$$

$$x + y \leqq 12,\ x \in \mathbb{Z}^1,\ y \in \mathbb{Z}^1. \tag{3.36e}$$

Feasible solutions of problem (3.36) are depicted by symbols ∘, • and ⊙ in Figure 3.11. In particular, solutions in the inducible region are denoted by •, and the Stackelberg solution is denoted by ⊙. If the integral condition is eliminated, the Stackelberg solution is a point represented by □ in Figure 3.11.

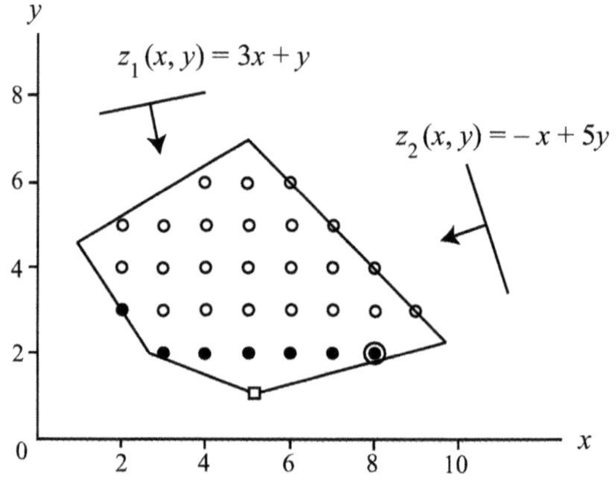

Fig. 3.11 Integer Stackelberg solution.

3.4.1 Computational methods based on genetic algorithms

In this subsection, we describe computational methods based on genetic algorithms for obtaining Stackelberg solutions to two-level integer programming problems.

Representation of individuals

A pair of decision variable vectors $(\boldsymbol{x},\boldsymbol{y})$ in problem (3.32) is represented as a binary string which is encoded as a Gray-coded integer in the artificial genetic systems. The Gray codes are represented by binary strings in a way similar to unsigned binary integers, but adjacent two Gray-coded integers differ by a single bit. From this property, a small change in the operation of mutation is implemented efficiently and easily. Let the length of binary strings be l. The relation between a Gray-coded integer $< g_{l-1} g_{l-2} \cdots g_0 >$ and an unsigned binary integer $< b_{l-1} b_{l-2} \cdots b_0 >$ is demonstrated as follows:

$$g_k = \begin{cases} b_{l-1} & \text{if } k = l-1 \\ b_{k+1} \oplus b_k & \text{if } k \leqq l-2, \end{cases} \quad k = 0, 1, \ldots, l-1, \tag{3.37}$$

where $\oplus$ means summation by modulo 2, i.e., the exclusive or. Transformation from a Gray-coded integer to the corresponding unsigned binary integer is represented by

$$b_k = \sum_{i=k}^{l-1} g_i \pmod 2, \; k = 0, 1, \ldots, l-1. \tag{3.38}$$

The length of a binary string is determined by the upper and the lower bounds of the decision variables. Namely, the bit length to be required for some decision variable is the least integer q such that a difference between the upper and the lower bounds is not larger than $2^q - 1$. When the upper bound of a decision variable x_i is not given explicitly, the upper bound is obtained by solving the following linear programming problem:

$$\text{maximize } x_i \tag{3.39a}$$

$$\text{subject to } \boldsymbol{Ax} + \boldsymbol{By} \leqq \boldsymbol{b} \tag{3.39b}$$

$$\boldsymbol{x} \geqq \boldsymbol{0}, \; \boldsymbol{y} \geqq \boldsymbol{0}. \tag{3.39c}$$

Penalties and fitness

It is desirable for each individual to satisfy the constraints of the given problem and to have a rational response of the follower with respect to a decision of the leader. Therefore, two types of penalties is introduced in proportion to the degrees of the violations of these conditions.

Penalty for the violations of the constraints: An individual which does not satisfy the constraints is penalized in proportion to the degree of the violation, and it is evaluated as follows:

$$d_i = \begin{cases} 0 & \text{if } A_i \boldsymbol{x} + B_i \boldsymbol{y} \leqq b_i \\ \dfrac{A_i \boldsymbol{x} + B_i \boldsymbol{y} - b_i}{|b_i|} & \text{if } A_i \boldsymbol{x} + B_i \boldsymbol{y} > b_i, \end{cases} \tag{3.40}$$

where A_i, B_i are the ith row vectors of A and B, respectively, and b_i is the ith element of $\boldsymbol{b}$. By using the value of d_i, the penalty of violations of the constraints is defined as

$$p_1 = \exp\left\{-\sum_{i=1}^{m} d_i\right\}. \tag{3.41}$$

Penalty for irrational responses: When it is assumed that there is no communication between the leader and the follower, or they do not make any binding agreement even if there exists such communication, the leader assumes that the follower takes a rational response, that is, the follower makes a decision so as to minimize the follower's objective function. Therefore, in the artificial genetic systems, an irrational response of the follower with respect to a decision of the leader is penalized in proportion to the degree of violations.

For an individual $(\boldsymbol{x},\boldsymbol{y})$ in the artificial genetic systems, let z_2^{IP} denote the objective function value $z_2(\boldsymbol{x},\boldsymbol{y})$ of the follower. To obtain the upper bound of $z_2(\boldsymbol{x},\boldsymbol{y})$, consider the following problem without the objective function of the leader:

$$\text{maximize } z_2(\boldsymbol{x},\boldsymbol{y}) = c_2\boldsymbol{x} + d_2\boldsymbol{y} \tag{3.42a}$$

$$\text{subject to } A\boldsymbol{x} + B\boldsymbol{y} \leqq \boldsymbol{b} \tag{3.42b}$$

$$\boldsymbol{x} \geqq \boldsymbol{0},\ \boldsymbol{y} \geqq \boldsymbol{0}. \tag{3.42c}$$

Let z_2^{worst} denote the optimal value of problem (3.42).

For a decision $\hat{\boldsymbol{x}}$ of the leader, the lower bound of $z_2(\hat{\boldsymbol{x}},\boldsymbol{y})$ can be calculated by solving the continuous relaxed problem:

$$\text{minimize } z_2(\hat{\boldsymbol{x}},\boldsymbol{y}) = c_2\hat{\boldsymbol{x}} + d_2\boldsymbol{y} \tag{3.43a}$$

$$\text{subject to } B\boldsymbol{y} \leqq \boldsymbol{b} - A\hat{\boldsymbol{x}} \tag{3.43b}$$

$$\boldsymbol{y} \geqq \boldsymbol{0}. \tag{3.43c}$$

Let $\bar{\boldsymbol{y}}$ denote an optimal solution to problem (3.43), and then the lower bound of $z_2(\hat{\boldsymbol{x}},\boldsymbol{y})$ is $z_2^{\mathrm{best}} \triangleq z_2(\hat{\boldsymbol{x}},\bar{\boldsymbol{y}})$. Then, the penalty for an irrational response is defined as:

$$p_2 = \exp\left\{-\frac{z_2^{\mathrm{IP}} - z_2^{\mathrm{best}}}{z_2^{\mathrm{worst}} - z_2^{\mathrm{best}}}\right\}. \tag{3.44}$$

Let $z_1^{\max}$ denote the optimal value of the following problem which is a continuous relaxed problem maximizing the objective function of the leader:

$$\text{maximize } z_1(\boldsymbol{x},\boldsymbol{y}) = \boldsymbol{c}_1\boldsymbol{x} + \boldsymbol{d}_1\boldsymbol{y} \tag{3.45a}$$

$$\text{subject to } A\boldsymbol{x} + B\boldsymbol{y} \leqq \boldsymbol{b} \tag{3.45b}$$

$$\boldsymbol{x} \geqq \boldsymbol{0},\ \boldsymbol{y} \geqq \boldsymbol{0}. \tag{3.45c}$$

The fitness of an individual with the objective function value z_1^{IP} is defined by using $z_1^{\max}$ and the two penalties p_1 and p_2 as follows:

$$f = p_1 p_2 \exp\{z_1^{\max} - z_1^{\mathrm{IP}}\}. \tag{3.46}$$

To save the computational time, the penalty p_2 for irrational responses is not computed for individuals which violate the constraints to the predefined extent, and then the fitness function is simply specified by

$$f = p_1. \tag{3.47}$$

Reproduction and genetic operations

As a reproduction operator, the elitist roulette wheel selection is adopted in the artificial genetic systems. The simple one-point crossover and bit-reverse type mutation are applied to reproduced individuals.

Algorithm

An procedure based on genetic algorithms for obtaining Stackelberg solutions to two-level integer programming problems is summarized as follows:

Step 1 After estimating the length of binary strings from the upper bounds of the decision variables, generate *pop_size* individuals at random, and then the initial population is formed.
Step 2 Calculate the fitness of each individual by (3.40)–(3.46), and reproduce the next generation in accordance with the elitist roulette wheel selection.
Step 3 With probabilities determined in advance, operations of the crossover and the mutation are performed.
Step 4 If the number of generation is over the maximum generation number *max_gen*, the algorithm stops, and an individual with the maximal fitness value is decoded as an approximate Stackelberg solution to the problem. Otherwise, return to Step 2.

Computation of a response of the follower

In the above mentioned algorithm, while the both decision variables $\boldsymbol{x}$ and $\boldsymbol{y}$ of the leader and the follower are combined and coded as a single binary string, substrings for $\boldsymbol{x}$ and $\boldsymbol{y}$ are generated independently. Because $\boldsymbol{y}$ must be a rational response with respect to $\boldsymbol{x}$ in Stackelberg solutions, for finding the Stackelberg solution efficiently, the random generation of the substring for $\boldsymbol{y}$ is not appropriate. As shown in definition (3.6), for a given decision $\hat{\boldsymbol{x}}$ of the leader, the rational response $\boldsymbol{y} \in R(\hat{\boldsymbol{x}})$ of the follower is an optimal solution to the following problem:

$$\text{minimize } z_2(\hat{\boldsymbol{x}},\boldsymbol{y}) = c_2\hat{\boldsymbol{x}} + d_2\boldsymbol{y} \tag{3.48a}$$
$$\text{subject to } B\boldsymbol{y} \leqq \boldsymbol{b} - A\hat{\boldsymbol{x}} \tag{3.48b}$$
$$\boldsymbol{y} \in \mathbb{Z}_+^{n_2}. \tag{3.48c}$$

On the other hand, in the above mentioned algorithm, the continuous relaxed problem (3.43) corresponding to problem (3.48) is solved. Therefore, to obtain a response of the follower which is supposed to be close to the exact integer rational response, we employ a round solution from an optimal solution to the continuous relaxed problem (3.43) for a given decision $\hat{\boldsymbol{x}}$ of the leader. By doing so, it is expected that the search of the Stackelberg solution based on the genetic algorithms proceeds efficiently.

To utilize the round solution efficiently, a new parameter *max_round* is introduced, and it means the number of maximal generations in which the round solution is employed as a response of the follower. Obviously, the parameter *max_round* is specified to be smaller than the maximum generation number *max_gen*. Until the generation number is smaller than *max_round*, the round solution is employed as a response of the follower, and the genetic operators are performed only to the substring corresponding to the decision $\boldsymbol{x}$ of the leader. Then, the fitness function f is defined as:

$$f = p_1 \exp\{z_1^{\max} - z_1^{\mathrm{IP}}\}. \tag{3.49}$$

After the generation number is over *max_round*, the genetic operators are performed to whole binary strings representing both the decisions $\boldsymbol{x}$ and $\boldsymbol{y}$ of the leader and the follower.

3.4.2 Computational Experiments

To demonstrate the feasibility and efficiency of the methods based on genetic algorithms shown in the previous subsection, computational experiments are carried out, and the methods based on genetic algorithms and the Moore and Bard method (1990) based on the branch-and-bound techniques are compared.

Two-level integer programming problems used in the computational experiments are prepared as follows. Each entry of the coefficient vectors $\boldsymbol{c}_1$, $\boldsymbol{c}_2$, $\boldsymbol{d}_1$ and $\boldsymbol{d}_2$ and the coefficient matrices A and B of the problems are randomly chosen from the set of integers in the interval $[-50, 50]$. Each entry of the right-hand side coefficient vector $\boldsymbol{b}$ of the constraints is determined by

$$b_i = (1-\gamma)\left(\sum_{j^a \in J_i^{a-}} \beta_{j^a} a_{ij^a} + \sum_{j^b \in J_i^{b-}} \beta_{j^b} b_{ij^b}\right) + \gamma\left(\sum_{j^a \in J_i^{a+}} \beta_{j^a} a_{ij^a} + \sum_{j^b \in J_i^{b+}} \beta_{j^b} b_{ij^b}\right), \tag{3.50}$$

where γ is a parameter which controls tightness of the constraints; β_{j^a} and β_{j^b} are the upper bounds of x_j and y_j, respectively; J_i^{a+}, J_i^{b+}, J_i^{a-} and J_i^{b-} are defined by $J_i^{a+} = \{j \mid a_{ij} \geqq 0, 1 \leqq j \leqq n_1\}$, $J_i^{b+} = \{j \mid b_{ij} \geqq 0, 1 \leqq j \leqq n_2\}$, $J_i^{a-} = \{j \mid a_{ij} < 0, 1 \leqq j \leqq n_1\}$, and $J_i^{b-} = \{j \mid b_{ij} < 0, 1 \leqq j \leqq n_2\}$. By determining coefficients as mentioned above, two-level integer programming problems with from 20 to 200 decision variables are generated. For each size of the problem, three problems are made by setting the parameter γ at 0.5, 0.7 and 0.9.

Parameters of the genetic algorithms are specified as follows: the population size $pop\_size = 100$, the maximum generation number $max\_gen = 1000$, the probability of crossover $p_c = 0.6$, the probability of mutation $p_m = 0.05$, the generation gap $gen\_gap = 0.5$. In particular, when the round solution is employed as a response of the follower, the maximum generation number is set at $max\_gen = 1100$ and the maximal number of generations for the round solution procedure is set at $max\_round = 1000$. If an individual satisfies 70% of the constraints, the fitness of the individual is calculated by the fitness function (3.46). Otherwise, the fitness is calculated by the simplified fitness function (3.47). In the computational experiment, the number of runs for the methods based on genetic algorithms is ten for each problem, i.e., each problem is solved ten times. The computational experiment is carried out on a PC with AMD k6-2 processor 450MHz, and computer programs are developed by using Microsoft Visual C++ 6.0.

To examine the effectiveness of introduction of the round solutions, the problems are solved by two methods with/without use of the round solutions. Moreover, the Moore and Bard method (1990) and the methods based on genetic algorithms are compared on the computational time (CPU time) and the accuracy of obtained solutions. In the Moore and Bard method, two-level linear programming problems continuously relaxed are solved iteratively, and for this end, the Hansen, Jaumard and Savard method (1992) is employed in this computational experiment.

Computational times of the three methods are shown in Table 3.8 and Figure 3.12, in which the method based on genetic algorithms with/without use of the round solutions are denoted by GA I and GA II, respectively, and the Moore and Bard method is denoted by M & B. The computational times in the columns of GA I and GA II are the mean values of ten runs of three problems.

Table 3.8 Computational time [sec.].

variables	M & B	GA I	GA II
20	496.74	2716.75	2981.27
40	1601.38	2834.69	3084.75
60	2893.21	3097.59	3299.67
80	5473.62	3233.80	3304.73
100	12762.07	3283.34	3407.21
120	19182.79	3407.27	3757.43
140	40633.91	3605.49	4208.62
160	—	3792.60	4456.82
180	—	3906.17	4589.06
200	—	4058.06	4692.79

From Table 3.8 and Figure 3.12, it is clear that, as to the computational time, GA I and GA II are superior to the Moore and Bard method for problems with 80 variables and over. The Moore and Bard method cannot find the exact Stackelberg

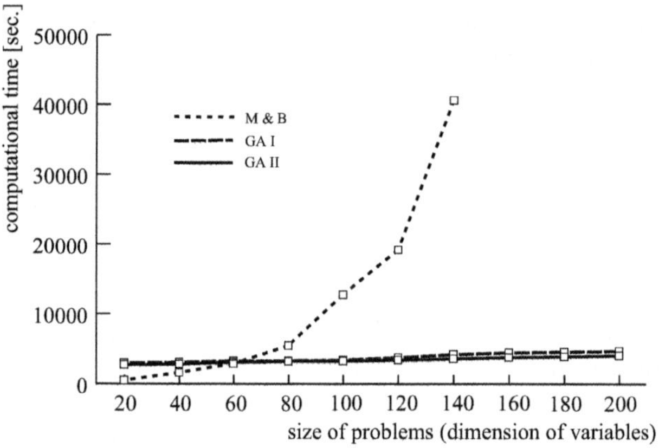

Fig. 3.12 Comparison of computational times.

solutions of problems with 160 variables and over in 12 hours at which we specify the upper bound of computational time for the experiments.

In Table 3.9, we show the ratios of the objective function value obtained by GA I and GA II to the objective function value of the exact Stackelberg solution computed by the Moore and Bard method for the problems with from 20 to 140 variables. Because, for all the problems, the objective function values are negative, we give the ratios as quotients which are the objective function values obtained by GA I and GA II divided by the objective function computed by the Moore and Bard method. Because the exact Stackelberg solutions to the problems with 160 and over cannot be found, in Table 3.9, we give the ratios with respect to incumbent solutions obtained by the Moore and Bard method at the upper bound of computational time.

Table 3.9 Accuracy of solutions [%].

variables	GA I	GA II
20	99.23	99.46
40	98.89	98.71
60	93.26	98.32
80	92.21	97.95
100	92.35	98.24
120	91.30	97.83
140	90.95	97.77
160	*86.29	*98.86
180	*95.53	*137.72
200	*101.30	*144.78

As seen in Table 3.9, while the method based on genetic algorithms without use of the round solutions, GA I, gradually reduces the accuracy of solutions from the problems with 60 variables, the method using the round solutions, GA II, finds better solutions with about 98 % of accuracy even for the problems with 140 variables. For the problem with 160 variables and over, it follows that we compare the solutions obtained by GA I and GA II with the incumbent solutions obtained by the Moore and Bard method at the upper bound of computational time and, to discriminate the ratios to the incumbent solutions from the ratios to the exact Stackelberg solutions, we mark asterisks * at the ratios to the incumbent solutions. The ratios of GA II for the problems with 180 and 200 variables are larger than 100%, and therefore GA II finds solutions which are more accurate than the incumbent solutions of the Moore and Bard method in a shorter computational time. Even GA I can find better solutions for the problems with 200 variables.

3.5 Multiobjective two-level linear programming

In this section, we consider a multiobjective two-level linear programming problem (Nishizaki and Sakawa, 1999). When the leader or the follower has multiple objectives, in order to employ traditional two-level programming techniques, the multiple objectives should be aggregated, and a single objective function or a scalar-valued utility function is required to be identified. Multiattribute utility analysis (Keeney and Raiffa, 1976) provides a heuristic device for constructing the scalar-valued utility function reflecting the preference of the decision maker.

In noncooperative environments, it is even more difficult for the leader to estimate or assess a single objective function or a utility function of the follower. From this observation, we consider multiobjective two-level linear programming problems.

It is often assumed that the set of rational responses of the follower is a singleton in a conventional two-level linear programming problem. In multiobjective environments, however, the assumption is not realistic, and consequently the leader takes into consideration multiple responses or an infinite number of responses of the followers with respect to the decision specified by the leader.

In this section, we take a prescriptive approach to the multiobjective two-level linear programming problem from a viewpoint of the leader. We assume that the leader has some subjective anticipation or belief. Such anticipation or belief can be considered three cases: the leader has the optimistic anticipation; the leader has the pessimistic anticipation; and the leader knows the preference of the follower. Stackelberg solutions for three cases are defined, and computational methods for obtaining the solutions are given.

A multiobjective two-level linear programming problem for obtaining a Stackelberg solution is expressed as

$$\underset{x}{\text{minimize}}\ z_{11}(\boldsymbol{x},\boldsymbol{y}) = \boldsymbol{c}_{11}\boldsymbol{x} + \boldsymbol{d}_{11}\boldsymbol{y} \tag{3.51a}$$

$$\cdots\cdots\cdots\cdots\cdots$$

$$\underset{x}{\text{minimize}}\ z_{1k_1}(\boldsymbol{x},\boldsymbol{y}) = \boldsymbol{c}_{1k_1}\boldsymbol{x} + \boldsymbol{d}_{1k_1}\boldsymbol{y} \tag{3.51b}$$

where $\boldsymbol{y}$ solves

$$\underset{y}{\text{minimize}}\ z_{21}(\boldsymbol{x},\boldsymbol{y}) = \boldsymbol{c}_{21}\boldsymbol{x} + \boldsymbol{d}_{21}\boldsymbol{y} \tag{3.51c}$$

$$\cdots\cdots\cdots\cdots\cdots$$

$$\underset{y}{\text{minimize}}\ z_{2k_2}(\boldsymbol{x},\boldsymbol{y}) = \boldsymbol{c}_{2k_2}\boldsymbol{x} + \boldsymbol{d}_{2k_2}\boldsymbol{y} \tag{3.51d}$$

$$\text{subject to}\ A\boldsymbol{x} + B\boldsymbol{y} \leqq \boldsymbol{b} \tag{3.51e}$$

$$\boldsymbol{x} \geqq \boldsymbol{0},\ \boldsymbol{y} \geqq \boldsymbol{0}, \tag{3.51f}$$

where $\boldsymbol{x}$ and $\boldsymbol{y}$ are the leader's n_1-dimensional and the follower's n_2-dimensional decision variable column vectors, respectively; $\boldsymbol{c}_{ij}$, $\boldsymbol{d}_{ij}$, $i = 1,2,\ j = 1,\dots,k_i$ are n_1-dimensional and n_2-dimensional coefficient row vectors, respectively; $\boldsymbol{b}$ is an m-dimensional constant column vector; A and B are $m \times n_1$ and $m \times n_2$ coefficient matrices, respectively; $z_{1j}(\boldsymbol{x},\boldsymbol{y})$, $j = 1,\dots,k_1$ and $z_{2j}(\boldsymbol{x},\boldsymbol{y})$, $j = 1,\dots,k_2$ are the leader's and the follower's linear objective functions, respectively.

In the case where the follower has a single objective, it is often assumed that the set of rational responses $R(\boldsymbol{x})$ for each of all the decisions $\boldsymbol{x}$ of the leader is a singleton, and most studies have aimed to develop computational methods for obtaining Stackelberg solution to such problems. However, when we consider a two-level linear programming problem with multiple objectives of the follower, the above assumption becomes meaningless, and hence we have to take account of the set of rational responses $R(\boldsymbol{x})$ which is not a singleton. It seems reasonable to suppose that the least requirement to be imposed on the concept corresponding to the rational responses $R(\boldsymbol{x})$ is that of the Pareto optimality.

Therefore, we introduce the following set of Pareto optimal responses $P(\boldsymbol{x})$ as a substitute for the set of rational responses $R(\boldsymbol{x})$ in a single objective case (3.6):

(iv) The set of Pareto optimal responses of the follower for $\boldsymbol{x}$ specified by the leader:

$$P(\boldsymbol{x}) = \{\boldsymbol{y} \in S(\boldsymbol{x}) \mid \text{there does not exist another}\ \boldsymbol{y}' \in S(\boldsymbol{x}) \\ \text{such that}\ \boldsymbol{z}_2(\boldsymbol{x},\boldsymbol{y}') \leq \boldsymbol{z}_2(\boldsymbol{x},\boldsymbol{y}).\}, \tag{3.52}$$

where $\boldsymbol{z}_2(\cdot,\cdot) = (z_{21}(\cdot,\cdot),\dots,z_{2k_2}(\cdot,\cdot))^T$ is the vector of the objective functions of the follower.

For a decision $\boldsymbol{x}$ specified by the leader, the leader can presume only that the response $\boldsymbol{y}$ of the follower belongs to the set of Pareto optimal responses $P(\boldsymbol{x})$.

If scalar-valued utility functions U_1 and U_2 of the leader and the follower are identified explicitly, problem (3.51) is expressed as

$$\underset{\boldsymbol{x}}{\text{maximize}} \quad U_1(z_{11}(\boldsymbol{x},\boldsymbol{y}),\ldots,z_{1k_1}(\boldsymbol{x},\boldsymbol{y})) \tag{3.53a}$$

$$\text{where } \boldsymbol{y} \text{ solves}$$

$$\underset{\boldsymbol{y}}{\text{maximize}} \quad U_2(z_{21}(\boldsymbol{x},\boldsymbol{y}),\ldots,z_{2k_2}(\boldsymbol{x},\boldsymbol{y})) \tag{3.53b}$$

$$\text{subject to} \quad A\boldsymbol{x}+B\boldsymbol{y} \leqq \boldsymbol{b} \tag{3.53c}$$

$$\boldsymbol{x} \geqq \boldsymbol{0},\ \boldsymbol{y} \geqq \boldsymbol{0}. \tag{3.53d}$$

For identifying such utility functions, multiattribute utility analysis provides a heuristic device when a decision maker assesses his own utility function according to his own preference and some independence assumptions on preference of the decision maker are satisfied (Keeney and Raiffa, 1976). In general, even if the leader can identify the utility function U_1 of himself, it is more difficult for the leader to know or to estimate the utility function U_2 of the follower.

As done in a single-objective two-level linear programming problem, after the leader has made a decision $\boldsymbol{x}$, the follower chooses a decision $\boldsymbol{y}$ which belongs to the set of Pareto optimal responses $P(\boldsymbol{x})$. In general, the leader does not know which response is chosen from the set of Pareto optimal responses by the follower. However, with some subjective anticipation or belief, we can provide reasonable decisions of the leader. Such anticipation or belief can be considered in three cases: (i) the leader anticipates that the follower will take a decision desirable for the leader (optimistic anticipation); (ii) the leader anticipates that the follower will take a decision undesirable for the leader (pessimistic anticipation); and (iii) the leader knows the preference of the follower. When the leader does not know the preference of the follower, the leader would make a decision by consulting outcomes yielded from the optimistic anticipation and the pessimistic anticipation. In contrast, when the leader and the follower have been often confronted with the decision making problem represented by (3.51), and they have experience in it, it seems to be natural that the leader knows or can estimate the preference of the follower.

3.5.1 Computational methods

We formulate multiobjective two-level linear programming problems with the anticipation or belief of the leader on condition that, for a decision $\boldsymbol{x}$ of the leader, the follower takes a decision $\boldsymbol{y}$ in the set of Pareto optimal responses $P(\boldsymbol{x})$.

Stackelberg solutions based on optimistic anticipation

After the leader has chosen a decision $\hat{\boldsymbol{x}}$, the follower chooses a decision $\boldsymbol{y}$ with full knowledge of the decision $\hat{\boldsymbol{x}}$ of the leader by solving the following multiobjective linear programming problem:

$$\underset{\boldsymbol{y}}{\text{minimize}}\ z_{21}(\hat{\boldsymbol{x}},\boldsymbol{y}) = \boldsymbol{d}_{21}\boldsymbol{y} + \boldsymbol{c}_{21}\hat{\boldsymbol{x}} \tag{3.54a}$$

$$\cdots\cdots\cdots\cdots\cdots$$

$$\underset{\boldsymbol{y}}{\text{minimize}}\ z_{2k_2}(\hat{\boldsymbol{x}},\boldsymbol{y}) = \boldsymbol{d}_{2k_2}\boldsymbol{y} + \boldsymbol{c}_{2k_2}\hat{\boldsymbol{x}} \tag{3.54b}$$

$$\text{subject to}\ B\boldsymbol{y} \leqq \boldsymbol{b} - A\hat{\boldsymbol{x}} \tag{3.54c}$$

$$\boldsymbol{y} \geqq \boldsymbol{0}. \tag{3.54d}$$

Let $P(\hat{\boldsymbol{x}})$ denote the set of Pareto optimal responses (solutions) of the multiobjective linear programming problem (3.54).

Hereafter, we call a Stackelberg solution based on the optimistic anticipation the optimistic Stackelberg solution simply. The optimistic Stackelberg solution is an optimal solution to the following problem if the utility function U_1 of the leader can be identified:

$$\underset{\boldsymbol{x}}{\text{maximize}}\ \max_{\boldsymbol{y}}\ U_1(z_{11}(\boldsymbol{x},\boldsymbol{y}),\ldots,z_{1k_1}(\boldsymbol{x},\boldsymbol{y})) \tag{3.55a}$$

$$\text{subject to}\ \boldsymbol{y} \in P(\boldsymbol{x}) \tag{3.55b}$$

$$A\boldsymbol{x} + B\boldsymbol{y} \leqq \boldsymbol{b} \tag{3.55c}$$

$$\boldsymbol{x} \geqq \boldsymbol{0},\ \boldsymbol{y} \geqq \boldsymbol{0}. \tag{3.55d}$$

Problem (3.55) expresses a situation where the leader chooses a decision $\boldsymbol{x}$ so as to maximize the leader's utility function $U_1(z_{11}(\boldsymbol{x},\boldsymbol{y}),\ldots,z_{1k_1}(\boldsymbol{x},\boldsymbol{y}))$ on the assumption (anticipation) that, for the given decision $\boldsymbol{x}$ of the leader, the follower takes a decision $\boldsymbol{y}$ in the set of Pareto optimal responses $P(\boldsymbol{x})$ such that the leader's utility function $U_1(z_{11}(\boldsymbol{x},\boldsymbol{y}),\ldots,z_{1k_1}(\boldsymbol{x},\boldsymbol{y}))$ is maximized.

It can be observed that, in most real-world situations, identifying the utility function U_1 even by the leader in person is difficult; even if possible, the function might be a nonlinear function. Because of the difficulty of identifying the utility function and the difficulty in computing solutions, it would be appropriate to use an interactive method in which a decision maker can learn and realize a local preference around solutions derived by solving some problem at each iteration. Here, we employ the reference point method by Wierzbicki (1979). Let an achievement function be

$$\max_{i=1,\ldots,k_1}\{z_{1i}(\boldsymbol{x},\boldsymbol{y}) - \bar{z}_{1i}\} + \rho\sum_{i=1}^{k_1}(z_{1i}(\boldsymbol{x},\boldsymbol{y}) - \bar{z}_{1i}), \tag{3.56}$$

where $(\bar{z}_{11},\ldots,\bar{z}_{1k_1})$ is a reference point specified by the leader, and ρ is a small positive scalar value. The function (3.56) is called the augmented Tchebyshev scalarizing function, which has some desirable properties.

Hence, the optimistic Stackelberg solutions can be obtained by interactively solving the following problem and updating the reference points:

$$\underset{x}{\text{minimize}}\ \underset{y}{\min}\left\{\max_{i=1,\dots,k_1}\{z_{1i}(\boldsymbol{x},\boldsymbol{y})-\bar{z}_{1i}\}+\rho\sum_{i=1}^{k_1}(z_{1i}(\boldsymbol{x},\boldsymbol{y})-\bar{z}_{1i})\right\} \quad (3.57a)$$

$$\text{subject to}\ \ \boldsymbol{y}\in P(\boldsymbol{x}) \quad (3.57b)$$

$$A\boldsymbol{x}+B\boldsymbol{y}\leqq\boldsymbol{b} \quad (3.57c)$$

$$\boldsymbol{x}\geqq\boldsymbol{0},\ \boldsymbol{y}\geqq\boldsymbol{0}. \quad (3.57d)$$

Problem (3.57) can be transformed into the following equivalent problem by introducing an auxiliary variable σ:

$$\underset{x,y,\sigma}{\text{minimize}}\ \ \sigma+\rho\sum_{i=1}^{k_1}(z_{1i}(\boldsymbol{x},\boldsymbol{y})-\bar{z}_{1i}) \quad (3.58a)$$

$$\text{subject to}\ \ \boldsymbol{y}\in P(\boldsymbol{x}) \quad (3.58b)$$

$$z_{1i}(\boldsymbol{x},\boldsymbol{y})-\bar{z}_{1i}\leqq\sigma,\ i=1,\dots,k_1 \quad (3.58c)$$

$$A\boldsymbol{x}+B\boldsymbol{y}\leqq\boldsymbol{b} \quad (3.58d)$$

$$\boldsymbol{x}\geqq\boldsymbol{0},\ \boldsymbol{y}\geqq\boldsymbol{0}. \quad (3.58e)$$

To solve problem (3.58), we apply the idea of the kth best method (Bialas and Karwan, 1984). We first start out by solving the following problem without the constraint $\boldsymbol{y}\in P(\boldsymbol{x})$ of the Pareto optimality:

$$\underset{x,y,\sigma}{\text{minimize}}\ \ \sigma+\rho\sum_{i=1}^{k_1}(z_{1i}(\boldsymbol{x},\boldsymbol{y})-\bar{z}_{1i}) \quad (3.59a)$$

$$\text{subject to}\ \ z_{1i}(\boldsymbol{x},\boldsymbol{y})-\bar{z}_{1i}\leqq\sigma,\ i=1,\dots,k_1 \quad (3.59b)$$

$$A\boldsymbol{x}+B\boldsymbol{y}\leqq\boldsymbol{b} \quad (3.59c)$$

$$\boldsymbol{x}\geqq\boldsymbol{0},\ \boldsymbol{y}\geqq\boldsymbol{0}. \quad (3.59d)$$

Let $(\hat{\boldsymbol{x}}^1,\hat{\boldsymbol{y}}^1,\hat{\sigma}^1)$ denote an optimal solution to problem (3.59), and let $(\hat{\boldsymbol{x}}^2,\hat{\boldsymbol{y}}^2,\hat{\sigma}^2),\dots,$ $(\hat{\boldsymbol{x}}^N,\hat{\boldsymbol{y}}^N,\hat{\sigma}^N)$ be the rest of $N-1$ basic feasible solutions such that

$$\sigma^j+\rho\sum_{i=1}^{k_1}(z_{1i}(\boldsymbol{x}^j,\boldsymbol{y}^j)-\bar{z}_{1i})\leqq\sigma^{j+1}+\rho\sum_{i=1}^{k_1}(z_{1i}(\boldsymbol{x}^{j+1},\boldsymbol{y}^{j+1})-\bar{z}_{1i}),$$
$$j=1,\dots,N-1.$$

To verify if the response $\hat{\boldsymbol{y}}^j$ of the follower belongs to the set of Pareto optimal solutions with respect to the decision $\hat{\boldsymbol{x}}^j$ of the leader, solve the linear programming problem:

$$\underset{\boldsymbol{y},\boldsymbol{\varepsilon}}{\text{minimize}} \quad v = -\sum_{i=1}^{k_2} \varepsilon_i \tag{3.60a}$$

$$\text{subject to} \quad \boldsymbol{d}_{2i}\boldsymbol{y} + \varepsilon_i = \boldsymbol{d}_{2i}\hat{\boldsymbol{y}}^j,\ i = 1,\ldots,k_2 \tag{3.60b}$$

$$\varepsilon_i \geqq 0,\ i = 1,\ldots,k_2 \tag{3.60c}$$

$$B\boldsymbol{y} \leqq \boldsymbol{b} - A\hat{\boldsymbol{x}}^j \tag{3.60d}$$

$$\boldsymbol{y} \geqq \boldsymbol{0}, \tag{3.60e}$$

where $\boldsymbol{\varepsilon} = (\varepsilon_1,\ldots,\varepsilon_{k_2})^T$. If an optimal value is equal to 0, i.e., $v = 0$, the response $\hat{\boldsymbol{y}}^j$ satisfies Pareto optimality for the decision $\hat{\boldsymbol{x}}^j$. Thus, solving problem (3.58) is equivalent to finding the minimal index j such that $v = 0$. Starting from the point $(\hat{\boldsymbol{x}}^1, \hat{\boldsymbol{y}}^1, \hat{\sigma}^1)$, we examine adjacent points in turn. A procedure for obtaining the optimistic Stackelberg solution is summarized as follows:

Step 0 The leader specifies an initial reference point $(\bar{z}_{11},\ldots,\bar{z}_{1k_1})$.

Step 1 Let $j := 1$. Solve the linear programming problem (3.59), and let $(\hat{\boldsymbol{x}}^1, \hat{\boldsymbol{y}}^1, \hat{\sigma}^1)$ denote an optimal solution to problem (3.59). Set $W := \{(\hat{\boldsymbol{x}}^1, \hat{\boldsymbol{y}}^1, \hat{\sigma}^1)\}$ and $T := \emptyset$.

Step 2 Solve the linear programming problem (3.60). If an optimal value is equal to 0, i.e., $v = 0$, then $(\hat{\boldsymbol{x}}^j, \hat{\boldsymbol{y}}^j)$ is the optimistic Stackelberg solution with respect to the reference point $(\bar{z}_{11},\ldots,\bar{z}_{1k_1})$. If the leader is satisfied with the solution, the interactive procedure stops. Otherwise, update the reference point $(\bar{z}_{11},\ldots,\bar{z}_{1k_1})$ and return to Step 1. If an optimal value to problem (3.60) is not equal to 0, i.e., $v \neq 0$, then go to Step 3.

Step 3 Let W^j be a set of extreme points of problem (3.59) which is adjacent to $(\hat{\boldsymbol{x}}^j, \hat{\boldsymbol{y}}^j, \hat{\sigma}^j)$ and satisfies

$$\sigma^j + \rho\sum_{i=1}^{k_1}(z_{1i}(\boldsymbol{x}^j,\boldsymbol{y}^j) - \bar{z}_{1i}) \leqq \sigma + \rho\sum_{i=1}^{k_1}(z_{1i}(\boldsymbol{x},\boldsymbol{y}) - \bar{z}_{1i}),$$

and let $T := T \cup (\hat{\boldsymbol{x}}^j, \hat{\boldsymbol{y}}^j, \hat{\sigma}^j)$ and $W := (W \cup W^j) \backslash T$.

Step 4 Let $j := j+1$. Choose an extreme point $(\hat{\boldsymbol{x}}^j, \hat{\boldsymbol{y}}^j, \hat{\sigma}^j)$ such that

$$\sigma^j + \rho\sum_{i=1}^{k_1}(z_{1i}(\boldsymbol{x}^j,\boldsymbol{y}^j) - \bar{z}_{1i}) = \min_{(\boldsymbol{x},\boldsymbol{y},\sigma)\in W} \{\sigma + \rho\sum_{i=1}^{k_1}(z_{1i}(\boldsymbol{x},\boldsymbol{y}) - \bar{z}_{1i})\},$$

and return to Step 2.

Stackelberg solutions based on pessimistic anticipation

Hereafter, we call a Stackelberg solutions based on the pessimistic anticipation as the pessimistic Stackelberg solutions. Let $(\bar{z}_{11},\ldots,\bar{z}_{1k_1})$ be the reference point, and let equation (3.56) be employed as an achievement function. Then, the pessimistic Stackelberg solution can be obtain by iteratively solving the following problem and updating the reference point:

$$\underset{x}{\text{minimize}}\ \underset{y}{\max}\ \left\{ \max_{i=1,\dots,k_1} \{z_{1i}(\boldsymbol{x},\boldsymbol{y}) - \bar{z}_{1i}\} + \rho \sum_{i=1}^{k_1} (z_{1i}(\boldsymbol{x},\boldsymbol{y}) - \bar{z}_{1i}) \right\} \tag{3.61a}$$

$$\text{subject to } \boldsymbol{y} \in P(\boldsymbol{x}) \tag{3.61b}$$

$$A\boldsymbol{x} + B\boldsymbol{y} \leqq \boldsymbol{b} \tag{3.61c}$$

$$\boldsymbol{x} \geqq \boldsymbol{0},\ \boldsymbol{y} \geqq \boldsymbol{0}. \tag{3.61d}$$

Problem (3.61) can be transformed into the following equivalent problem by introducing an auxiliary variable σ:

$$\underset{x}{\text{minimize}}\ \underset{y,\sigma}{\max}\ \sigma + \rho \sum_{i=1}^{k_1} (z_{1i}(\boldsymbol{x},\boldsymbol{y}) - \bar{z}_{1i}) \tag{3.62a}$$

$$\text{subject to } \boldsymbol{y} \in P(\boldsymbol{x}) \tag{3.62b}$$

$$z_{1i}(\boldsymbol{x},\boldsymbol{y}) - \bar{z}_{1i} \leqq \sigma,\ i = 1,\dots,k_1 \tag{3.62c}$$

$$A\boldsymbol{x} + B\boldsymbol{y} \leqq \boldsymbol{b} \tag{3.62d}$$

$$\boldsymbol{x} \geqq \boldsymbol{0},\ \boldsymbol{y} \geqq \boldsymbol{0}. \tag{3.62e}$$

Suppose that the leader makes a decision $\hat{\boldsymbol{x}}^j$ and anticipates that the follower chooses a decision undesirable for the leader in the set of Pareto optimal responses. Namely, we assume that the leader judges that the follower takes an optimal solution $y(\hat{\boldsymbol{x}}^j)$ to the following problem as a response to the decision $\hat{\boldsymbol{x}}^j$:

$$\underset{y,\sigma}{\text{maximize}}\ \sigma + \rho \sum_{i=1}^{k_1} (z_{1i}(\hat{\boldsymbol{x}}^j,\boldsymbol{y}) - \bar{z}_{1i}) \tag{3.63a}$$

$$\text{subject to } \boldsymbol{y} \in P(\hat{\boldsymbol{x}}^j) \tag{3.63b}$$

$$z_{1i}(\hat{\boldsymbol{x}}^j,\boldsymbol{y}) - \bar{z}_{1i} \leqq \sigma,\ i = 1,\dots,k_1 \tag{3.63c}$$

$$B\boldsymbol{y} \leqq \boldsymbol{b} - A\hat{\boldsymbol{x}}^j \tag{3.63d}$$

$$\boldsymbol{y} \geqq \boldsymbol{0}. \tag{3.63e}$$

Turning our attention to the similarity between problems (3.63) and (3.58), we find that problem (3.63) can be solved by a procedure similar to that for solving problem (3.58). Let $(\hat{\boldsymbol{y}}^l(\hat{\boldsymbol{x}}^j), \sigma^l(\hat{\boldsymbol{x}}^j))$ be a basic feasible solution which yields the lth largest value of the objective function of the following linear programming problem:

$$\underset{y,\sigma}{\text{maximize}}\ \sigma + \rho \sum_{i=1}^{k_1} (z_{1i}(\hat{\boldsymbol{x}}^j,\boldsymbol{y}) - \bar{z}_{1i}) \tag{3.64a}$$

$$\text{subject to } z_{1i}(\hat{\boldsymbol{x}}^j,\boldsymbol{y}) - \bar{z}_{1i} \leqq \sigma,\ i = 1,\dots,k_1 \tag{3.64b}$$

$$B\boldsymbol{y} \leqq \boldsymbol{b} - A\hat{\boldsymbol{x}}^j \tag{3.64c}$$

$$\boldsymbol{y} \geqq \boldsymbol{0}. \tag{3.64d}$$

To examine if the response $\hat{\boldsymbol{y}}^l(\hat{\boldsymbol{x}}^j)$ of the follower belongs to the set of Pareto optimal responses $P(\hat{\boldsymbol{x}}^j)$ of problem (3.54), we solve the following linear programming problem:

$$\underset{\boldsymbol{y},\boldsymbol{\varepsilon}}{\text{minimize}} \ w = -\sum_{i=1}^{k_2} \varepsilon_i \tag{3.65a}$$

$$\text{subject to } \boldsymbol{d}_{2i}\boldsymbol{y} + \varepsilon_i = \boldsymbol{d}_{2i}\hat{\boldsymbol{y}}^l(\hat{\boldsymbol{x}}^j), \ i = 1,\ldots,k_2 \tag{3.65b}$$

$$\varepsilon_i \geqq 0, \ i = 1,\ldots,k_2 \tag{3.65c}$$

$$B\boldsymbol{y} \leqq \boldsymbol{b} - A\hat{\boldsymbol{x}}^j \tag{3.65d}$$

$$\boldsymbol{y} \geqq \boldsymbol{0}. \tag{3.65e}$$

If an optimal value is equal to 0, i.e., $w = 0$, the response $\hat{\boldsymbol{y}}^l(\hat{\boldsymbol{x}}^j)$ satisfies the Pareto optimality for the decision $\hat{\boldsymbol{x}}^j$. Thus, solving problem (3.63) is equivalent to finding the minimal index l such that $w = 0$. Let l^* be such a minimal index. Then, the response $\hat{\boldsymbol{y}}^{l^*}(\hat{\boldsymbol{x}}^j)$ which is not desirable for the leader in the set of Pareto optimal responses is determined. Namely, solving problem (3.62) is equivalent to finding the minimal index j such that $\hat{\boldsymbol{y}}^j$ is equal to $\hat{\boldsymbol{y}}^{l^*}(\hat{\boldsymbol{x}}^j)$. First, pick up an extreme point of the following problem (3.66) in nondecreasing order of the value of the objective function (3.66a) from $(\hat{\boldsymbol{x}}^1,\hat{\boldsymbol{y}}^1)$ to $(\hat{\boldsymbol{x}}^N,\hat{\boldsymbol{y}}^N)$; these points satisfy $\hat{\sigma}^j + \rho\sum_{i=1}^{k_1}(z_{1i}(\hat{\boldsymbol{x}}^j,\hat{\boldsymbol{y}}^j) - \bar{z}_{1i}) \leqq \hat{\sigma}^{j+1} + \rho\sum_{i=1}^{k_1}(z_{1i}(\hat{\boldsymbol{x}}^{j+1},\hat{\boldsymbol{y}}^{j+1}) - \bar{z}_{1i})$, $j = 1,\ldots,N-1$:

$$\underset{\boldsymbol{x},\boldsymbol{y},\sigma}{\text{minimize}} \ \sigma + \rho\sum_{i=1}^{k_1}(z_{1i}(\boldsymbol{x},\boldsymbol{y}) - \bar{z}_{1i}) \tag{3.66a}$$

$$\text{subject to } z_{1i}(\boldsymbol{x},\boldsymbol{y}) - \bar{z}_{1i} \leqq \sigma, \ i = 1,\ldots,k_1 \tag{3.66b}$$

$$A\boldsymbol{x} + B\boldsymbol{y} \leqq \boldsymbol{b} \tag{3.66c}$$

$$\boldsymbol{x} \geqq \boldsymbol{0}, \ \boldsymbol{y} \geqq \boldsymbol{0}. \tag{3.66d}$$

Second, find $\hat{\boldsymbol{y}}^{l^*}(\hat{\boldsymbol{x}}^j)$ by the above mentioned procedure. Third, check whether $\hat{\boldsymbol{y}}^j$ is equal to $\hat{\boldsymbol{y}}^{l^*}(\hat{\boldsymbol{x}}^j)$. A procedure for obtaining the pessimistic Stackelberg solution is summarized as follows:

Step 0 The leader specifies an initial reference point $(\bar{z}_{11},\ldots,\bar{z}_{1k_1})$.

Step 1 Let $j := 1$. Solve the linear programming problem (3.66), and let $(\hat{\boldsymbol{x}}^1,\hat{\boldsymbol{y}}^1,\hat{\sigma}^1)$ denote an optimal solution to problem (3.66). Let $W^1 := \{(\hat{\boldsymbol{x}}^1,\hat{\boldsymbol{y}}^1,\hat{\sigma}^1)\}$ and $T^1 := \emptyset$.

Step 2 Let $l := 1$. Solve the linear programming problem (3.64), and let $\hat{\boldsymbol{y}}^1(\hat{\boldsymbol{x}}^j)$ denote an optimal solution to problem (3.64). Let $W^2 := \{(\hat{\boldsymbol{y}}^1(\hat{\boldsymbol{x}}^j),\hat{\sigma}^1)\}$ and $T^2 := \emptyset$.

Step 3 Solve the linear programming problem (3.65). If an optimal value is equal to 0, i.e., $w = 0$ and $\hat{\boldsymbol{y}}^j$ is equal to $\hat{\boldsymbol{y}}^{l^*}(\hat{\boldsymbol{x}}^j)$, then $(\hat{\boldsymbol{x}}^j,\hat{\boldsymbol{y}}^j)$ is the pessimistic Stackelberg solution with respect to the reference point $(\bar{z}_{11},\ldots,\bar{z}_{1k_1})$. If the leader is satisfied with the solution, the interactive procedure stops. Otherwise, update the reference point $(\bar{z}_{11},\ldots,\bar{z}_{1k_1})$, and return to Step 1. If an optimal value to problem

(3.65) is not equal to 0, i.e., $w \neq 0$, then go to Step 4. If $w = 0$ and $\hat{\boldsymbol{y}}^j \neq \hat{\boldsymbol{y}}^{l^*}(\hat{\boldsymbol{x}}^j)$, then go to Step 6.

Step 4 Let W^{2l} be a set of extreme points of problem (3.64) which is adjacent to $\hat{\boldsymbol{y}}^l(\hat{\boldsymbol{x}}^j)$ and satisfies

$$\hat{\sigma}^l + \rho \sum_{i=1}^{k_1} (z_{1i}(\hat{\boldsymbol{x}}^j, \hat{\boldsymbol{y}}^l(\hat{\boldsymbol{x}}^j)) - \bar{z}_{1i}) \geqq \sigma + \rho \sum_{i=1}^{k_1} (z_{1i}(\boldsymbol{x}, \boldsymbol{y}) - \bar{z}_{1i}),$$

and let $T^2 := T^2 \cup (\hat{\boldsymbol{y}}^l(\hat{\boldsymbol{x}}^j), \hat{\sigma}^l)$ and $W^2 := (W^2 \cup W^{2l}) \backslash T^2$.

Step 5 Let $l := l+1$. Choose an extreme point $(\hat{\boldsymbol{y}}^l(\hat{\boldsymbol{x}}^j), \hat{\sigma}^l)$ such that

$$\hat{\sigma}^l + \rho \sum_{i=1}^{k_1} (z_{1i}(\hat{\boldsymbol{x}}^j, \hat{\boldsymbol{y}}^l(\hat{\boldsymbol{x}}^j)) - \bar{z}_{1i}) = \max_{(\boldsymbol{y}, \sigma) \in W^2} \{\sigma + \rho \sum_{i=1}^{k_1} (z_{1i}(\boldsymbol{x}, \boldsymbol{y}) - \bar{z}_{1i})\},$$

and return to Step 3.

Step 6 Let W^{1j} be a set of extreme points of problem (3.66) which is adjacent to $(\hat{\boldsymbol{x}}^j, \hat{\boldsymbol{y}}^j, \hat{\sigma}^j)$ and satisfies

$$\hat{\sigma}^j + \rho \sum_{i=1}^{k_1} (z_{1i}(\hat{\boldsymbol{x}}^j, \hat{\boldsymbol{y}}^j) - \bar{z}_{1i}) \leqq \sigma + \rho \sum_{i=1}^{k_1} (z_{1i}(\boldsymbol{x}, \boldsymbol{y}) - \bar{z}_{1i}),$$

and let $T^1 := T^1 \cup (\hat{\boldsymbol{x}}^j, \hat{\boldsymbol{y}}^j, \hat{\sigma}^j)$ and $W^1 := (W^1 \cup W^{1j}) \backslash T^1$.

Step 7 Let $j := j+1$. Choose an extreme point $(\hat{\boldsymbol{x}}^j, \hat{\boldsymbol{y}}^j, \hat{\sigma}^j)$ such that

$$\hat{\sigma}^j + \rho \sum_{i=1}^{k_1} (z_{1i}(\hat{\boldsymbol{x}}^j, \hat{\boldsymbol{y}}^j) - \bar{z}_{1i}) = \min_{(\boldsymbol{x}, \boldsymbol{y}, \sigma) \in W^1} \{\sigma + \rho \sum_{i=1}^{k_1} (z_{1i}(\boldsymbol{x}, \boldsymbol{y}) - \bar{z}_{1i})\},$$

and return to Step 2.

Stackelberg solutions based on follower preference

We consider an algorithm based on an interactive method similar to the methods shown in the above because it can be expected that the decision maker derives the satisfactory solutions by learning and recognizing local preferences during an interactive process.

Suppose situations where the leader and the follower have been confronted often with the decision making problem represented by (3.51); in such a case, it is natural to consider that the leader knows or can estimate the preference of the follower. We assume that the leader can learn and recognize the local preference of the leader by an interactive process, but cannot learn and recognize that of the follower; therefore, the follower's reference point is estimated only once by the leader.

Let $(\bar{z}_{21}, \ldots, \bar{z}_{2k_2})$ denote the follower's reference point specified by the leader. Then, it follows that the follower makes a decision $\hat{\boldsymbol{y}}$ by solving the following problem with respect to a decision $\hat{\boldsymbol{x}}$ of the leader:

$$\underset{y}{\text{minimize}} \left\{ \max_{i=1,\dots,k_2} \{z_{2i}(\hat{\boldsymbol{x}},\boldsymbol{y}) - \bar{z}_{2i}\} + \rho \sum_{i=1}^{k_2} (z_{2i}(\hat{\boldsymbol{x}},\boldsymbol{y}) - \bar{z}_{2i}) \right\} \tag{3.67a}$$

$$\text{subject to } \boldsymbol{B}\boldsymbol{y} \leqq \boldsymbol{b} - \boldsymbol{A}\hat{\boldsymbol{x}} \tag{3.67b}$$

$$\boldsymbol{y} \geqq \boldsymbol{0}. \tag{3.67c}$$

Problem (3.67) can be transformed into the following equivalent problem by introducing an auxiliary variable η:

$$\underset{\boldsymbol{y},\eta}{\text{minimize}} \ \eta + \rho \sum_{i=1}^{k_2} (z_{2i}(\hat{\boldsymbol{x}},\boldsymbol{y}) - \bar{z}_{2i}) \tag{3.68a}$$

$$\text{subject to } z_{2i}(\hat{\boldsymbol{x}},\boldsymbol{y}) - \bar{z}_{2i} \leqq \eta,\ i = 1,\dots,k_2 \tag{3.68b}$$

$$\boldsymbol{B}\boldsymbol{y} \leqq \boldsymbol{b} - \boldsymbol{A}\hat{\boldsymbol{x}} \tag{3.68c}$$

$$\boldsymbol{y} \geqq \boldsymbol{0}. \tag{3.68d}$$

Given a decision $\hat{\boldsymbol{x}}$ of the leader, a set of rational responses $R(\hat{\boldsymbol{x}})$ of the follower is a set of optimal solutions to problem (3.68). We assume that the set of rational responses $R(\hat{\boldsymbol{x}})$ is a singleton for any $\hat{\boldsymbol{x}}$, i.e., $R(\hat{\boldsymbol{x}})$ is a function of $\hat{\boldsymbol{x}}$.

We use the reference point method for the leader in a way similar to the cases where the leader has optimistic or pessimistic anticipation. Stackelberg solutions based on the preference of the follower can be obtained by updating the reference point of the leader and iteratively solving the problem:

$$\underset{\boldsymbol{x}}{\text{minimize}} \left\{ \max_{i=1,\dots,k_1} \{z_{1i}(\boldsymbol{x},\boldsymbol{y}) - \bar{z}_{1i}\} + \rho \sum_{i=1}^{k_1} (z_{1i}(\boldsymbol{x},\boldsymbol{y}) - \bar{z}_{1i}) \right\} \tag{3.69a}$$

$$\text{subject to } \boldsymbol{y} = R(\boldsymbol{x}) \tag{3.69b}$$

$$\boldsymbol{A}\boldsymbol{x} + \boldsymbol{B}\boldsymbol{y} \leqq \boldsymbol{b} \tag{3.69c}$$

$$\boldsymbol{x} \geqq \boldsymbol{0},\ \boldsymbol{y} \geqq \boldsymbol{0}. \tag{3.69d}$$

By introducing an auxiliary variable σ, problem (3.69) can be transformed into the following equivalent problem:

$$\underset{\boldsymbol{x},\sigma}{\text{minimize}} \ \sigma + \rho \sum_{i=1}^{k_1} (z_{1i}(\boldsymbol{x},\boldsymbol{y}) - \bar{z}_{1i}) \tag{3.70a}$$

$$\text{subject to } \boldsymbol{y} = R(\boldsymbol{x}) \tag{3.70b}$$

$$z_{1i}(\boldsymbol{x},\boldsymbol{y}) - \bar{z}_{1i} \leqq \sigma,\ i = 1,\dots,k_1 \tag{3.70c}$$

$$\boldsymbol{A}\boldsymbol{x} + \boldsymbol{B}\boldsymbol{y} \leqq \boldsymbol{b} \tag{3.70d}$$

$$\boldsymbol{x} \geqq \boldsymbol{0},\ \boldsymbol{y} \geqq \boldsymbol{0}. \tag{3.70e}$$

Because problem (3.68) is a linear programming problem and problem (3.70) is also linear with the exception of the constraint $\boldsymbol{y} = R(\boldsymbol{x})$, problem (3.70) can be reduced to a conventional two-level linear programming problem, and it can be solved by using one of the developed algorithms (Bialas and Karwan, 1984; Bard and Falk, 1982; Bard, 1983a; Fortuny-Amat and McCarl, 1981; Bard and Moore, 1990a; White and Anandalingam, 1993).

Kuhn-Tucker approach

The Kuhn-Tucker optimality conditions of the lower level's problem (3.2) are used in order to solve the two-level linear programming problem (3.1); as shown in the subsection 3.1, such a computational method is called the Kuhn-Tucker approach (Bialas and Karwan, 1984; Bard and Falk, 1982; Bard, 1983a; Fortuny-Amat and McCarl, 1981; Bard and Moore, 1990a).

The Kuhn-Tucker optimality conditions of the follower's problem (3.2) are represented by

$$\boldsymbol{u}B - \boldsymbol{v} = -\boldsymbol{d}_2 \tag{3.71}$$

$$\boldsymbol{u}(A\boldsymbol{x} + B\boldsymbol{y} - \boldsymbol{b}) - \boldsymbol{v}\boldsymbol{y} = 0 \tag{3.72}$$

$$A\boldsymbol{x} + B\boldsymbol{y} \leqq \boldsymbol{b} \tag{3.73}$$

$$\boldsymbol{y} \geqq \boldsymbol{0},\ \boldsymbol{u}^T \geqq \boldsymbol{0},\ \boldsymbol{v}^T \geqq \boldsymbol{0}, \tag{3.74}$$

where $\boldsymbol{u}$ and $\boldsymbol{v}$ are an m-dimensional and an n_2-dimensional variable row vectors. Thus, the Stackelberg solution is derived by solving the problem

$$\begin{aligned}
&\text{minimize } z_1(\boldsymbol{x},\boldsymbol{y}) = \boldsymbol{c}_1\boldsymbol{x} + \boldsymbol{d}_1\boldsymbol{y} && (3.75\text{a})\\
&\text{subject to } \boldsymbol{u}B - \boldsymbol{v} = -\boldsymbol{d}_2 && (3.75\text{b})\\
&\qquad \boldsymbol{u}(A\boldsymbol{x} + B\boldsymbol{y} - \boldsymbol{b}) - \boldsymbol{v}\boldsymbol{y} = 0 && (3.75\text{c})\\
&\qquad A\boldsymbol{x} + B\boldsymbol{y} \leqq \boldsymbol{b} && (3.75\text{d})\\
&\qquad \boldsymbol{x} \geqq \boldsymbol{0},\ \boldsymbol{y} \geqq \boldsymbol{0}, \boldsymbol{u}^T \geqq \boldsymbol{0},\ \boldsymbol{v}^T \geqq \boldsymbol{0}. && (3.75\text{e})
\end{aligned}$$

We consider the multiobjective two-level linear programming problem (3.51) through the Kuhn-Tucker approach. According to Kuhn and Tucker (1951), the Pareto optimality conditions of the lower level's multiobjective linear programming problem (3.54) are

$$\boldsymbol{w}D_2 + \boldsymbol{u}A_2 - \boldsymbol{v} = \boldsymbol{0} \tag{3.76}$$

$$B\boldsymbol{y} + A\boldsymbol{x} - \boldsymbol{b} \leqq \boldsymbol{0} \tag{3.77}$$

$$\boldsymbol{u}(\boldsymbol{b} - A\boldsymbol{x} - B\boldsymbol{y}) + \boldsymbol{v}\boldsymbol{y} = 0 \tag{3.78}$$

$$\boldsymbol{y} \geqq \boldsymbol{0},\ \boldsymbol{u}^T \geqq \boldsymbol{0},\ \boldsymbol{v}^T \geqq \boldsymbol{0} \tag{3.79}$$

$$\boldsymbol{w}^T > \boldsymbol{0}, \tag{3.80}$$

where $\boldsymbol{w}$ is a k_2-dimensional variable row vector and $D_2 = [\boldsymbol{d}_{21} \cdots \boldsymbol{d}_{2k_2}]^T$. Necessary and sufficient conditions that a response $\boldsymbol{y}$ of the follower to a decision $\hat{\boldsymbol{x}} \in S(X)$ of the leader be Pareto optimal are that the response $\boldsymbol{y}$ satisfies the above conditions (3.76)–(3.80).

Introducing the Kuhn-Tucker conditions (3.76)–(3.80) to Pareto optimality, we consider the following problem for obtaining the optimistic Stackelberg solution:

$$\text{minimize} \quad \sigma + \rho \sum_{i=1}^{k_1} (z_{1i}(\boldsymbol{x},\boldsymbol{y}) - \bar{z}_{1i}) \tag{3.81a}$$

$$\text{subject to} \quad z_{1i}(\boldsymbol{x},\boldsymbol{y}) - \bar{z}_{1i} \leqq \sigma, \; i = 1,\dots,k_1 \tag{3.81b}$$

$$A\boldsymbol{x} + B\boldsymbol{y} \leqq \boldsymbol{b} \tag{3.81c}$$

$$\boldsymbol{w}D_2 + \boldsymbol{u}B - \boldsymbol{v} = \boldsymbol{0} \tag{3.81d}$$

$$\boldsymbol{u}(\boldsymbol{b} - A\boldsymbol{x} - B\boldsymbol{y}) + \boldsymbol{v}\boldsymbol{y} = 0 \tag{3.81e}$$

$$\boldsymbol{x} \geqq \boldsymbol{0}, \; \boldsymbol{y} \geqq \boldsymbol{0}, \; \boldsymbol{u}^T \geqq \boldsymbol{0}, \; \boldsymbol{v}^T \geqq \boldsymbol{0} \tag{3.81f}$$

$$\boldsymbol{w}^T > \boldsymbol{0}. \tag{3.81g}$$

To solve problem (3.81), we introduce an m-dimensional zero-one decision column vector $\boldsymbol{\lambda}$ and an n_2-dimensional zero-one decision column vector $\boldsymbol{\mu}$ to the complementarity constraint in a way similar to the method by Fortuny-Amat and McCarl (1981), and a mixed zero-one programming problem can be formulated by appending the following constraints in place of the complementarity constraint (3.81e):

$$\boldsymbol{u}^T \leqq M\boldsymbol{\lambda} \tag{3.82a}$$

$$\boldsymbol{b} - A\boldsymbol{x} - B\boldsymbol{y} \leqq M(\boldsymbol{1}^m - \boldsymbol{\lambda}) \tag{3.82b}$$

$$\boldsymbol{v}^T \leqq M\boldsymbol{\mu} \tag{3.82c}$$

$$\boldsymbol{y} \leqq M(\boldsymbol{1}^{n_2} - \boldsymbol{\mu}) \tag{3.82d}$$

$$\boldsymbol{\lambda} \in \{0,1\}^m, \; \boldsymbol{\mu} \in \{0,1\}^{n_2}, \tag{3.82e}$$

where M is some large positive number, and $\boldsymbol{1}^m$ and $\boldsymbol{1}^{n_2}$ are m-dimensional and n_2-dimensional column vectors of ones.

Therefore, we can obtain an optimal solution to the mixed zero-one programming problem by using the branch-and-bound technique. Bard and Moore (1990a) also present another algorithm for obtaining an optimal solution to problem (3.75). In their method, the branch-and-bound enumeration is employed, and linear programming problems, which are yielded from problem (3.75) by eliminating the complementarity constraint (3.75c), are repeatedly solved. By doing so, it is verified whether or not the complementarity condition holds.

We can apply these methods to solving problem (3.81), but it must be noted that there exist strict inequality constraints $\boldsymbol{w}^T > \boldsymbol{0}$ in the relaxed linear programming problem. The linear programming problem with the constraints $\boldsymbol{w}^T > \boldsymbol{0}$ cannot be solved via the usual Simplex method. Fortunately, this difficulty is resolved easily by revising the pivot procedure so as to let the variables $\boldsymbol{w}$ always remain in the basis.

3.5.2 Numerical examples

To understand the geometric properties of the optimistic and the pessimistic Stackelberg solutions to the multiobjective two-level linear programming problem, and to demonstrate the feasibility of the algorithms, two illustrative numerical examples are given. In the first example, we examine rational responses of the follower.

Example 1

We consider the following two-level multiobjective linear programming problem, which has a single objective function of the leader and two objective functions of the follower:

$$
\begin{aligned}
&\underset{x}{\text{minimize}} \ z_{11}(x,y_1,y_2) = -x-2y_1+4y_2\\
&\quad \text{where } (y_1,y_2) \text{ solves}\\
&\underset{y_1,y_2}{\text{minimize}} \ z_{21}(x,y_1,y_2) = x+2y_1-y_2\\
&\underset{y_1,y_2}{\text{minimize}} \ z_{22}(x,y_1,y_2) = 2x-2y_1-y_2\\
&\text{subject to } \begin{array}{lll}
y_2 \leqq 100, & x+y_2 \leqq 170, & x+y_1+y_2 \leqq 240\\
y_1+y_2 \leqq 170, & -x+y_1+y_2 \leqq 130, & -x+y_2 \leqq 60\\
-x-y_1+y_2 \leqq 20, & -y_1+y_2 \leqq 60, & x+y_1-y_2 \leqq 130\\
x \leqq 100, & x+y_1 \leqq 170, & y_1 \leqq 100\\
-x+y_1 \leqq 60, & -x \leqq -10, & x+y_1 \geqq 50\\
-y_1 \leqq -10, & x-y_1 \leqq 60, & x-y_2 \leqq 60\\
x+y_1-y_2 \leqq 130, & y_1-y_2 \leqq 60, & -x+y_1-y_2 \leqq 20\\
-x-y_2 \leqq -50, & -x-y_1-y_2 \leqq -90, & -y_1-y_2 \leqq -170\\
x-y_1-y_2 \leqq 20, & -y_2 \leqq -10. &
\end{array}
\end{aligned}
$$

The decision variable of the leader is x and those of the follower are y_1 and y_2. The leader minimizes the single objective function $z_{11}(x,y_1,y_2) = -x-2y_1+4y_2$ and the follower minimizes the two objective functions $z_{21}(x,y_1,y_2) = x+2y_1-y_2$ and $z_{22}(x,y_1,y_2) = 2x-2y_1-y_2$. We can compute the optimistic and the pessimistic Stackelberg solutions by applying the algorithms described above or by using the graphical solution procedure.

Table 3.10 Stackelberg solutions and values of objective functions.

	optimistic solution	pessimistic solution
(x,y_1,y_2)	$(70,100,70)$	$(100,40,70)$
z_{11}	10	100
z_{21}	200	110
z_{22}	-130	50

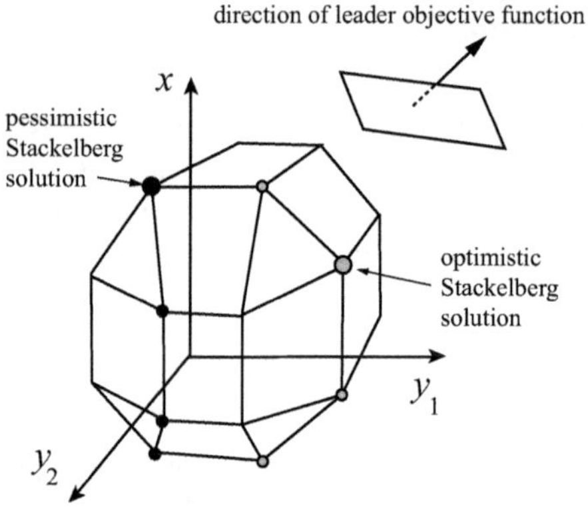

Fig. 3.13 Multiobjective Stackelberg solutions.

Table 3.11 Search sequence of Stackelberg solutions.

search sequence	(x, y_1, y_2)	z_1	optimistic solution	pessimistic solution
1	(70, 70, 10)	−170	(70, 100, 70)	(70, 40, 100)
2	(40, 70, 10)	−140	(40, 100, 70)	(40, 40, 70)
3	(70, 100, 40)	−110	(70, 100, 70)	(70, 40, 100)
4	(70, 40, 10)	−110	(70, 100, 70)	(70, 40, 100)
5	(100, 70, 40)	−80	(100, 70, 70)	(100, 40, 70)
6	(40, 40, 10)	−80	(40, 100, 70)	(40, 40, 70)
7	(40, 100, 40)	−80	(40, 100, 70)	(40, 40, 100)
8	(100, 40, 40)	−20	(100, 70, 70)	(100, 40, 70)
9	(10, 70, 40)	10	(10, 70, 70)	(10, 40, 70)
10	(70, 100, 70)	10	∗(70, 100, 70)	(70, 40, 100)
11	(40, 100, 70)	40	(40, 100, 70)	(40, 40, 70)
12	(100, 70, 70)	40	(100, 70, 70)	(100, 40, 70)
13	(10, 40, 40)	70	(10, 70, 70)	(10, 40, 70)
14	(70, 10, 40)	70	(70, 100, 70)	(70, 40, 100)
15	(100, 40, 70)	100	(100, 70, 70)	∗(100, 40, 70)
16	(40, 10, 40)	100	(40, 100, 70)	(40, 40, 70)
17	(10, 70, 70)	130	(10, 70, 70)	(10, 40, 70)
18	(10, 40, 70)	190	(10, 70, 70)	(10, 40, 70)
19	(70, 70, 100)	190	(70, 100, 70)	(70, 40, 100)
20	(70, 10, 70)	190	(70, 100, 70)	(70, 40, 100)
21	(40, 70, 100)	220	(40, 100, 70)	(40, 40, 70)
22	(40, 10, 70)	220	(40, 100, 70)	(40, 40, 70)
23	(70, 40, 100)	250	(70, 100, 70)	(70, 40, 100)
24	(40, 40, 100)	280	(40, 100, 70)	(40, 40, 70)

The optimistic Stackelberg solution $(x, y_1, y_2) = (70, 100, 70)$ and the pessimistic Stackelberg solution $(x, y_1, y_2) = (100, 40, 70)$ are depicted in Figure 3.13. The value of the objective function of the leader and the values of the two objective functions of the follower are shown in Table 3.10. In Table 3.11, all of the extreme points of the problem are enumerated in ascending order of the objective function z_1 of the leader, and the optimistic Stackelberg solution is found at the 10th extreme point, and the pessimistic Stackelberg solution is also found at the 15th extreme point.

Example 2

We consider the following two-level multiobjective linear programming problem, which has a single objective function and three objective functions of the follower:

$$\begin{aligned}
\underset{x_1,x_2,x_3}{\text{minimize}}\ & z_{11} = -14x_1 + 11x_2 + 8x_3 - 15y_1 - 3y_2 + 4y_3 \\
& \text{where } (y_1, y_2, y_3) \text{ solves} \\
\underset{y_1,y_2,y_3}{\text{minimize}}\ & z_{21} = 6x_1 - 2x_2 + 4x_3 - 4y_1 + 7y_2 - 7y_3 \\
\underset{y_1,y_2,y_3}{\text{minimize}}\ & z_{22} = -1x_1 - 13x_2 - 3x_3 + 4y_1 + 2y_2 + 4y_3 \\
\underset{y_1,y_2,y_3}{\text{minimize}}\ & z_{23} = -1x_1 - 2x_2 - 18x_3 + 3y_1 - 9y_2 + 8y_3 \\
\text{subject to}\ & 15x_1 - 7x_2 + 3x_3 + 2y_1 - 7y_2 + 2y_3 \leqq 200 \\
& 7x_1 + 7x_2 + 6x_3 + 1y_1 + 13y_2 + 1y_3 \leqq 140 \\
& 2x_1 + 2x_2 - 1x_3 + 14y_1 + 2y_2 + 2y_3 \leqq 240 \\
& -3x_1 + 6x_2 + 12x_3 + 4y_1 - 8y_2 + 1y_3 \leqq 140 \\
& 4x_1 - 7x_2 + 7x_3 + 2y_1 + 4y_2 - 7y_3 \leqq 45 \\
& 4x_1 + 5x_2 + 1x_3 - 7y_1 - 6y_2 + 1y_3 \leqq 800 \\
& x_1,\ x_2,\ x_3 \geqq 0,\ y_1,\ y_2,\ y_3 \geqq 0.
\end{aligned}$$

The decision variable vectors (x_1,x_2,x_3) and (y_1,y_2,y_3) of the leader and the follower, respectively, are three-dimensional vectors. This problem is larger than that in Example 1 and cannot be solved by the graphical solution procedure. Thus, we must use the algorithms described above to obtain the optimistic and the pessimistic Stackelberg solutions to this problem. The solutions are shown in Table 3.12.

Table 3.12 Stackelberg solutions and objective function values for the six-dimensional problem.

	optimistic solution	pessimistic solution
(x_1,x_2,x_3)	(11.938397, 0.000000, 0.000000)	(14.221800, 0.000000, 0.000000)
(y_1,y_2,y_3)	(14.177088, 2.786012, 6.035973)	(0.000000, 2.855206, 3.329718)
z_{11}	−364.008028	−194.351952
z_{21}	−7.827698	82.009219
z_{22}	74.485873	4.807484
z_{23}	53.806546	−13.280911

Discussion

To examine rational responses of the follower, suppose that the leader has chosen a decision $x = 70$ in the problem of Example 1. Then, the feasible region of the follower $S(\boldsymbol{x})$ is an octagonal area in the y_1-y_2 plane as shown in Figure 3.14. The follower can choose a decision in the region $S(\boldsymbol{x})$ and must choose it in the set of Pareto optimal responses $P(\boldsymbol{x})$, which is depicted as thick lines and is derived by considering the preference cone of the follower. Because the objective functions of the follower are $z_{21}(x, y_1, y_2) = x + 2y_1 - y_2$ and $z_{22}(x, y_1, y_2) = 2x - 2y_1 - y_2$, the preference cone can be sketched as shown in Figure 3.14.

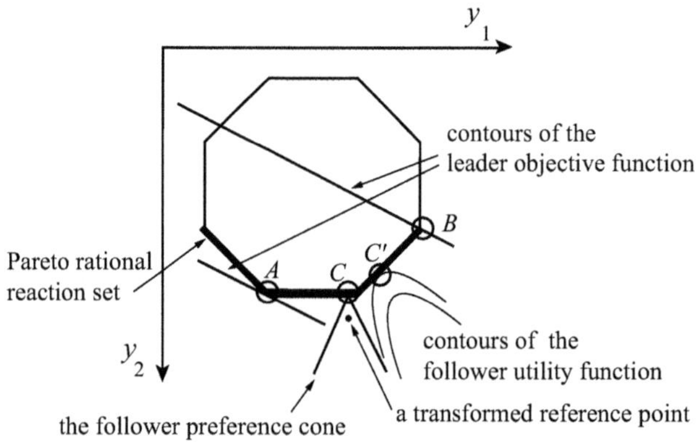

Fig. 3.14 Pareto optimal response set.

In general, the set of rational responses is not a singleton, but a set including multiple points or an infinite number of points in multiobjective environments; therefore, it is difficult for the leader to predict which point the follower is going to choose. From a prescriptive viewpoint, it is significant for the leader confronting such a problem to estimate the follower's decision on the basis of some subjective anticipation or belief.

Some candidates for a response of the follower are shown in Figure 3.14. Point A denotes the follower's rational response to the leader, which is a most pessimistic one in the set of Pareto optimal responses $P(\boldsymbol{x})$; Point B denotes a most optimistic response. When the preference of the follower can be expressed as a reference point or the leader can estimate a reference point which reflects the preference of the follower, Point C becomes a rational response of the follower. In Figure 3.14, the reference point transformed from the z_1-z_2 objective space into the y_1-y_2 plane is depicted. If the leader could assess a utility function of the follower as seen in Figure 3.14, Point C' would be a rational response of the follower.

When the follower chooses one of Points A, B and C as a rational response, we can compute Stackelberg solutions to the multiobjective two-level linear programming problem by using algorithms presented above.

3.6 Stochastic two-level linear programming

In this section, we treat two-level programming problems with random variable coefficients. Stackelberg solutions are defined in the E-model and the V-model, and the corresponding computational methods are given. Such formulations can be interpreted as a prescriptive approach from a viewpoint of the leader.

We consider the following two-level programming problem with random variable coefficients:

$$\begin{aligned}
&\underset{\boldsymbol{x}}{\text{minimize}} \;\; z_1(\boldsymbol{x},\boldsymbol{y}) = \tilde{\boldsymbol{c}}_1\boldsymbol{x} + \tilde{\boldsymbol{d}}_1\boldsymbol{y} && (3.83\text{a})\\
&\quad \text{where } \boldsymbol{y} \text{ solves}\\
&\underset{\boldsymbol{y}}{\text{minimize}} \;\; z_2(\boldsymbol{x},\boldsymbol{y}) = \tilde{\boldsymbol{c}}_2\boldsymbol{x} + \tilde{\boldsymbol{d}}_2\boldsymbol{y} && (3.83\text{b})\\
&\text{subject to } \; A\boldsymbol{x} + B\boldsymbol{y} \leqq \tilde{\boldsymbol{b}} && (3.83\text{c})\\
&\qquad\qquad\quad \boldsymbol{x} \geqq \boldsymbol{0},\; \boldsymbol{y} \geqq \boldsymbol{0}, && (3.83\text{d})
\end{aligned}$$

where $\boldsymbol{x}$ and $\boldsymbol{y}$ are the leader's n_1-dimensional and the follower's n_2-dimensional decision variable column vectors, respectively; $\tilde{\boldsymbol{c}}_i$, $\tilde{\boldsymbol{d}}_i$, $i = 1,2$ are n_1-dimensional and n_2-dimensional random variable coefficient row vectors, respectively; $\tilde{\boldsymbol{b}}$ is an m-dimensional random variable column vector; A and B are $m \times n_1$ and $m \times n_2$ constant matrices, respectively; $z_1(\boldsymbol{x},\boldsymbol{y})$ and $z_2(\boldsymbol{x},\boldsymbol{y})$ are the leader's and the follower's linear objective functions, respectively.

Because problem (3.83) includes random variables in the right-hand side of the constraints, the chance constraint programming approach (Charnes and Cooper, 1963) is helpful, and in this approach, constraint violations are permitted up to the specified probability limits. Let α_i be a probability of the extent to which the ith constraint violation is admitted. Then, the chance constraint conditions are represented as follows:

$$P\left[A^i\boldsymbol{x} + B^i\boldsymbol{y} \leqq \tilde{b}_i\right] \geqq \alpha_i,\; i = 1,\dots,m, \tag{3.84}$$

where P means a probability measure, and A^i, B^i and $\tilde{b}_i$ are the coefficients of the ith constraint. Inequality (3.84) means that the ith constraint may be violated, but at most $1-\alpha_i$ proportion of the time.

Let $F_i(\tau)$ be a distribution function of the random variable $\tilde{b}_i$. Then, because of the fact that

$$P\left[A^i\boldsymbol{x} + B^i\boldsymbol{y} \leqq \tilde{b}_i\right] = 1 - F(A^i\boldsymbol{x} + B^i\boldsymbol{y}),$$

inequality (3.84) is rewritten as

$$F(A^i\boldsymbol{x} + B^i\boldsymbol{y}) \leqq 1 - \alpha_i.$$

Let $K_{1-\alpha_i}$ be the maximum value of τ satisfying $\tau = F^{-1}(1-\alpha_i)$. Then, from the monotonicity of the distribution function $F(\tau)$, inequality (3.84) is rewritten as

$$A^i\boldsymbol{x} + B^i\boldsymbol{y} \leqq K_{1-\alpha_i},\; i = 1,\dots,m, \tag{3.85}$$

and it is represented equivalently by

$$Ax + By \leqq \boldsymbol{K}_{1-\alpha}, \tag{3.86}$$

where $\boldsymbol{K}_{1-\alpha} = (K_{1-\alpha_1}, \ldots, K_{1-\alpha_m})^T$.

With the chance constraint formulation, problem (3.83) is rewritten as the following problem with deterministic constraints:

$$\begin{aligned}
&\underset{\boldsymbol{x}}{\text{minimize}} \;\; z_1(\boldsymbol{x},\boldsymbol{y}) = \tilde{\boldsymbol{c}}_1\boldsymbol{x} + \tilde{\boldsymbol{d}}_1\boldsymbol{y} && \text{(3.87a)}\\
&\quad \text{where } \boldsymbol{y} \text{ solves}\\
&\underset{\boldsymbol{y}}{\text{minimize}} \;\; z_2(\boldsymbol{x},\boldsymbol{y}) = \tilde{\boldsymbol{c}}_2\boldsymbol{x} + \tilde{\boldsymbol{d}}_2\boldsymbol{y} && \text{(3.87b)}\\
&\text{subject to } \; A\boldsymbol{x} + B\boldsymbol{y} \leqq \boldsymbol{K}_{1-\alpha} && \text{(3.87c)}\\
&\qquad\qquad\quad \boldsymbol{x} \geqq \boldsymbol{0},\; \boldsymbol{y} \geqq \boldsymbol{0}. && \text{(3.87d)}
\end{aligned}$$

3.6.1 Stochastic two-level linear programming models

To deal with the objective functions with random variables in two-level linear programming problems, the concepts of the minimum expected value model (E-model) and the minimum variance model (V-model) are applied (Charnes and Cooper, 1963).

E-model for two-level linear programming problems

In the E-model, means of the objective functions of the leader and the follower are minimized, and the deterministic problem corresponding to problem (3.87) is formulated as

$$\begin{aligned}
&\underset{\boldsymbol{x}}{\text{minimize}} \;\; E[\tilde{\boldsymbol{c}}_1\boldsymbol{x} + \tilde{\boldsymbol{d}}_1\boldsymbol{y}] = \boldsymbol{m}_1^c\boldsymbol{x} + \boldsymbol{m}_1^d\boldsymbol{y} && \text{(3.88a)}\\
&\quad \text{where } \boldsymbol{y} \text{ solves}\\
&\underset{\boldsymbol{x}}{\text{minimize}} \;\; E[\tilde{\boldsymbol{c}}_2\boldsymbol{x} + \tilde{\boldsymbol{d}}_2\boldsymbol{y}] = \boldsymbol{m}_2^c\boldsymbol{x} + \boldsymbol{m}_2^d\boldsymbol{y} && \text{(3.88b)}\\
&\text{subject to } \; A\boldsymbol{x} + B\boldsymbol{y} \leqq \boldsymbol{K}_{1-\alpha} && \text{(3.88c)}\\
&\qquad\qquad\quad \boldsymbol{x} \geqq \boldsymbol{0},\; \boldsymbol{y} \geqq \boldsymbol{0}, && \text{(3.88d)}
\end{aligned}$$

where $E[\tilde{f}]$ denotes the mean of $\tilde{f}$, and $(\boldsymbol{m}_1^c, \boldsymbol{m}_1^d)$ and $(\boldsymbol{m}_2^c, \boldsymbol{m}_2^d)$ are vectors of the means of $(\tilde{\boldsymbol{c}}_1, \tilde{\boldsymbol{d}}_1)$ and $(\tilde{\boldsymbol{c}}_2, \tilde{\boldsymbol{d}}_2)$, respectively.

A Stackelberg solution in the E-model, i.e., an optimal solution to problem (3.88) is interpreted as follows: the leader makes a decision so as to minimize the mean of the leader's objective function under the assumption that the follower also makes a decision so as to minimize the mean of the follower's objective function for a given decision of the leader.

Because (3.88) is regarded as a conventional two-level linear programming problem, the Stackelberg solution in the E-model can be obtained by using existing solution methods (Fortuny-Amat and McCarl, 1981; Bard and Falk, 1982; Bard, 1983a; Bialas and Karwan, 1984; Bard and Moore, 1990a; White and Anandalingam, 1993).

V-model for two-level linear programming problems

Although a Stackelberg solution by the E-model minimizes the mean of the objective function, it might be undesirable for the decision maker if the dispersion of the objective function value is large. In such a situation, it is reasonable for the decision maker to employ the V-model which minimizes the variance of the objective function. In the V-model, problems with coefficients characterized by the elements governed by stochastic events are often formulated such that the variance of the objective function value is minimized under the condition that the mean of the objective function value is smaller than a certain level specified by the decision maker. Let β_1 and β_2 denote such specified levels for the means of the objective function values for the leader and the follower, respectively, and then a two-level linear programming problem with random variables is formulated as

$$\underset{\boldsymbol{x}}{\text{minimize}} \;\; Var[\tilde{\boldsymbol{c}}_1\boldsymbol{x}+\tilde{\boldsymbol{d}}_1\boldsymbol{y}] = \begin{bmatrix}\boldsymbol{x}\\ \boldsymbol{y}\end{bmatrix}^T V_1 \begin{bmatrix}\boldsymbol{x}\\ \boldsymbol{y}\end{bmatrix} \tag{3.89a}$$

where $\boldsymbol{y}$ solves

$$\underset{\boldsymbol{y}}{\text{minimize}} \;\; Var[\tilde{\boldsymbol{c}}_2\boldsymbol{x}+\tilde{\boldsymbol{d}}_2\boldsymbol{y}] = \begin{bmatrix}\boldsymbol{x}\\ \boldsymbol{y}\end{bmatrix}^T V_2 \begin{bmatrix}\boldsymbol{x}\\ \boldsymbol{y}\end{bmatrix} \tag{3.89b}$$

$$\text{subject to} \;\; A\boldsymbol{x}+B\boldsymbol{y} \leqq \boldsymbol{K}_{1-\alpha} \tag{3.89c}$$

$$\boldsymbol{m}_1^c\boldsymbol{x}+\boldsymbol{m}_1^d\boldsymbol{y} \leqq \beta_1 \tag{3.89d}$$

$$\boldsymbol{m}_2^c\boldsymbol{x}+\boldsymbol{m}_2^d\boldsymbol{y} \leqq \beta_2 \tag{3.89e}$$

$$\boldsymbol{x} \geqq \boldsymbol{0},\; \boldsymbol{y} \geqq \boldsymbol{0}, \tag{3.89f}$$

where $Var[\tilde{f}]$ denotes the variance of $\tilde{f}$, and V_1 and V_2 are the variance-covariance matrices of $(\tilde{\boldsymbol{c}}_1, \tilde{\boldsymbol{d}}_1)$ and $(\tilde{\boldsymbol{c}}_2, \tilde{\boldsymbol{d}}_2)$, respectively. We assume that V_1 and V_2 are positive-definite without loss of generality.

A Stackelberg solution in the V-model, i.e., an optimal solution to problem (3.89) is interpreted as follows: the leader makes a decision so as to minimize the variance of the leader's objective function under the assumption that the follower also makes a decision so as to minimize the variance of the follower's objective function for a given decision of the leader.

3.6.2 Computational method for V-model

For a decision $\hat{\boldsymbol{x}}$ specified by the leader, the rational response of the follower is an optimal solution to the following problem:

$$\underset{\boldsymbol{y}}{\text{minimize}} \quad Var[\tilde{\boldsymbol{c}}_2\hat{\boldsymbol{x}} + \tilde{\boldsymbol{d}}_2\boldsymbol{y}] = \begin{bmatrix}\hat{\boldsymbol{x}}\\ \boldsymbol{y}\end{bmatrix}^T V_2 \begin{bmatrix}\hat{\boldsymbol{x}}\\ \boldsymbol{y}\end{bmatrix} \tag{3.90a}$$

$$\text{subject to} \quad A\hat{\boldsymbol{x}} + B\boldsymbol{y} \leqq \boldsymbol{K}_{1-\alpha} \tag{3.90b}$$

$$\boldsymbol{m}_1^c\hat{\boldsymbol{x}} + \boldsymbol{m}_1^d\boldsymbol{y} \leqq \beta_1 \tag{3.90c}$$

$$\boldsymbol{m}_2^c\hat{\boldsymbol{x}} + \boldsymbol{m}_2^d\boldsymbol{y} \leqq \beta_2 \tag{3.90d}$$

$$\boldsymbol{y} \geqq \boldsymbol{0}, \tag{3.90e}$$

Let $R(\boldsymbol{x})$ be the set of rational response of the follower, i.e., the set of optimal solutions to problem (3.90). Then, the Stackelberg solution in the V-model is an optimal solution to the following problem:

$$\text{minimize} \quad Var[\tilde{\boldsymbol{c}}_1\boldsymbol{x} + \tilde{\boldsymbol{d}}_1\boldsymbol{y}] = \begin{bmatrix}\boldsymbol{x}\\ \boldsymbol{y}\end{bmatrix}^T V_1 \begin{bmatrix}\boldsymbol{x}\\ \boldsymbol{y}\end{bmatrix} \tag{3.91a}$$

$$\text{subject to} \quad \boldsymbol{y} \in R(\boldsymbol{x}) \tag{3.91b}$$

$$A_1'\boldsymbol{x} + A_2'\boldsymbol{y} \leqq \boldsymbol{b}' \tag{3.91c}$$

$$\boldsymbol{x} \geqq \boldsymbol{0},\ \boldsymbol{y} \geqq \boldsymbol{0}, \tag{3.91d}$$

where, for the sake of simplicity, we use coefficient matrices A_1', A_2' and a coefficient vector $\boldsymbol{b}'$ in place of A_1, A_2, $\boldsymbol{m}_1^c$, $\boldsymbol{m}_1^d$, $\boldsymbol{m}_2^c$, $\boldsymbol{m}_2^d$, $\boldsymbol{K}_{1-\alpha}$, β_1 and β_2 in problem (3.90).

Because the variance-covariance matrix V_2 is positive-definite, the objective function (3.90a) is strictly convex. It should be noted that from this fact, one finds that problem (3.90) is a convex programming problem with a strict convex objective function, and therefore, the set $R(\boldsymbol{x})$ of rational responses is a singleton. Moreover, the constraint (3.91b) in problem (3.91) can be replaced by the Kuhn-Tucker conditions for problem (3.90).

Let $\boldsymbol{x}^*$ be a decision of the leader. The Lagrange function corresponding to problem (3.90) is defined as

$$L(\boldsymbol{y}, \boldsymbol{\lambda}, \boldsymbol{\omega}; \boldsymbol{x}^*) = \begin{bmatrix}\boldsymbol{x}^*\\ \boldsymbol{y}\end{bmatrix}^T V_2 \begin{bmatrix}\boldsymbol{x}^*\\ \boldsymbol{y}\end{bmatrix} + \boldsymbol{\lambda}(A_1'\boldsymbol{x}^* + A_2'\boldsymbol{y} - \boldsymbol{b}') - \boldsymbol{\omega}\boldsymbol{y}, \tag{3.92}$$

where $\boldsymbol{\lambda}$ and $\boldsymbol{\omega}$ are the Lagrange multiplier vectors. Then, the Kuhn-Tucker conditions are given as follows:

$$2\sum_{j=1}^{n_1} v_{2(n_1+i)j}x_j^* + 2\sum_{j=n_1+1}^{n_1+n_2} v_{2(n_1+i)j}y_{j-n_1} + \boldsymbol{\lambda}A'_{2\cdot i} - \omega_i = 0,\ i=1,\ldots,n_2 \quad (3.93)$$

$$A'_1\boldsymbol{x}^* + A'_2\boldsymbol{y} - \boldsymbol{b}' \leqq \boldsymbol{0} \quad (3.94)$$

$$\boldsymbol{\lambda}(A'_1\boldsymbol{x}^* + A'_2\boldsymbol{y} - \boldsymbol{b}') = 0,\ \boldsymbol{\omega}\boldsymbol{y} = 0 \quad (3.95)$$

$$\boldsymbol{y} \geqq \boldsymbol{0},\ \boldsymbol{\lambda} \geqq \boldsymbol{0},\ \boldsymbol{\omega} \geqq \boldsymbol{0}, \quad (3.96)$$

where v_{2ij} is the ij-element of V_2, and $A'_{2\cdot i}$ is an ith column vector of A'_2.

Substituting the Kuhn-Tucker conditions (3.93)–(3.96) for the constraint (3.91b) in problem (3.91), problem (3.91) can be transformed into the following single-level quadratic programming problem with linear complementarity constraints:

$$\text{minimize } Var[\tilde{\boldsymbol{c}}_1\boldsymbol{x} + \tilde{\boldsymbol{d}}_1\boldsymbol{y}] = \begin{bmatrix}\boldsymbol{x}\\ \boldsymbol{y}\end{bmatrix}^T V_1 \begin{bmatrix}\boldsymbol{x}\\ \boldsymbol{y}\end{bmatrix} \quad (3.97a)$$

$$\text{subject to } 2\sum_{j=1}^{n_1} v_{2(n_1+i)j}x_j + 2\sum_{j=n_1+1}^{n_1+n_2} v_{2(n_1+i)j}y_{j-n_1} + \boldsymbol{\lambda}A'_{2\cdot i} - \omega_i = 0,\ i=1,\ldots,n_2 \quad (3.97b)$$

$$A'_1\boldsymbol{x} + A'_2\boldsymbol{y} - \boldsymbol{b}' \leqq \boldsymbol{0} \quad (3.97c)$$

$$\boldsymbol{\lambda}(\boldsymbol{b}' - A'_1\boldsymbol{x} - A'_2\boldsymbol{y}) + \boldsymbol{\omega}\boldsymbol{y} = 0 \quad (3.97d)$$

$$\boldsymbol{x} \geqq \boldsymbol{0},\ \boldsymbol{y} \geqq \boldsymbol{0},\ \boldsymbol{\lambda} \geqq \boldsymbol{0},\ \boldsymbol{\omega} \geqq \boldsymbol{0}. \quad (3.97e)$$

Although problem (3.97) is not convex and therefore an optimal solution to the problem is not obtained by using conventional convex programming techniques, it becomes a usual quadratic programming problem with linear constraints by eliminating the linear complementarity constraints. Applying the branching technique with respect to the linear complementarity constraints in the Bard and Moore method (1990a) for solving two-level linear programming problems, problem (3.97) can be solved. Namely, after branching with respect to the linear complementarity constraints, subproblems generated through the branching operation are solved by conventional convex programming methods such as Wolfe (1959) and Lemke (1965), and finally we can obtain a Stackelberg solution in the V-model, i.e., an optimal solution to the deterministic problem (3.97). For the sake of simplicity in the description of the algorithm, let $\sum_{i=1}^{m+n_2+2} u_i g_i = 0$ denote the constraint (3.97d). The algorithm for obtaining Stackelberg solutions in the V-model is summarized as follows:

Step 0 Set $k := 0$, $S_k^+ := \emptyset$, $S_k^- := \emptyset$, $S_k^0 := W$ and $\bar{V} := \infty$.

Step 1 Let $u_i = 0$ for $i \in S_k^+$ and $g_i = 0$ for $i \in S_k^-$. Solve problem (3.97) without the constraint (3.97d). If the problem is infeasible, go to Step 5. Otherwise, set $k := k+1$ and let $(\boldsymbol{x}^k, \boldsymbol{y}^k, \boldsymbol{\lambda}^k, \boldsymbol{\omega}^k)$ denote the obtained solution.

Step 2 If $Var[\tilde{\boldsymbol{c}}_1\boldsymbol{x}^k + \tilde{\boldsymbol{d}}_1\boldsymbol{y}^k] \geqq \bar{V}$, then go to Step 5.

Step 3 If the linear complementarity constraint is satisfied, i.e., $u_i g_i = 0$, $i = 1,\ldots,m+n_2+2$, then go to Step 4. Otherwise, find i^* such that the amount of violation of the linear complementarity constraint $u_{i^*} g_{i^*}$ is the largest, and let $S_k^+ := S_k^+ \cup i^*$, $S_k^0 := S_k^0 \backslash i^*$, append i^* to P_k, and then, return to Step 1.

Step 4 Set $\bar{V} := Var[\tilde{\boldsymbol{c}}_1 \boldsymbol{x}^k + \tilde{\boldsymbol{d}}_1 \boldsymbol{y}^k]$.

Step 5 If there exists no unexplored node, then go to Step 6. Otherwise, branch to the newest unexplored node, and update S_k^+, S_k^-, S_k^0 and P_k. Then, return to Step 1.

Step 6 Terminate the algorithm. If $\bar{V} = \infty$, there is no feasible solution. Otherwise, the solution corresponding to the current $\bar{V}$ is a Stackelberg solution.

3.6.3 Numerical example

Consider a simple two-level linear programming problem with random variable coefficients, in which each of the leader and the follower has one decision variable and there are five coefficients. Stackelberg solutions in the V-model are derived by the computational method described above. Consider the following problem:

$$\underset{x}{\text{minimize}} \ z_1(x,y) = \tilde{c}_1 x + \tilde{d}_1 y \tag{3.98a}$$

$$\text{where } y \text{ solves}$$

$$\underset{y}{\text{minimize}} \ z_2(x,y) = \tilde{c}_2 x + \tilde{d}_2 y \tag{3.98b}$$

$$\text{subject to} \ -x + 3y \leqq \tilde{b}_1, \ 10x - y \leqq \tilde{b}_2 \tag{3.98c}$$

$$3x + y \geqq \tilde{b}_3, \ x + 2y \geqq \tilde{b}_4 \tag{3.98d}$$

$$3x + 2y \geqq \tilde{b}_5, \ x \geqq 0, y \geqq 0. \tag{3.98e}$$

The means of random variables $\tilde{c}_1$, $\tilde{d}_1$, $\tilde{c}_2$, and $\tilde{d}_2$ are given as

$$m_1^c = -2.0, \ m_1^d = -3.0, \ m_2^c = 2.0, \ m_2^d = 1.0,$$

and the variance-covariance matrices of $(\tilde{c}_1, \tilde{d}_1)$ and $(\tilde{c}_2, \tilde{d}_2)$ are given as

$$V_1 = \begin{bmatrix} 2 & 1 \\ 1 & 3 \end{bmatrix}, \ V_2 = \begin{bmatrix} 1 & -1 \\ -1 & 6 \end{bmatrix}.$$

It is assumed that all the right-hand side values of the constraints are represented by normal random variables. Table 3.13 shows the means and the variances of the random variables, and the acceptable probabilities for the constraint violations.

The formulation of the two-level linear programming problem with random variable coefficients in the V-model including the chance constraint conditions is given as follows:

Table 3.13 Means and variances of random variables in the constraints and acceptable probabilities for the constraint violations.

coefficient	$\tilde{b}_1$	$\tilde{b}_2$	$\tilde{b}_3$	$\tilde{b}_4$	$\tilde{b}_5$
mean	50.11	113.15	15.16	13.16	25.63
variance	9.0	36.0	9.0	4.0	16.0
probability	0.85	0.70	0.90	0.70	0.80

$$\underset{x}{\text{minimize}} \quad [x\ y]\begin{bmatrix}2 & 1\\ 1 & 3\end{bmatrix}\begin{bmatrix}x\\ y\end{bmatrix} \tag{3.99a}$$

where y solves

$$\underset{y}{\text{minimize}} \quad [x\ y]\begin{bmatrix}1 & -1\\ -1 & 6\end{bmatrix}\begin{bmatrix}x\\ y\end{bmatrix} \tag{3.99b}$$

$$\text{subject to} \quad -x+3y \leqq 47,\ 10x-y \leqq 110,\ -3x-y \leqq -19 \tag{3.99c}$$

$$-x-2y \leqq -15,\ -3x-2y \leqq -29 \tag{3.99d}$$

$$-2x-3y \leqq -31,\ 2x+y \leqq 33 \tag{3.99e}$$

$$x \geqq 0,\ y \geqq 0, \tag{3.99f}$$

where the constraints (3.99c) and (3.99d) are the chance constraint conditions, and the constraints (3.99e) are the conditions for the means of the objective function values.

Problem (3.99) is transformed into the following single-level quadratic programming problem with linear complementarity constraints:

$$\text{minimize} \quad [x\ y]\begin{bmatrix}2 & 1\\ 1 & 3\end{bmatrix}\begin{bmatrix}x\\ y\end{bmatrix} \tag{3.100a}$$

$$\text{subject to} \quad -x+3y \leqq 47,\ 10x-y \leqq 110,\ -3x-y \leqq -19 \tag{3.100b}$$

$$-x-2y \leqq -15,\ -3x-2y \leqq -29 \tag{3.100c}$$

$$-2x-3y \leqq -31,\ 2x+y \leqq 33 \tag{3.100d}$$

$$2x+12y+3\lambda_1-\lambda_2-\lambda_3-2\lambda_4-2\lambda_5-3\lambda_6+\lambda_7-\omega=0 \tag{3.100e}$$

$$\begin{aligned}&\lambda_1(x-3y+47)+\lambda_2(-10x+y+110)+\lambda_3(3x+y-19)\\ &+\lambda_4(x+2y-15)+\lambda_5(3x+2y-29)+\lambda_6(2x+3y-31)\\ &+\lambda_7(-2x-y+33)+\omega y=0\end{aligned} \tag{3.100f}$$

$$x \geqq 0,\ y \geqq 0,\ \lambda_i \geqq 0,\ i=1,\ldots,7,\ \omega \geqq 0. \tag{3.100g}$$

The feasible region of problem (3.99) which is formulated by the V-model is depicted in Figure 3.15. The Stackelberg solution to problem (3.99) or the optimal solution to problem (3.100) is denoted by the white circle at the point $(5.17, 6.89)$, and when the constraints for the means of the objective function values in problem (3.99) are eliminated, the corresponding Stackelberg solution is denoted by the black circle at the point $(7,4)$. The black square at the point $(1,16)$ shows the Stackelberg solution to the problem formulated by the E-model. For these Stackelberg solutions, the means and the variances of the objective functions of the leader and the follower are also given in Table 3.14.

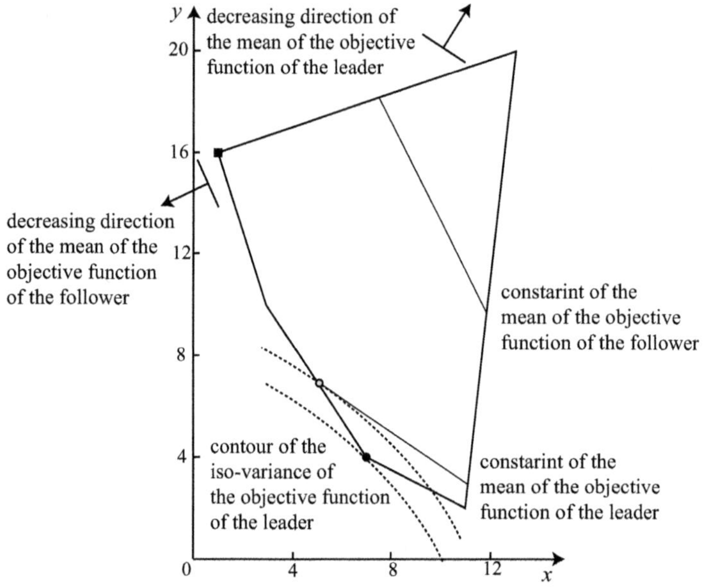

Fig. 3.15 Feasible region and the Stackelberg solutions.

Table 3.14 Stackelberg solutions in V-model and E-model.

model	solution	mean of the leader	mean of the follower	variance of the leader	variance of the follower
V-model	$(5.17, 6.89)$	-31	17	266.94	240.25
V-model without means	$(7, 4)$	-26	18	202	89
E-model	$(1, 16)$	-50	18	802	1505

As seen in Table 3.14, the mean of the leader's objective function in the E-model is small compared to that of the V-model as a matter of course, but one finds that the variances of the E-model are considerably large. In particular, although the means of the follower in the V-model and the E-model are almost the same, the variances of the V-model are reduced substantially. It is found that the variances of the V-model with the constraints for the means are reduced adequately with the suitable levels of the means specified by the leader.

Chapter 4
Cooperative Decision Making in Hierarchical Organizations

Although the previous chapter deals with decision making problems with two decision makers under noncooperative environments and Stackelberg solutions are employed, if the decision makers reach an agreement to make decisions cooperatively, it is not always appropriate to employ the concept of Stackelberg solutions as a working plan. In this chapter, we give solution methods for decision making problems in hierarchical organizations, assuming cooperative behavior of the decision makers. Taking into account vagueness of judgements of the decision makers, we present interactive fuzzy programming for two-level linear programming problems. In the interactive method, after determining the fuzzy goals of the decision makers at both levels, a satisfactory solution is derived by updating some reference points with respect to the satisfactory level. From the viewpoint of experts' imprecise or fuzzy understanding of the nature of parameters in a problem-formulation process, the method is extended to interactive fuzzy programming for two-level linear programming problems with fuzzy parameters. These results are extended to deal with two-level linear fractional programming problems and decentralized two-level linear programming problems. Furthermore, applying genetic algorithms, we deal with 0-1 and nonlinear nonconvex programming problems. Finally, multiobjective two-level linear programming and stochastic two-level linear programming are dealt with.

4.1 Solution concept for cooperative decision making

We consider a linear programming problem with two decision makers. For the sake of simplicity, we call the two decision makers DM1 and DM2 in this chapter. Let $\boldsymbol{x}_1$ and $\boldsymbol{x}_2$ denote the column vectors of the decision variables of DM1 and DM2, respectively, and let $z_1(\boldsymbol{x}_1,\boldsymbol{x}_2) = \boldsymbol{c}_{11}\boldsymbol{x}_1 + \boldsymbol{c}_{12}\boldsymbol{x}_2$ and $z_2(\boldsymbol{x}_1,\boldsymbol{x}_2) = \boldsymbol{c}_{21}\boldsymbol{x}_1 + \boldsymbol{c}_{22}\boldsymbol{x}_2$ denote the objective functions of DM1 and DM2, respectively, where $\boldsymbol{c}_{i1}$, $i = 1,2$ are n_1-dimensional coefficient row vector, $\boldsymbol{c}_{i2}$, $i = 1,2$ are n_2-dimensional coefficient row vector. Assume that $A_1\boldsymbol{x}_1 + A_2\boldsymbol{x}_2 \leq \boldsymbol{b}$ be the common constraints between DM1 and

M. Sakawa and I. Nishizaki (eds.), *Cooperative and Noncooperative Multi-Level Programming*, 83
Operations Research/Computer Science Interfaces Series 48, DOI: 10.1007/978-1-4419-0676-2_4,

DM2, where A_1 is an $m \times n_1$ coefficient matrix, A_2 is an $m \times n_2$ coefficient matrix, $\boldsymbol{b}$ is an m-dimensional constant column vector. Then, for a given $\boldsymbol{x}_2$, DM1 deals with the following linear programming problem:

$$\underset{\boldsymbol{x}_1}{\text{minimize}} \quad z_1(\boldsymbol{x}_1, \boldsymbol{x}_2) = \boldsymbol{c}_{11}\boldsymbol{x}_1 + \boldsymbol{c}_{12}\boldsymbol{x}_2 \tag{4.1a}$$

$$\text{subject to} \quad A_1\boldsymbol{x}_1 \leqq \boldsymbol{b} - A_2\boldsymbol{x}_2 \tag{4.1b}$$

$$\boldsymbol{x}_1 \geqq \boldsymbol{0}. \tag{4.1c}$$

Similarly, for a given $\boldsymbol{x}_1$, the linear programming problem for DM2 is formulated as

$$\underset{\boldsymbol{x}_2}{\text{minimize}} \quad z_2(\boldsymbol{x}_1, \boldsymbol{x}_2) = \boldsymbol{c}_{21}\boldsymbol{x}_1 + \boldsymbol{c}_{22}\boldsymbol{x}_2 \tag{4.2a}$$

$$\text{subject to} \quad A_2\boldsymbol{x}_2 \leqq \boldsymbol{b} - A_1\boldsymbol{x}_1 \tag{4.2b}$$

$$\boldsymbol{x}_2 \geqq \boldsymbol{0}. \tag{4.2c}$$

Under noncooperative environments, such situations can be considered two cases: DM1 and DM2 make decisions simultaneously; and either of DM1 or DM2 first makes a decision, and then the rest of them makes a decision with knowledge of the decision of the leader.

The first case where DM1 and DM2 make decisions simultaneously is equivalent to the situation where even if either of them first makes a decision and then the other one makes a decision later, the follower does not know the decision of the leader. For such situations, the concept of Nash equilibrium solutions is considered to be appropriate because, by a unilateral deviation of one of the decision makers from a Nash equilibrium, the decision maker cannot decrease his own objective function value. The set of Nash equilibrium solutions is a set of solutions satisfying the optimality conditions of the two problems (4.1) and (4.2) simultaneously. For the later case, we already considered in the previous chapter.

To understand the difference between the two solution concepts and introduce the other types of solution methods under cooperative environments, consider the following linear programming problem with two decision makers which is the same example as that given in the previous chapter:

$$\begin{aligned}
&\underset{x_1}{\text{minimize}} \quad z_1(x_1, x_2) = -x_1 - 8x_2 \\
&\underset{x_2}{\text{minimize}} \quad z_2(x_1, x_2) = -4x_1 + x_2 \\
&\text{subject to} \quad (x_1, x_2) \in S \triangleq \{-x_1 + 2x_2 \leqq 13,\ 2x_1 + 3x_2 \leqq 37,\ 2x_1 - x_2 \leqq 17, \\
&\qquad\qquad\qquad 2x_1 - 3x_2 \leqq 11,\ x_1 + 4x_2 \geqq 11,\ 5x_1 + 2x_2 \geqq 19\}.
\end{aligned}$$

The feasible region S of the two-level linear programming problem together with directions decreasing the objective function values of the two decision makers is depicted in Figure 4.1.

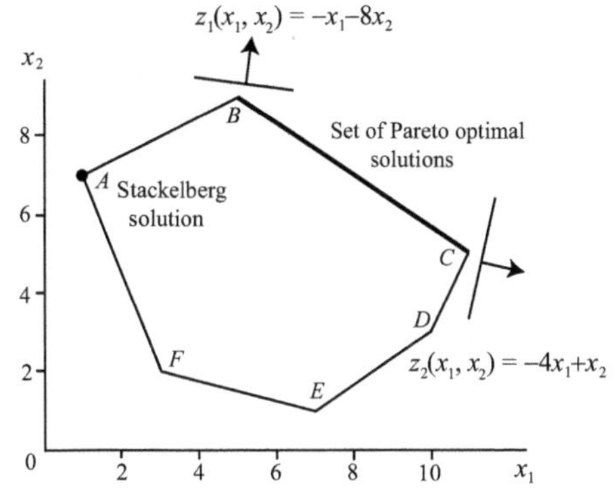

Fig. 4.1 Feasible region of the linear programming problem with two decision makers.

First, consider the cases where the two decision makers choose their actions in turn. If DM1 and DM2 are the leader and the follower, respectively, the inducible region *IR* is the piecewise-linear boundary *C*-*D*-*E*-*F*-*A* and then the Stackelberg solution is $(x_1,x_2) = (1,7)$ at point *A*. Conversely, If DM2 and DM1 are the leader and the follower, respectively, the inducible region *IR* is the piecewise-linear boundary *B*-*C*-*D*-*E* and then the Stackelberg solution is $(x_1,x_2) = (11,5)$ at point *C*.

Next, assuming that DM1 and DM2 make decisions simultaneously, we consider Nash equilibrium solutions. In this situation, the set of Nash equilibrium solutions is the intersection *C*-*D*-*E* of the above mentioned *IR*s, *C*-*D*-*E*-*F*-*A* and *B*-*C*-*D*-*E*, where a unilateral deviation cannot improve the objective function value.

Unlike the noncooperative situations described above, if DM1 and DM2 make a binding agreement to select actions cooperatively, it is predicted that the selected actions should be in the set of Pareto optimal solutions which is the boundary *B*-*C*; this does not depend on whether they make decisions in turn or simultaneously. The set of Pareto optimal solutions can be easily verified by the preference cone which consists of the two vectors of coefficients of the objective functions of DM1 and DM2. Although Pareto optimal solutions in Definition 2.1 are defined for multiple objectives held by one decision maker, Pareto optimal solutions here are thought to be defined for objective functions z_1 and z_2 of the two decision maker, DM1 and DM2.

Assume that DM1 first makes a decision and then DM2 makes a decision later. If they do not have a motivation to cooperate with each other and there is no communication between them, the outcome of the problem can be explained by the Stackelberg solution $(x_1,x_2) = (1,7)$ at point *A*, and then the objective function values of them are $z_1 = -57$ and $z_2 = 3$. However, if they make decisions cooperatively, the Stackelberg solution seems to be not reasonable because there exist solutions dominating the Stackelberg solution. The set of Pareto optimal solutions which are not dominated by any feasible solution is the linear boundary *B*-*C*, and it is justifiable for the two decision makers to make decisions in a coordinated manner so as to yield a certain Pareto optimal solution in the linear boundary *B*-*C*.

Even if the two decision makers reach an agreement to select actions cooperatively, a subsequent controversial aspect is which point in the set of Pareto optimal solutions should be selected. If a utility function which reflects and aggregates the preferences of the two decision makers is identified ingeniously, a Pareto optimal solution maximizing the utility function should be selected. Studies on assessing such a utility function which is called a social welfare function have been accumulated from various perspectives (Arrow, 1963; Fishburn, 1969; Harsanyi, 1955; Keeney and Kirkwood, 1975; Keeney and Raiffa, 1976; Dyer and Sarin, 1979; Baucells and Shapley, 2006).

Let N denote the number of decision makers, and then a social welfare function can be represented by $U(\boldsymbol{x}) = u(u_1(\boldsymbol{x}), \ldots, u_N(\boldsymbol{x}))$, where $u_j(\boldsymbol{x})$ is a utility of the jth individual or group and u is a unspecified function. If some conditions hold, a social welfare function is the additive or the multiplicative form (Harsanyi, 1955; Keeney and Kirkwood, 1975). From the viewpoint of ways to specify the function u, group decision making problems can be classified into two classes. In the first type of problems, a single decision maker identifies the function u, while in the second type of problems, all of the decision makers as a whole decide the form of the function u. In the first case, the single supra decision maker is interpreted as a "benevolent dictator" who takes responsibility of the decisions in the problem. In fuzzy two-level linear programming and its extensions provided in the subsequent sections, one of the decision makers would be considered as some kind of benevolent dictator.

4.2 Fuzzy two- and multi-level linear programming

As shown in chapter 3, the concept of Stackelberg solutions has been usually employed as a solution concept to two-level programming problems. When Stackelberg solutions are employed, it is assumed that there is no communication between two decision makers, or they do not make any binding agreement even if there exists such communication. However, the above assumption is not always reasonable when we deal with decision making problems in a decentralized firm as two-level programming problems in which top management is the leader and an operation division of the firm is the follower because it is supposed that there exists cooperative relationship between them. Namely, top management or an executive board is interested in overall management policy such as long-term corporate growth or market share. In contrast, operation divisions of the firm are concerned with coordination of daily activities. After the top management chooses a strategy in accordance with the overall management policy, each division determines goals which are relevant to the strategy chosen by the top management, and it tries to achieve them. Although decision making problems in a decentralized firm can be formulated as two-level programming problems, there is essentially cooperative relationship between the leader and the follower. Moreover, as for a computational aspect of Stackelberg solutions, even if the objective functions of both decision makers and the common constraint functions are linear, it is known that the mathematical programming problem for obtaining Stackelberg solutions is a nonconvex programming problem with special structure as shown in chapter 3. Although a large number of algorithms for obtaining Stackelberg solutions have been developed, it is also known that solving the mathematical programming problem for obtaining Stackelberg solutions is NP-hard (Shimizu, Ishizuka and Bard, 1997). From such difficulties, a new solution concept which is easy to compute and reflects structure of two-level programming problems had been expected.

Lai (1996) and Shih, Lai and Lee (1996) propose a solution method, which is different from the concept of Stackelberg solutions, for two-level linear programming problems with cooperative relationship between decision makers. Their method is based on an idea that the decision maker at the lower level optimizes her own objective function, taking a goal or preference of the decision maker at the upper level into consideration. The decision makers identify membership functions of the fuzzy goals for their objective functions, and in particular, the decision maker at the upper level also specifies those of the fuzzy goals for the decision variables. The decision maker at the lower level solves a fuzzy programming problem with a constraint with respect to a satisfactory degree of the decision maker at the upper level. Unfortunately, there is a possibility that their method leads a final solution to an undesirable one because of inconsistency between the fuzzy goals of the objective function and those of the decision variables.

Sakawa, Nishizaki and Uemura (1998) present interactive fuzzy programming for two- or multi-level linear programming problems. In order to overcome the problem in the methods of Lai (1996) and Shih, Lai and Lee (1996), after eliminating the fuzzy goals for decision variables, they formulate the two- or multi-level linear programming problem. In their interactive method, after determining the fuzzy goals of the decision makers at all the levels, a satisfactory solution is derived efficiently by updating the satisfactory degree of the decision maker at the upper level with considerations of overall satisfactory balance among all the levels.

4.2.1 Interactive fuzzy programming for two-level problem

As shown in chapter 3, a two-level linear programming problem for obtaining a Stackelberg solution is formulated as:

$$\underset{\boldsymbol{x}_1}{\text{minimize}} \quad z_1(\boldsymbol{x}_1, \boldsymbol{x}_2) = \boldsymbol{c}_{11}\boldsymbol{x}_1 + \boldsymbol{c}_{12}\boldsymbol{x}_2 \tag{4.3a}$$

$$\text{where } y \text{ solves}$$

$$\underset{\boldsymbol{x}_2}{\text{minimize}} \quad z_2(\boldsymbol{x}_1, \boldsymbol{x}_2) = \boldsymbol{c}_{21}\boldsymbol{x}_1 + \boldsymbol{c}_{22}\boldsymbol{x}_2 \tag{4.3b}$$

$$\text{subject to} \quad A_1\boldsymbol{x}_1 + A_2\boldsymbol{x}_2 \leqq \boldsymbol{b} \tag{4.3c}$$

$$\boldsymbol{x}_1 \geqq \boldsymbol{0},\ \boldsymbol{x}_2 \geqq \boldsymbol{0}, \tag{4.3d}$$

where $\boldsymbol{x}_i$, $i = 1,2$ is an n_i-dimensional decision variable column vector, $\boldsymbol{c}_{i1}$, $i = 1,2$ is an n_1-dimensional coefficient row vector, $\boldsymbol{c}_{i2}$, $i = 1,2$ is an n_2-dimensional coefficient row vector, $\boldsymbol{b}$ is an m-dimensional constant column vector, and A_i, $i = 1,2$ is an $m \times n_i$ coefficient matrix. For the sake of simplicity, we use the following notations: $\boldsymbol{x} = (\boldsymbol{x}_1^T, \boldsymbol{x}_2^T)^T \in \mathbb{R}^{n_1+n_2}$, $\boldsymbol{c}_i = (\boldsymbol{c}_{i1}, \boldsymbol{c}_{i2})$, $i = 1,2$, and $A = [A_1\ A_2]$, and let DM1 denote the decision maker at the upper level and let DM2 denote the decision maker at the lower level in this chapter. In the two-level linear programming problem (4.3), $z_1(\boldsymbol{x}_1, \boldsymbol{x}_2)$ and $z_2(\boldsymbol{x}_1, \boldsymbol{x}_2)$ represent the objective functions of DM1 and DM2, respectively, and $\boldsymbol{x}_1$ and $\boldsymbol{x}_2$ represent the decision variables of DM1 and DM2, respectively.

In contrast to the above formulation, in fuzzy linear programming, DM1 specifies a fuzzy goal and a minimal satisfactory level and evaluates a solution proposed by DM2, and DM2 solves an optimization problem with constraints for the fuzzy goal and the minimal satisfactory level of DM1. Combining the two problems (4.1) and (4.2) into one problem, we formulate the following linear programming problem with DM1 and DM2:

$$\underset{\text{for DM1}}{\text{minimize}}\ z_1(\boldsymbol{x}_1, \boldsymbol{x}_2) = \boldsymbol{c}_{11}\boldsymbol{x}_1 + \boldsymbol{c}_{12}\boldsymbol{x}_2 \tag{4.4a}$$

$$\underset{\text{for DM2}}{\text{minimize}}\ z_2(\boldsymbol{x}_1, \boldsymbol{x}_2) = \boldsymbol{c}_{21}\boldsymbol{x}_1 + \boldsymbol{c}_{22}\boldsymbol{x}_2 \tag{4.4b}$$

$$\text{subject to}\ A_1\boldsymbol{x}_1 + A_2\boldsymbol{x}_2 \leqq \boldsymbol{b} \tag{4.4c}$$

$$\boldsymbol{x}_2 \geqq \boldsymbol{0},\ \boldsymbol{x}_1 \geqq \boldsymbol{0}, \tag{4.4d}$$

where the two objective functions z_1 and z_2 are those of DM1 and DM2, respectively, and "$\underset{\text{for DM1}}{\text{minimize}}$" and "$\underset{\text{for DM2}}{\text{minimize}}$" mean that DM1 and DM2 are minimizers for their objective functions.

It is natural that the decision makers have fuzzy goals for their objective functions when they take fuzziness of human judgments into consideration. For each of the objective functions $z_i(\boldsymbol{x})$, $i = 1,2$ of problem (4.4), assume that the decision makers have fuzzy goals such as "the objective function $z_i(\boldsymbol{x})$ should be substantially less than or equal to some specific value p_i."

Let S denote the feasible region of problem (4.4), and then, the individual minimum of the objective function for DMi is

$$z_i^{\min} = z_i(\boldsymbol{x}^{io}) = \min_{\boldsymbol{x} \in S} z_i(\boldsymbol{x}),\ i = 1,2, \tag{4.5}$$

and the individual maximum of the objective function for DMi is

$$z_i^{\max} = \max_{\boldsymbol{x} \in S} z_i(\boldsymbol{x}), i = 1,2. \tag{4.6}$$

The individual maximum and the individual maximum are helpful for DMi to identify a membership function prescribing the fuzzy goal for the objective function $z_i(\boldsymbol{x})$, $i = 1,2$. DMi determines the membership function $\mu_i(z_i(\boldsymbol{x}))$, which is strictly monotone decreasing with $z_i(\boldsymbol{x})$, based on the variation ratio of degree of satisfaction in the interval between the individual minimum (4.5) and the individual maximum (4.6). The domain of the membership function is the interval $[z_i^{\min}, z_i^{\max}]$, $i = 1,2$, and DMi specifies the value z_i^0 of the objective function such that the degree of satisfaction is 0, $\mu_i(z_i^0) = 0$, and the value z_i^1 of the objective function such that the degree of satisfaction is 1, $\mu_i(z_i^1) = 1$. For the value undesired (larger) than z_i^0, it is defined that $\mu_i(z_i(\boldsymbol{x})) = 0$, and for the value desired (smaller) than z_i^1, it is defined that $\mu_i(z_i(\boldsymbol{x})) = 1$.

Although the membership function does not always need to be linear, for the sake of simplicity, we adopt a linear membership function which characterizes the fuzzy goal of each decision maker.

The linear membership function $\mu_i(z_i)$ is defined as:

$$\mu_i(z_i(\boldsymbol{x})) = \begin{cases} 0 & \text{if } z_i(\boldsymbol{x}) > z_i^0 \\ \dfrac{z_i(\boldsymbol{x}) - z_i^0}{z_i^1 - z_i^0} & \text{if } z_i^1 < z_i(\boldsymbol{x}) \leqq z_i^0 \\ 1 & \text{if } z_i(\boldsymbol{x}) \leqq z_i^1, \end{cases} \tag{4.7}$$

and it is depicted in Figure 4.2.

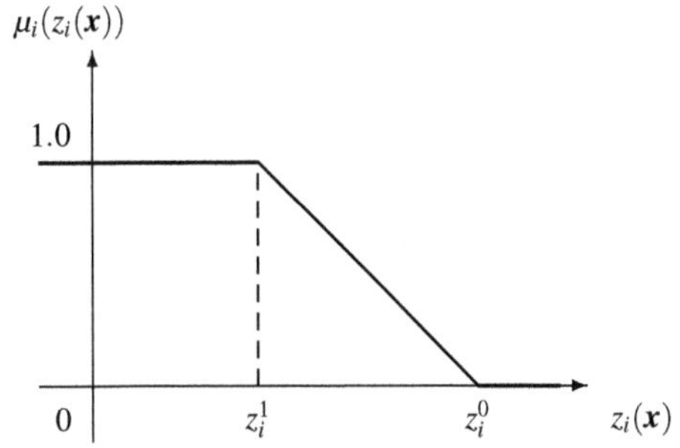

Fig. 4.2 Linear membership function.

Zimmermann (1978) suggests a method for assessing the parameters of the membership function. In his method, the parameter z_i^1 is determined as

$$z_i^1 = z_i^{\min}, \tag{4.8}$$

and the parameter z_i^0 is specified as

$$z_i^0 = z_i^{\mathrm{m}} = z_i(\boldsymbol{x}^{jo}),\ i \neq j, \tag{4.9}$$

where $\boldsymbol{x}^{jo}$ is a feasible solution minimizing $z_j(\boldsymbol{x})$, i.e., $\boldsymbol{x}^{jo} \in \arg\min_{\boldsymbol{x}\in S} z_j(\boldsymbol{x})$. Then, by setting the parameters as described above, the linear membership functions (4.7) is identified.

By identifying the membership functions $\mu_1(z_1(\boldsymbol{x}))$ and $\mu_2(z_2(\boldsymbol{x}))$ for the objective functions $z_1(\boldsymbol{x})$ and $z_2(\boldsymbol{x})$, the original two-level linear programming problem (4.4) can be interpreted as the membership function maximization problem defined by:

$$\begin{aligned}
&\underset{\text{for DM1}}{\text{minimize}} \quad \mu_1(z_1(\boldsymbol{x})) && (4.10a)\\
&\underset{\text{for DM2}}{\text{minimize}} \quad \mu_2(z_2(\boldsymbol{x})) && (4.10b)\\
&\text{subject to} \quad A\boldsymbol{x} \leqq \boldsymbol{b} && (4.10c)\\
&\qquad\qquad\quad \boldsymbol{x} \geqq \boldsymbol{0}. && (4.10d)
\end{aligned}$$

In problem (4.10), $\boldsymbol{x} \in \mathbb{R}^n$ is an n-dimensional decision variable vector, and it is divided into two vectors $\boldsymbol{x}_1$ and $\boldsymbol{x}_2$ which are n_1- and n_2-dimensional decision variable vectors of DM1 and DM2, respectively, i.e., $n = n_1 + n_2$. However, because the two decision makers make decisions cooperatively, the decision variable vector is represented simply by $\boldsymbol{x}$ without partition.

To derive an overall satisfactory solution to the membership function maximization problem (4.10), we first find the maximizing decision of the fuzzy decision proposed by Bellman and Zadeh (1970). Namely, the following problem is solved for obtaining a solution which maximizes the smaller degree of satisfaction between those of the two decision makers:

$$\begin{aligned}
&\text{maximize} \quad \min\{\mu_1(z_1(\boldsymbol{x})), \mu_2(z_2(\boldsymbol{x}))\} && (4.11a)\\
&\text{subject to} \quad A\boldsymbol{x} \leqq \boldsymbol{b} && (4.11b)\\
&\qquad\qquad\quad \boldsymbol{x} \geqq \boldsymbol{0}. && (4.11c)
\end{aligned}$$

By introducing an auxiliary variable λ, this problem can be transformed into the following equivalent problem:

$$\begin{aligned}
&\text{maximize} \quad \lambda && (4.12a)\\
&\text{subject to} \quad \mu_1(z_1(\boldsymbol{x})) \geqq \lambda && (4.12b)\\
&\qquad\qquad\quad \mu_2(z_2(\boldsymbol{x}))\} \geqq \lambda && (4.12c)\\
&\qquad\qquad\quad A\boldsymbol{x} \leqq \boldsymbol{b} && (4.12d)\\
&\qquad\qquad\quad \boldsymbol{x} \geqq \boldsymbol{0}. && (4.12e)
\end{aligned}$$

Solving problem (4.12), we can obtain a solution which maximizes the smaller satisfactory degree between those of both decision makers. It should be noted that if the membership functions $\mu_i(z_i(\boldsymbol{x}))$, $i = 1, 2$ are linear membership functions such as (4.7), problem (4.12) becomes a linear programming problem. Let $\boldsymbol{x}^*$ denote an optimal solution to problem (4.12). Then, we define the satisfactory degree of both decision makers under the constraints as

$$\lambda^* = \min\{\mu_1(z_1(\boldsymbol{x}^*)), \mu_2(z_2(\boldsymbol{x}^*))\}. \tag{4.13}$$

If DM1 is satisfied with the optimal solution $\boldsymbol{x}^*$, it follows that the optimal solution $\boldsymbol{x}^*$ becomes a satisfactory solution; however, DM1 is not always satisfied with the solution $\boldsymbol{x}^*$. It is quite natural to assume that DM1 specifies the minimal satisfactory level $\hat{\delta} \in [0,1]$ for the membership function $\mu_1(z_1(\boldsymbol{x}))$ subjectively.

Consequently, if DM1 is not satisfied with the solution $\boldsymbol{x}^*$ to problem (4.12), the following problem is formulated:

$$\text{maximize } \mu_2(z_2(\boldsymbol{x})) \tag{4.14a}$$

$$\text{subject to } \mu_1(z_1(\boldsymbol{x})) \geqq \hat{\delta} \tag{4.14b}$$

$$A\boldsymbol{x} \leqq \boldsymbol{b} \tag{4.14c}$$

$$\boldsymbol{x} \geqq \boldsymbol{0}, \tag{4.14d}$$

where DM2's membership function is maximized under the condition that DM1's membership function $\mu_1(z_1(\boldsymbol{x}))$ is larger than or equal to the minimal satisfactory level $\hat{\delta}$ specified by DM1. It should be also noted that if the membership functions $\mu_i(z_i(\boldsymbol{x}))$, $i = 1,2$ are linear membership functions such as (4.7), problem (4.14) becomes a linear programming problem.

If there exists an optimal solution to problem (4.14), it follows that DM1 obtains a satisfactory solution having a satisfactory degree larger than or equal to the minimal satisfactory level specified by DM1's self. However, the larger the minimal satisfactory level $\hat{\delta}$ is assessed, the smaller the DM2's satisfactory degree becomes when the objective functions of DM1 and DM2 conflict with each other. Consequently, a relative difference between the satisfactory degrees of DM1 and DM2 becomes larger, and it follows that the overall satisfactory balance between both decision makers is not appropriate.

In order to take account of the overall satisfactory balance between both decision makers, DM1 needs to compromise with DM2 on DM1's own minimal satisfactory level. To do so, the following ratio of the satisfactory degree of DM2 to that of DM1 is helpful:

$$\Delta = \frac{\mu_2(z_2(\boldsymbol{x}))}{\mu_1(z_1(\boldsymbol{x}))}, \tag{4.15}$$

which is originally introduced by Lai (1996).

DM1 is guaranteed to have a satisfactory degree larger than or equal to the minimal satisfactory level for the fuzzy goal because the corresponding constraint (4.14b) is involved in problem (4.14). To take into account the overall satisfactory balance between both decision makers, we provide two methods for evaluating the ratio Δ of satisfactory degrees. In the first method, DM1 specifies the lower bound $\Delta_{\min}$ and the upper bound $\Delta_{\max}$ of the ratio, and the ratio Δ is evaluated by verifying whether or not it is in the interval $[\Delta_{\min}, \Delta_{\max}]$. The condition that the overall satisfactory balance is appropriate is represented by

$$\Delta \in [\Delta_{\min}, \Delta_{\max}]. \tag{4.16}$$

In the second method, DM1 identifies a fuzzy goal $\tilde{R}$ for the ratio Δ of satisfactory degrees, which is expressed in words such as "the ratio Δ should be in the vicinity of a certain value m," and he gives the permissible level $\hat{\delta}_{\tilde{\Delta}}$ to the membership value of the fuzzy goal. The condition that the overall satisfactory balance is appropriate is represented by

$$\mu^{\tilde{R}}(\Delta) \geqq \hat{\delta}_{\tilde{\Delta}}, \tag{4.17}$$

where $\mu^{\tilde{R}}$ denotes a membership function of the fuzzy goal $\tilde{R}$. In particular, the extended version of the second method is employed in multiobjective environments as we will show later.

The two methods have relevance to each other by relating the α-level set $\{\Delta \mid \mu^{\tilde{R}}(\Delta) \geqq \hat{\delta}_{\tilde{\Delta}}\}$ to the interval $[\Delta_{\min}, \Delta_{\max}]$. Namely, by regarding the interval of the α-level set as the interval $[\Delta_{\min}, \Delta_{\max}]$ for the appropriate ratio, the two methods can be treated in the same way. Therefore, we explain only interactive procedure with the first method.

At the iteration l, let $\mu_1(z_1^l)$, $\mu_2(z_2^l)$, λ^l and $\Delta^l = \mu_2(z_2^l)/\mu_1(z_1^l)$ denote DM1's and DM2's satisfactory degrees, a satisfactory degree of both decision makers, and the ratio of satisfactory degrees of the two decision makers, respectively. Let the solution be $\boldsymbol{x}^l$ at the iteration l. The interactive process terminates if the following two conditions are satisfied and DM1 concludes the solution as an overall satisfactory solution.

[Termination conditions of the interactive process]

Condition 1 DM1's satisfactory degree is larger than or equal to the minimal satisfactory level $\hat{\delta}$ specified by DM1's self, i.e., $\mu_1(z_1^l) \geqq \hat{\delta}$.

Condition 2 The ratio Δ^l of satisfactory degrees lies in the closed interval between the lower and the upper bounds specified by DM1, i.e., $\Delta^l \in [\Delta_{\min}, \Delta_{\max}]$.

Condition 1 ensures the minimal satisfaction to DM1 in the sense of the attainment of the fuzzy goal, and condition 2 is provided in order to keep overall satisfactory balance between both decision makers. If these two conditions are not satisfied simultaneously, DM1 needs to update the minimal satisfactory level $\hat{\delta}$. The updating procedures are summarized as follows.

[Procedure for updating the minimal satisfactory level $\hat{\delta}$]

Case 1 If condition 1 is not satisfied, then DM1 decreases the minimal satisfactory level $\hat{\delta}$.

Case 2 If the ratio Δ^l exceeds its upper bound, then DM1 increases the minimal satisfactory level $\hat{\delta}$. Conversely, if the ratio Δ^l is below its lower bound, then DM1 decreases the minimal satisfactory level $\hat{\delta}$.

Case 3 Although conditions 1 and 2 are satisfied, if DM1 is not satisfied with the obtained solution and judges that it is desirable to increase the satisfactory degree of DM1 at the expense of the satisfactory degree of DM2, then DM1 increases the minimal satisfactory level $\hat{\delta}$. Conversely, if DM1 judges that it is desirable to increase the satisfactory degree of DM2 at the expense of the satisfactory degree of DM1, then DM1 decreases the minimal satisfactory level $\hat{\delta}$.

In particular, if condition 1 is not satisfied, there does not exist any feasible solution for problem (4.14), and therefore DM1 has to moderate the minimal satisfactory level.

We are now ready to give a procedure of interactive fuzzy programming for deriving an overall satisfactory solution to problem (4.4), which is summarized in the following, and it is illustrated with a flowchart in Figure 4.3:

[Algorithm of interactive fuzzy programming]

Step 0 Calculate the individual minimum (4.5) and maximum (4.6) for each decision maker.

Step 1 Ask DM1 to identify the membership function $\mu_1(z_1)$ of the fuzzy goal for the objective function $z_1(\boldsymbol{x})$. Similarly, ask DM2 to identify the membership function $\mu_2(z_2)$ of the fuzzy goal for the objective function $z_2(\boldsymbol{x})$.

Step 2 Set $l := 1$ and solve problem (4.12), in which a smaller satisfactory degree between those of DM1 and DM2 is maximized. If DM1 is satisfied with the obtained optimal solution, the solution becomes a satisfactory solution. Otherwise, ask DM1 to specify the minimal satisfactory level $\hat{\delta}$ together with the lower and the upper bounds $[\Delta_{\min}, \Delta_{\max}]$ of the ratio of satisfactory degrees Δ^l with the satisfactory degree λ^* of both decision makers and the related information about the solution in mind.

Step 3 Set $l := l+1$. Solve problem (4.14), in which the satisfactory degree of DM2 is maximized under the condition that the satisfactory degree of DM1 is larger than or equal to the minimal satisfactory level $\hat{\delta}$, and then an optimal solution $\boldsymbol{x}^l$ to problem (4.14) is proposed to DM1 together with λ^l, $\mu_1(z_1^l)$, $\mu_2(z_2^l)$ and Δ^l.

Step 4 If the solution $\boldsymbol{x}^l$ satisfies the termination conditions and DM1 accepts it, then the procedure stops, and the solution $\boldsymbol{x}^l$ is determined to be a satisfactory solution.

Step 5 Ask DM1 to revise the minimal satisfactory level $\hat{\delta}$ in accordance with the procedure of updating minimal satisfactory level. Return to Step 4.

4.2.2 Numerical example for two-level problem

To illustrate the procedure of interactive fuzzy programming described above, consider the following two-level linear programming problem:

$$\underset{\text{for DM1}}{\text{minimize}} \quad z_1 = \boldsymbol{c}_1\boldsymbol{x}_1 + \boldsymbol{c}_2\boldsymbol{x}_2 \tag{4.18a}$$

$$\underset{\text{for DM2}}{\text{minimize}} \quad z_2 = \boldsymbol{c}_3\boldsymbol{x}_1 + \boldsymbol{c}_4\boldsymbol{x}_2 \tag{4.18b}$$

$$\text{subject to} \quad A_1\boldsymbol{x}_1 + A_2\boldsymbol{x}_2 \leqq \boldsymbol{b} \tag{4.18c}$$

$$x_{1j} \geqq 0,\ j = 1,\ldots,7,\ x_{2j} \geqq 0,\ j = 1,\ldots,8, \tag{4.18d}$$

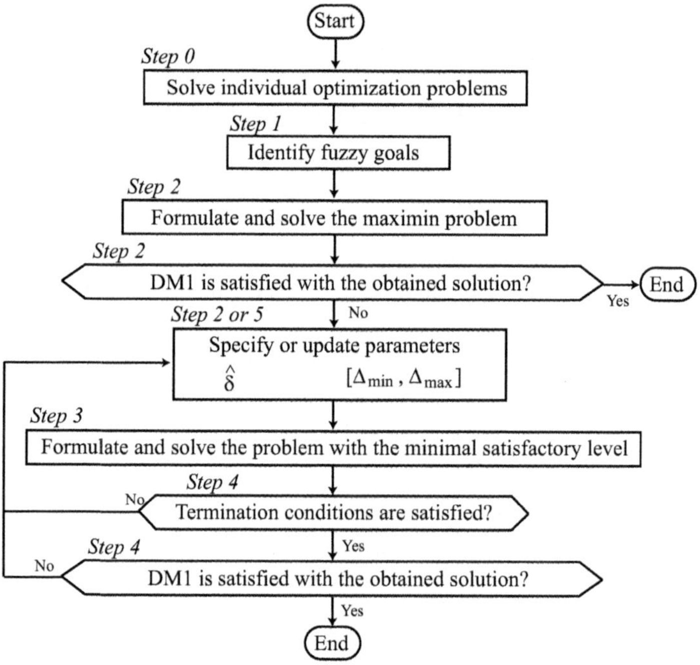

Fig. 4.3 Flowchart of interactive fuzzy programming.

where $\boldsymbol{x}_1 = (x_{11},\dots,x_{17})^T$ and $\boldsymbol{x}_2 = (x_{21},\dots,x_{28})^T$; each entry of coefficient vectors $\boldsymbol{c}_i$, $i = 1,2,3,4$, 16×7 coefficient matrix A_1 and 16×8 coefficient matrix A_2 is a randomly selected number from the interval $[-50,50]$; each entry of the right hand side constant column vector $\boldsymbol{b}$ is a sum of entries of the corresponding row vector of A_1 and A_2 multiplied by 0.6. These coefficients are shown in Table 4.1.

To identify membership functions of the fuzzy goals for the objective functions, the two individual minimization problems of both decision makers are solved at the beginning of the procedure. The individual minima and the corresponding optimal solutions are shown in Table 4.2.

Suppose that the decision makers employ the linear membership function (4.7) whose parameters are determined by the Zimmermann method (1978). Then, one finds that $z_1^0 = z_1^{\rm m} = -431.707$ and $z_2^0 = z_2^{\rm m} = -364.147$, and the maximin problem (4.12) for this numerical example can be written as

$$\begin{aligned}
&\text{maximize } \lambda && (4.19\text{a})\\
&\text{subject to } (z_1(\boldsymbol{x}) + 431.707)/(-530.681 + 431.707) \geqq \lambda && (4.19\text{b})\\
&\qquad (z_2(\boldsymbol{x}) + 364.147)/(-374.497 + 364.147) \geqq \lambda && (4.19\text{c})\\
&\qquad \boldsymbol{x} \in S, && (4.19\text{d})
\end{aligned}$$

Table 4.1 Coefficients of problem (4.18).

c_1	−45	−5	−45	−14	−9	−26	−47	c_2	−19	−27	−27	−33	−7	−28	−21	−12		
c_3	−45	−17	−28	−49	−24	−4	−5	c_4	−49	−38	−32	−2	−5	−33	−4	−9		
A_1	46	−29	−48	31	21	−47	−37	A_2	37	32	19	21	2	−11	−35	−2	b	0
	0	39	12	−14	29	−42	−26		42	31	−19	−25	6	4	2	5		26
	38	−27	5	−31	14	−38	−29		−5	−47	49	−45	−45	5	10	−40		−111
	−26	16	44	6	19	17	27		32	17	−6	−27	1	18	−6	15		88
	−48	13	2	−33	19	22	−35		27	−35	−35	−26	−16	37	47	−2		−37
	9	−6	12	−17	−32	−8	24		−24	45	−31	16	−9	−19	17	44		12
	−9	2	−16	8	32	−6	−25		−25	−8	4	23	41	30	36	−11		45
	24	30	42	−26	16	19	−18		−18	9	−34	−46	30	3	−1	−45		−8
	−26	−8	0	41	−42	−19	13		−42	49	−27	4	−2	−12	24	−33		−47
	−29	16	−16	−4	18	45	−8		21	6	47	43	46	26	22	−5		136
	−7	1	−3	38	18	−43	−15		31	−34	23	−35	−34	20	−15	−26		−48
	−12	−4	47	0	−4	−18	−19		28	47	−36	−45	20	40	3	−15		19
	−12	−46	11	−47	−47	19	30		50	12	−24	13	20	−43	−8	20		−31
	5	−2	37	38	0	12	−34		34	28	−40	−18	33	39	14	2		88
	49	41	3	12	−48	15	12		32	31	−28	−25	−23	−6	−25	−15		14
	−17	−6	34	21	11	5	−28		−46	−15	9	12	49	4	−17	−47		−18

Table 4.2 Optimal solutions to individual problems.

$z_1^{\min}$	−530.681							
x_1	2.723	0.000	0.000	2.224	2.565	1.642	4.417	
x_2	0.000	0.000	1.051	1.290	0.000	0.000	1.557	0.000
$z_2^{\min}$	−374.497							
x_1	2.152	0.000	1.058	2.857	1.365	2.005	1.582	
x_2	0.000	0.000	1.288	0.854	0.000	0.000	1.507	1.156

where S denotes the feasible region of problem (4.18). The result of the first iteration including an optimal solution to problem (4.19) is shown in Table 4.3.

Table 4.3 Iteration 1.

x_1^1	2.525	0.000	0.979	2.359	2.046	1.718	3.007	
x_2^1	0.000	0.000	1.054	1.562	0.000	0.000	1.304	0.232
(z_1^1, z_2^1)	$(-505.332, -371.846)$							
$(\mu_1(z_1^1), \mu_2(z_2^1))$	$(0.744, 0.744)$							
λ^1	0.744							
Δ^1	1.000							

Suppose that DM1 is not satisfied with the solution obtained in Iteration 1, and then he specifies the minimal satisfactory level at $\hat{\delta} = 0.8$ and the bounds of the ratio at the interval $[\Delta_{\min}, \Delta_{\max}] = [0.8, 0.9]$, taking account of the result of the first iteration. Then, the problem with the minimal satisfactory level (4.14) is written as

$$\text{maximize} \quad \mu_2(z_2(\boldsymbol{x})) \tag{4.20a}$$
$$\text{subject to} \quad (z_1(\boldsymbol{x})+431.707)/(-530.681+431.707) \geqq 0.8 \tag{4.20b}$$
$$\boldsymbol{x} \in S. \tag{4.20c}$$

The result of the second iteration including an optimal solution to problem (4.20) is shown in Table 4.4.

Table 4.4 Iteration 2.

$\boldsymbol{x}_1^2$	2.554	0.000	0.974	2.322	2.097	1.697	3.114	
$\boldsymbol{x}_2^2$	0.000	0.000	1.037	1.616	0.000	0.000	1.288	0.162
(z_1^2, z_2^2)	$(-510.886, -371.646)$							
$(\mu_1(z_1^2), \mu_2(z_2^2))$	$(0.800, 0.725)$							
λ^2	0.725							
Δ^2	0.906							

At the second iteration, the satisfactory degree $\mu_1(z_1^2) = 0.800$ of DM1 becomes equal to the minimal satisfactory level $\hat{\delta} = 0.800$, but the ratio $\Delta^2 = 0.906$ of satisfactory degrees is not in the valid interval $[0.8, 0.9]$ of the ratio. Therefore, this solution does not satisfy the second condition of termination of the interactive process. Suppose that DM1 updates the minimal satisfactory level at $\hat{\delta} = 0.9$. Then, the problem with the revised minimal satisfactory level (4.14) is solved, and the result of the third iteration is shown in Table 4.5.

Table 4.5 Iteration 3.

$\boldsymbol{x}_1^3$	2.604	0.000	0.963	2.255	2.189	1.658	3.306	
$\boldsymbol{x}_2^3$	0.000	0.000	1.005	1.711	0.000	0.000	1.261	0.038
(z_1^3, z_2^3)	$(-520.783, -371.290)$							
$(\mu_1(z_1^3), \mu_2(z_2^3))$	$(0.900, 0.690)$							
λ^3	0.690							
Δ^3	0.767							

At the third iteration, the ratio $\Delta^3 = 0.767$ of satisfactory degrees is not in the valid interval $[0.8, 0.9]$ of the ratio yet. Therefore, this solution does not satisfy the second condition of termination of the interactive process. Suppose that DM1 updates the minimal satisfactory level at $\hat{\delta} = 0.85$. Then, the problem with the revised minimal satisfactory level (4.14) is solved, and the result of the fourth iteration is given in Table 4.6.

At the fourth iteration, the satisfactory degree $\mu_1(z_1^4) = 0.850$ of DM1 becomes equal to the minimal satisfactory level $\hat{\delta} = 0.85$, and the ratio $\Delta^4 = 0.832$ of satisfactory degrees is in the valid interval $[0.8, 0.9]$ of the ratio. Therefore, this solution

Table 4.6 Iteration 4.

x_1^4	2.579	0.000	0.968	2.288	2.143	1.677	3.210	
x_2^4	0.000	0.000	1.021	1.663	0.000	0.000	1.275	0.100
(z_1^4, z_2^4)	$(-515.834, -371.468)$							
$(\mu_1(z_1^4), \mu_2(z_2^4))$	$(0.850, 0.707)$							
λ^4	0.707							
Δ^4	0.832							

satisfies the termination conditions of the interactive process, and it becomes a satisfactory solution for both decision makers if DM1 accepts the solution.

4.2.3 Interactive fuzzy programming for multi-level problem

In this subsection, we extend interactive fuzzy programming for two-level linear programming problems to that for multi-level linear programming problems. Suppose that there is a single decision maker at each level in an organization with t levels, which is depicted in Figure 4.4.

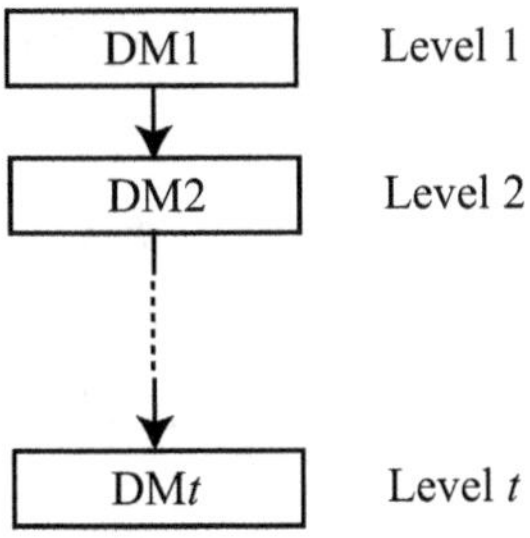

Fig. 4.4 Multi-level organization.

A multi-level linear programming problem is formally represented as:

$$\underset{\text{for DM1}}{\text{minimize}} \quad z_1(\boldsymbol{x}_1, \ldots, \boldsymbol{x}_t) = \boldsymbol{c}_{11}\boldsymbol{x}_1 + \cdots + \boldsymbol{c}_{1t}\boldsymbol{x}_t \tag{4.21a}$$

$$\cdots\cdots\cdots\cdots\cdots\cdots$$

$$\underset{\text{for DM}t}{\text{minimize}} \quad z_t(\boldsymbol{x}_1, \ldots, \boldsymbol{x}_t) = \boldsymbol{c}_{t1}\boldsymbol{x}_1 + \cdots + \boldsymbol{c}_{tt}\boldsymbol{x}_t \tag{4.21b}$$

$$\text{subject to} \quad A_1\boldsymbol{x}_1 + \cdots + A_t\boldsymbol{x}_t \leqq \boldsymbol{b} \tag{4.21c}$$

$$\boldsymbol{x}_1 \geqq \boldsymbol{0}, \ldots, \boldsymbol{x}_t \geqq \boldsymbol{0}, \tag{4.21d}$$

where $\boldsymbol{x}_i$, $i = 1,\ldots,t$ is an n_i-dimensional decision variable column vector; $\boldsymbol{c}_{ij}$, $i,j = 1,\ldots,t$ is an n_j-dimensional coefficient row vector; A_j, $j = 1,\ldots,t$ is an $m \times n_j$ coefficient matrix; $\boldsymbol{x}_i$ and $z_i(\boldsymbol{x}_1,\ldots,\boldsymbol{x}_t)$, $i = 1,\ldots,t$ are the decision variable vector and the objective function of the decision maker at the ith level, respectively. In the multi-level linear programming problem (4.21) with cooperative t decision makers, "$\underset{\text{for DM1}}{\text{minimize}}$", ..., "$\underset{\text{for DM}t}{\text{minimize}}$" mean that t decision makers in t levels are all minimizers for their objective functions.

For the sake of simplicity, we use the following notations: $\boldsymbol{x}^T = (\boldsymbol{x}_1^T,\ldots,\boldsymbol{x}_t^T)^T \in \mathbb{R}^{n_1+\cdots+n_t}$, $\boldsymbol{c}_1 = (\boldsymbol{c}_{11},\ldots,\boldsymbol{c}_{1t})$, ..., $\boldsymbol{c}_t = (\boldsymbol{c}_{t1},\ldots,\boldsymbol{c}_{tt})$, and $A = [A_1 \cdots A_t]$, and let DMi, $i = 1,\ldots,t$ denote the decision maker at the ith level.

As in the two-level linear programming problems considered before, it is natural that each of the decision makers has a fuzzy goal for the objective function when the decision maker takes fuzziness of human judgments into consideration. To identify membership functions of the fuzzy goals for the objective functions, we solve t individual single-objective problems for $i = 1,\ldots,t$:

$$\text{minimize} \quad z_i(\boldsymbol{x}_1,\ldots,\boldsymbol{x}_t) = \boldsymbol{c}_{i1}\boldsymbol{x}_1 + \cdots + \boldsymbol{c}_{it}\boldsymbol{x}_t \tag{4.22a}$$

$$\text{subject to} \quad A_1\boldsymbol{x}_1 + \cdots + A_t\boldsymbol{x}_t \leqq \boldsymbol{b} \tag{4.22b}$$

$$\boldsymbol{x}_1 \geqq \boldsymbol{0},\ldots,\boldsymbol{x}_t \geqq \boldsymbol{0}. \tag{4.22c}$$

Suppose that, for the multi-level programming problems, the decision makers employ the linear membership function (4.7) whose parameters are determined by the Zimmermann method (1978). The individual minimum is

$$\begin{aligned} z_i^{\min} &= z_i(\boldsymbol{x}^{io}) \\ &= \min\{z_i(\boldsymbol{x}) \mid A_1\boldsymbol{x}_1 + \cdots + A_t\boldsymbol{x}_t \leqq \boldsymbol{b}, \boldsymbol{x}_1 \geqq \boldsymbol{0},\ldots,\boldsymbol{x}_t \geqq \boldsymbol{0}\}, \end{aligned} \tag{4.23}$$

and z_i^{m} is defined by

$$z_i^{\mathrm{m}} = \max_{j=1,\ldots,t,\, j\neq i} \{z_i(\boldsymbol{x}^{jo})\}. \tag{4.24}$$

Then, the parameters z_i^0 and z_i^1 of the linear membership function are determined by choosing $z_i^1 = z_i^{\min}$, $z_i^0 = z_i^{\mathrm{m}}$, $i = 1,\ldots,t$.

After identifying the membership functions $\mu_i(z_i(\boldsymbol{x}))$, $i = 1,\ldots,t$, to derive an overall satisfactory solution to the formulated problem (4.21), we first solve the following maximin problem for obtaining a solution which maximizes the smallest degree of satisfaction among those of the t decision makers:

$$\text{maximize} \quad \min\{\mu_1(z_1(\boldsymbol{x})),\ldots,\mu_t(z_t(\boldsymbol{x}))\} \tag{4.25a}$$

$$\text{subject to} \quad \boldsymbol{x} \in S, \tag{4.25b}$$

where S denotes the feasible region of problem (4.21).

By introducing an auxiliary variable λ, this problem can be transformed into the following equivalent maximization problem:

$$\begin{aligned} & \text{maximize } \lambda && (4.26a) \\ & \text{subject to } \mu_1(z_1(\boldsymbol{x})) \geqq \lambda && (4.26b) \\ & \qquad \cdots\cdots \\ & \qquad \mu_t(z_t(\boldsymbol{x})) \geqq \lambda && (4.26c) \\ & \qquad \boldsymbol{x} \in S. && (4.26d) \end{aligned}$$

Solving problem (4.26), we can obtain a solution which maximizes the smallest satisfactory degree among those of the t decision makers. Let $\boldsymbol{x}^*$ denote an optimal solution to problem (4.26). Then, the t decision makers try to derive a satisfactory solution for all of them, taking account of the satisfactory degree of the t decision makers:

$$\lambda^* = \min\{\mu_1(z_1(\boldsymbol{x}^*)), \ldots, \mu_t(z_t(\boldsymbol{x}^*))\}, \tag{4.27}$$

and the ratio of satisfactory degrees of decision makers in adjacent two levels:

$$\Delta_i = \frac{\mu_{i+1}(z_{i+1}(\boldsymbol{x}))}{\mu_i(z_i(\boldsymbol{x}))}, \; i = 1, \ldots, t-1. \tag{4.28}$$

If the decision makers are not satisfied with the optimal solution $\boldsymbol{x}^*$ to problem (4.26), from the bottom level to the top level, problems for deriving satisfactory solutions are formulated and solved one after another. First, we take up the two-level problem between the $(t-1)$th level and the tth level, and derive a satisfactory solution for the two decision makers by using a procedure similar to that of two-level problem described before. Next, we deal with the problem for the three decision makers at the $(t-2)$th, $(t-1)$th and tth levels. Through such sequential procedures, we can derive a satisfactory solution for all the t decision makers.

Because it is difficult to adjust appropriate ratios of satisfactory degrees for multi-level programming problems in comparison with two-level problems, we employ the second method for taking the overall satisfactory balance between two levels, and formulate a problem where not only the condition of the minimal satisfactory level $\hat{\delta}$ of the decision maker at the upper level but also the condition of the fuzzy goal $\tilde{R}$ for the ratio Δ of satisfactory degrees are included in the constraints.

Namely, we formulate the following subproblem for the two decision makers at the $(t-1)$th level and the tth level:

$$\begin{aligned} & \text{maximize } \lambda && (4.29a) \\ & \text{subject to } \mu_{t-1}(z_{t-1}(\boldsymbol{x})) \geqq \hat{\delta}_{t-1} && (4.29b) \\ & \qquad \mu_{t-1}^{\tilde{R}}(\Delta_{t-1}(\boldsymbol{x})) \geqq \hat{\delta}_{\Delta_{t-1}} && (4.29c) \\ & \qquad \mu_t(z_t(\boldsymbol{x})) \geqq \lambda && (4.29d) \\ & \qquad \boldsymbol{x} \in S. && (4.29e) \end{aligned}$$

The interactive process for DM$(t-1)$ and DMt terminates if an optimal solution to problem (4.29) can be obtained and DM$(t-1)$ concludes the solution as an overall satisfactory solution between DM$(t-1)$ and DMt.

Because problem (4.29) includes the condition of DM$(t-1)$'s satisfactory degree, $\mu_{t-1}(z_{t-1}(\boldsymbol{x})) \geqq \hat{\delta}_{t-1}$, and the condition of the ratio of satisfactory degrees, $\mu_{t-1}^{\tilde{R}}(\Delta_{t-1}(\boldsymbol{x})) \geqq \hat{\delta}_{\Delta_{t-1}}$, in the constraints, if the minimal satisfactory level $\hat{\delta}_{t-1}$ is assessed too large and/or the permissible level $\delta_{\Delta_{t-1}}$ is also assessed too large, there may not exist any feasible solution. Therefore, DM$(t-1)$ has to decrease either of the minimal satisfactory level $\hat{\delta}_{t-1}$ or the permissible level $\delta_{\Delta_{t-1}}$, or both of them if an optimal solution to problem (4.29) cannot be obtained.

Suppose that an optimal solution to problem (4.29) is obtained, and DM$(t-1)$ agrees that the solution is an overall satisfactory solution between DM$(t-1)$ and DMt. Let $\hat{\Delta}_{t-1}$ denote the ratio of satisfactory degrees between DM$(t-1)$ and DMt in problem (4.29). To keep the ratio in the next subproblem, we introduce the constraint, $(1/\hat{\Delta}_{t-1})\mu_t(z_t(\boldsymbol{x})) \geqq \lambda$, and formulate the following problem for the three decision makers at the $(t-2)$th, $(t-1)$th and tth levels:

$$\text{maximize } \lambda \tag{4.30a}$$
$$\text{subject to } \mu_{t-2}(z_{t-2}(\boldsymbol{x})) \geqq \hat{\delta}_{t-2} \tag{4.30b}$$
$$\mu_{t-2}^{\tilde{R}}(\Delta_{t-2}(\boldsymbol{x})) \geqq \hat{\delta}_{\Delta_{t-2}} \tag{4.30c}$$
$$\mu_{t-1}(z_{t-1}(\boldsymbol{x})) \geqq \lambda \tag{4.30d}$$
$$(1/\hat{\Delta}_{t-1})\mu_t(z_t(\boldsymbol{x})) \geqq \lambda \tag{4.30e}$$
$$\boldsymbol{x} \in S. \tag{4.30f}$$

In general, the problem for decision makers from the qth level to the bottom tth level can be represented as follows:

$$\text{maximize } \lambda \tag{4.31a}$$
$$\text{subject to } \mu_q(z_q(\boldsymbol{x})) \geqq \hat{\delta}_q \tag{4.31b}$$
$$\mu_q^{\tilde{R}}(\Delta_q(\boldsymbol{x})) \geqq \hat{\delta}_{\Delta_q} \tag{4.31c}$$
$$\mu_{q+1}(z_{q+1}(\boldsymbol{x})) \geqq \lambda \tag{4.31d}$$
$$(1/\hat{\Delta}_{q+1})\mu_{q+2}(z_{q+2}(\boldsymbol{x})) \geqq \lambda \tag{4.31e}$$
$$\cdots\cdots\cdots$$
$$(1/\hat{\Delta}_{q+1})\cdots(1/\hat{\Delta}_{t-1})\mu_t(z_t(\boldsymbol{x})) \geqq \lambda \tag{4.31f}$$
$$\boldsymbol{x} \in S, \tag{4.31g}$$

where, for $i = q+1,\ldots,t-1$, $\hat{\Delta}_i$ denote the ratio of satisfactory degrees between DMi and DM$(i+1)$ obtained in the previous iteration.

We present an interactive algorithm for deriving an overall satisfactory solution to the multi-level linear programming problem (4.21), which is summarized in the following and is illustrated with a flowchart in Figure 4.5:

[Algorithm of interactive fuzzy programming for multi-level problem]

Step 0 Solve the individual problems (4.22) for the t decision makers.

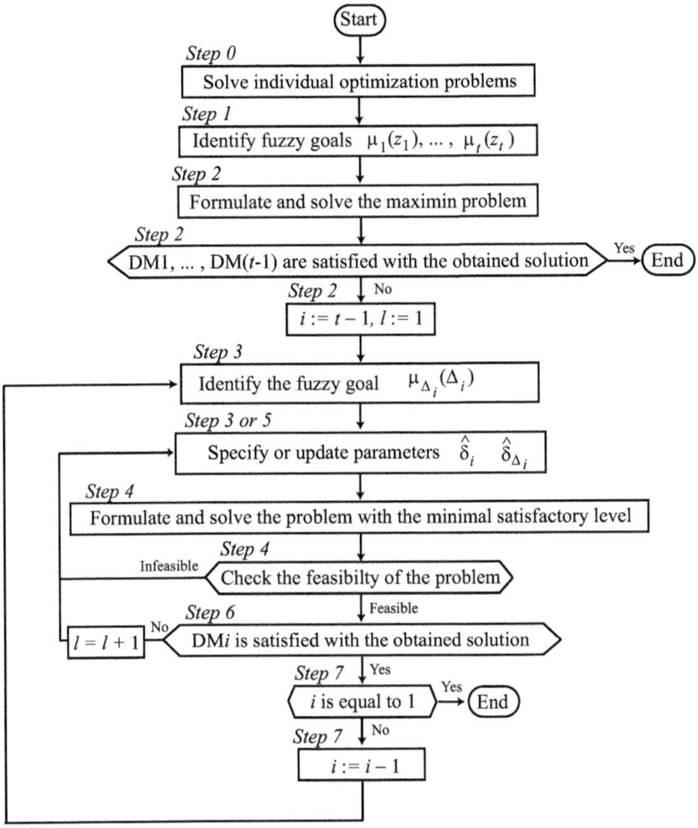

Fig. 4.5 Flowchart of interactive fuzzy programming for multi-level problem.

Step 1 Ask all of DM1, ..., DMt to identify the membership functions $\mu_1(z_1), \ldots, \mu_t(z_t)$ of the fuzzy goals for the objective functions.

Step 2 Set $i := t-1$ and $l := 1$. Solve problem (4.26) where the smallest degree among the satisfactory degrees of DM1, ..., DMt is maximized. If DM1, ..., DM$(t-1)$ are satisfied with the obtained optimal solution, the solution becomes a satisfactory solution. Otherwise, go to Step3.

Step 3 Ask DMi to specify the minimal satisfactory level $\hat{\delta}_i$. Moreover, ask DMi to identify the membership function of the fuzzy goal for the ratio of satisfactory degrees between DMi and DM$(i+1)$ and to specify the permissible level δ_{Δ_i} by taking account of the satisfactory degrees the related information about the optimal solutions to problems (4.26) or (4.31).

Step 4 Solve problem (4.31), and if an optimal solution can be obtained, then an optimal solution $\boldsymbol{x}^l$ of problem (4.31) is proposed to DMi together with λ^l, $\mu_i(z_i^l)$, $\ldots$, $\mu_t(z_t^l)$ and $\Delta_i^l, \ldots, \Delta_{t-1}^l$, and go to Step 6.

Step 5 Ask DMi to update the minimal satisfactory level $\hat{\delta}_i$ and/or the permissible level δ_{Δ_i}, and return to Step 4.
Step 6 If DMi is satisfied with the solution obtained in Step 4, go to Step 7. Otherwise, let $l := l+1$ and return to Step 5.
Step 7 If the counter i is equal to 1, the algorithm stops. Otherwise, let $i := i-1$, and return to Step3.

4.2.4 Numerical example for multi-level problem

To illustrate the procedure of interactive fuzzy programming described above, consider the following multi-level linear programming problem:

$$\underset{\text{for DM1}}{\text{minimize}} \quad z_1 = \boldsymbol{c}_1\boldsymbol{x}_1 + \boldsymbol{c}_2\boldsymbol{x}_2 + \boldsymbol{c}_3\boldsymbol{x}_3 \tag{4.32a}$$

$$\underset{\text{for DM2}}{\text{minimize}} \quad z_2 = \boldsymbol{c}_4\boldsymbol{x}_1 + \boldsymbol{c}_5\boldsymbol{x}_2 + \boldsymbol{c}_6\boldsymbol{x}_3 \tag{4.32b}$$

$$\underset{\text{for DM3}}{\text{minimize}} \quad z_3 = \boldsymbol{c}_7\boldsymbol{x}_1 + \boldsymbol{c}_8\boldsymbol{x}_2 + \boldsymbol{c}_9\boldsymbol{x}_3 \tag{4.32c}$$

$$\text{subject to} \quad A_1\boldsymbol{x}_1 + A_2\boldsymbol{x}_2 + A_3\boldsymbol{x}_3 \leqq \boldsymbol{b} \tag{4.32d}$$

$$x_{ij} \geqq 0,\ i = 1,2,3,\ j = 1,\ldots,5, \tag{4.32e}$$

where $\boldsymbol{x}_1 = (x_{11},\ldots,x_{15})^T$, $\boldsymbol{x}_2 = (x_{21},\ldots,x_{25})^T$ and $\boldsymbol{x}_3 = (x_{31},\ldots,x_{35})^T$; each entry of coefficient vectors $\boldsymbol{c}_i$, $i = 1,\ldots,9$ and 16×5 coefficient matrices A_i, $i = 1,2,3$ is a randomly selected number from the interval $[-50,50]$; each entry of the right-hand side constant column vector $\boldsymbol{b}$ is a sum of entries of the corresponding row vector of A_1, A_2 and A_3 multiplied by 0.6. These coefficients are shown in Table 4.7.

To identify membership functions of the fuzzy goals for the objective functions, three individual minimization problems of the three decision makers are solved at the beginning of the procedure. The individual minima and the corresponding optimal solutions are shown in Table 4.8.

Suppose that the decision makers employ the linear membership function (4.7) whose parameters are determined by the Zimmermann method (1978). Then, one finds that $(z_1^0, z_1^1) = (z_1^{\mathrm{m}}, z_1^{\min}) = (-381.245, -429.733)$, $(z_2^0, z_2^1) = (z_2^{\mathrm{m}}, z_2^{\min}) = (-269.466, -344.445)$ and $(z_3^0, z_3^1) = (z_3^{\mathrm{m}}, z_3^{\min}) = (-279.084, -327.455)$, and the maximin problem (4.26) for this problem can be written as

$$\text{maximize} \quad \lambda \tag{4.33a}$$

$$\text{subject to} \quad (z_1(\boldsymbol{x}) + 381.245)/(-429.733 + 381.245) \geqq \lambda \tag{4.33b}$$

$$(z_2(\boldsymbol{x}) + 269.466)/(-344.445 + 269.466) \geqq \lambda \tag{4.33c}$$

$$(z_3(\boldsymbol{x}) + 279.084)/(-327.455 + 279.084) \geqq \lambda \tag{4.33d}$$

$$\boldsymbol{x} \in S, \tag{4.33e}$$

Table 4.7 Coefficients of problem (4.32).

c_1	−49	−5	−45	−5	−34	c_2	−16	−33	−19	−12	−21	c_3	−7	−23	−12	−8	−11		
c_4	−12	−10	−4	−19	−10	c_5	−30	−25	−49	−18	−37	c_6	−17	−13	−24	−15	−10		
c_7	−25	−7	−31	−12	−24	c_8	−12	−5	−4	−33	−6	c_9	−41	−22	−37	−14	−23		
A_1	46	−29	−48	31	21	A_2	−47	−37	37	32	19	A_3	21	2	−11	−35	−2	$\boldsymbol{b}$	0
	0	39	12	−14	29		−42	−26	42	31	−19		−25	6	4	2	5		26
	38	−27	5	−31	14		−38	−29	−5	−47	49		−45	−45	5	10	−40		−111
	−26	16	44	6	19		17	27	32	17	−6		−27	1	18	−6	15		88
	−48	13	2	−33	19		22	−35	27	−35	−35		−26	−16	37	47	−2		−37
	9	−6	12	−17	−32		−8	24	−24	45	−31		16	−9	−19	17	44		12
	−9	2	−16	8	32		−6	−25	−25	−8	4		23	41	30	36	−11		45
	24	30	42	−26	16		19	−18	−18	9	−34		−46	30	3	−1	−45		−8
	−26	−8	0	41	−42		−19	13	−42	49	−27		4	−2	−12	24	−33		−47
	−29	16	−16	−4	18		45	−8	21	6	47		43	46	26	22	−5		136
	−7	1	−3	38	18		−43	−15	31	−34	23		−35	−34	20	−15	−26		−48
	−12	−4	47	0	−4		−18	−19	28	47	−36		−45	20	40	3	−15		19
	−12	−46	11	−47	−47		19	30	50	12	−24		13	20	−43	−8	20		−31
	5	−2	37	38	0		12	−34	34	28	−40		−18	33	39	14	2		88
	49	41	3	12	−48		15	12	32	31	−28		−25	−23	−6	−25	−15		14
	−17	−6	34	21	11		5	−28	−46	−15	9		12	49	4	−17	−47		−18

Table 4.8 Optimal solutions to individual problems.

$z_1^{\min}$	−429.733				
$\boldsymbol{x}_1$	3.18955	0	0	1.00390	3.46578
$\boldsymbol{x}_2$	2.24532	4.08649	0	0	0.307015
$\boldsymbol{x}_3$	1.55737	0	0	0.926562	0
$z_2^{\min}$	−344.445				
$\boldsymbol{x}_1$	2.72265	0	0	2.22359	2.56515
$\boldsymbol{x}_2$	1.64204	4.41723	0	0	1.05148
$\boldsymbol{x}_3$	1.28966	0	0	1.55682	0
$z_3^{\min}$	−327.455				
$\boldsymbol{x}_1$	2.63502	0	0.888399	0.616815	2.41906
$\boldsymbol{x}_2$	1.61707	3.01909	0	0	0.262904
$\boldsymbol{x}_3$	2.21810	0	1.12324	0	0

where S denotes the feasible region of problem (4.32). The result of the first iteration including an optimal solution to problem (4.33) is shown in Table 4.9.

Suppose that DM2 is not satisfied with the solution obtained in Iteration 1, and then taking account of the result of the first iteration, she specifies the minimal satisfactory level at $\hat{\delta}_2 = 0.7$. Moreover, suppose that she identifies the membership function of the fuzzy goal $\tilde{R}$ for the ratio Δ_2 of satisfactory degrees as

$$\mu_2^{\tilde{R}}(\Delta_2(\boldsymbol{x})) = (\Delta_2(\boldsymbol{x}) - 0.5)/(1.0 - 0.5),$$

Table 4.9 Iteration 1.

x_1^1	2.77721	0	0.778426	1.87060	2.54963
x_2^1	1.82450	3.46482	0	0	0.785458
x_3^1	1.73425	0	0	1.12318	0
(z_1^1, z_2^1, z_3^1)	$(-410.191, -314.225, -307.959)$				
$(\mu_1(z_1^1), \mu_2(z_2^1), \mu_3(z_3^1))$	(0.596958, 0.596958, 0.596960)				
λ^1	0.596958				
(Δ_1^1, Δ_2^1)	(1.00000, 1.00000)				

and she also specifies the permissible level at $\hat{\delta}_{\Delta_2} = 0.4$. Then, problem (4.29) with the minimal satisfactory level $\hat{\delta}_2$ and the permissible level $\hat{\delta}_{\Delta_2}$ for the fuzzy goals of DM2's satisfactory degree and the ratio between the satisfactory degrees of DM2 and DM3 is written as

$$\begin{aligned}
&\text{maximize } \lambda && (4.34\text{a})\\
&\text{subject to } (z_2(\boldsymbol{x}) + 269.466)/(-344.445 + 269.466) \geqq 0.70 && (4.34\text{b})\\
&\qquad (z_3(\boldsymbol{x})/z_2(\boldsymbol{x}) - 0.5)/(1.0 - 0.5) \geqq 0.40 && (4.34\text{c})\\
&\qquad (z_3(\boldsymbol{x}) + 279.084)/(-327.455 + 279.084) \geqq \lambda && (4.34\text{d})\\
&\qquad \boldsymbol{x} \in S. && (4.34\text{e})
\end{aligned}$$

The result of the second iteration including an optimal solution to problem (4.34) is shown in Table 4.10.

Table 4.10 Iteration 2.

x_1^2	2.63263	0	0.834957	2.23300	2.26198
x_2^2	1.64582	3.50088	0	0	1.00298
x_3^2	1.68136	0	0	1.29220	0
(z_1^2, z_2^2, z_3^2)	$(-401.166, -321.951, -303.082)$				
$(\mu_1(z_1^2), \mu_2(z_2^2), \mu_3(z_3^2))$	(0.410842, 0.700000, 0.496118)				
(Δ_1^2, Δ_2^2)	(1.70382, 0.70874)				
$\mu_2^{\tilde{R}}(\Delta_2^2)$	0.417481				

At the second iteration, suppose that DM2 judges that it is desirable to increase the satisfactory degree of DM3 at the expense of the satisfactory degree of self and she updates the minimal satisfactory level at $\hat{\delta}_2 = 0.65$. Then, the problem with the minimal satisfactory levels $\hat{\delta}_2$ and $\hat{\delta}_{\Delta_2}$ is solved, and the result of the third iteration is shown in Table 4.5.

Suppose that DM2 is satisfied with the solution obtained in the third iteration, and DM1 specifies the minimal satisfactory level at $\hat{\delta}_1 = 0.7$. Moreover, suppose that DM1 identifies the membership function of the fuzzy goal $\tilde{R}$ for the ratio Δ_1 of

Table 4.11 Iteration 3.

x_1^3	2.62462	0	0.960431	2.21999	2.22707
x_2^3	1.65352	3.35926	0	0	0.986894
x_3^3	1.74339	0	0	1.24498	0
(z_1^3, z_2^3, z_3^3)	$(-401.894, -318.202, -306.947)$				
$(\mu_1(z_1^3), \mu_2(z_2^3), \mu_3(z_3^3))$	$(0.425857, 0.649999, 0.576030)$				
(Δ_1^3, Δ_2^3)	(1.526331,0.886202)				
$\mu_2^{\tilde{R}}(\Delta_2^3)$	0.772404				

satisfactory degrees as

$$\mu_1^{\tilde{R}}(\Delta_1(\boldsymbol{x})) = (\Delta_1(\boldsymbol{x}) - 0.5)/(1.0 - 0.5),$$

and he also specifies the permissible level at $\hat{\delta}_{\Delta_1} = 0.3$. Then, the problem (4.30) with the minimal satisfactory level $\hat{\delta}_1$ and the permissible level $\hat{\delta}_{\Delta_1}$ for the fuzzy goals of DM1's satisfactory degree and the ratio between the satisfactory degrees of DM1 and DM2 is written as

$$\begin{aligned}
&\text{maximize } \lambda && (4.35\text{a})\\
&\text{subject to } (z_1(\boldsymbol{x}) + 381.245)/(-429.733 + 381.245) \geqq 0.70 && (4.35\text{b})\\
&\qquad (z_2(\boldsymbol{x})/z_1(\boldsymbol{x}) - 0.5)/(1.0 - 0.5) \geqq 0.30 && (4.35\text{c})\\
&\qquad (z_2(\boldsymbol{x}) + 269.466)/(-344.445 + 269.466) \geqq \lambda && (4.35\text{d})\\
&\qquad (1/\Delta_2^3)(z_3(\boldsymbol{x}) + 279.084)/(-327.455 + 279.084) \geqq \lambda && (4.35\text{e})\\
&\qquad \boldsymbol{x} \in S. && (4.35\text{f})
\end{aligned}$$

The result of the fourth iteration including an optimal solution to problem (4.35) is shown in Table 4.12.

Table 4.12 Iteration 4.

x_1^4	2.88880	0	0.538626	1.64392	2.80209
x_2^4	1.93419	3.66726	0	0	0.662677
x_3^4	1.67091	0	0	1.08344	0
(z_1^4, z_2^4, z_3^4)	$(-415.187, -314.959, -305.093)$				
$(\mu_1(z_1^4), \mu_2(z_2^4), \mu_3(z_3^4))$	$(0.700000, 0.606742, 0.537695)$				
(Δ_1^4, Δ_2^4)	(0.866777,0.886201)				
$\mu_1^{\tilde{R}}(\Delta_1^4)$	0.733553				

At the fourth iteration, suppose that DM1 judges that it is desirable to increase the satisfactory degree of self at the expense of the satisfactory degree of DM2 and he updates the minimal satisfactory level at $\hat{\delta}_1 = 0.75$. Then, the problem with the minimal satisfactory levels $\hat{\delta}_1$ and $\hat{\delta}_{\Delta_1}$ is solved, and the result of the fifth iteration is shown in Table 4.13.

Table 4.13 Iteration 5.

x_1^5	2.93698	0	0.461694	1.53885	2.90697
x_2^5	1.98538	3.72344	0	0	0.603544
x_3^5	1.65769	0	0	1.05398	0
(z_1^5, z_2^5, z_3^5)	$(-416.611, -314.367, -304.754)$				
$(\mu_1(z_1^5), \mu_2(z_2^5), \mu_3(z_3^5))$	(0.750000,0.598853,0.530703)				
(Δ_1^5, Δ_2^5)	(0.798472,0.886199)				
$\mu_1^{\tilde{R}}(\Delta_1^5)$	0.596944				

At the fifth iteration, the satisfactory degree $\mu_1(z_1^5) = 0.75$ of DM1 becomes equal to the minimal satisfactory level $\hat{\delta}_1 = 0.75$, and the satisfactory degree $\mu_1^{\tilde{R}}(\Delta_1^5) = 0.596944$ of DM1 is larger than the permissible level of $\hat{\delta}_{\Delta_1} = 0.3$. Therefore, this solution satisfies the termination conditions of the interactive process, and it becomes a satisfactory solution for the three decision makers if DM1 accepts the solution.

4.3 Fuzzy two-level linear programming with fuzzy parameters

In this section, we consider two-level linear programming problem with fuzzy parameters from the viewpoint of experts' imprecise or fuzzy understanding of the nature of parameters in a problem-formulation process (Sakawa, Nishizaki and Uemura, 2000a), and give a fuzzy programming method for the problems.

4.3.1 Interactive fuzzy programming

When a real-world decision situation is formulated as a mathematical programming problem, various factors of the real-world system should be reflected in the description of objective functions and constraints in the mathematical programming problem. Naturally, these objective functions and constraints involve many parameters whose possible values may be estimated by experts. In the conventional approaches, such parameters are required to be fixed at some values in an experimental and/or subjective manner through the experts' understanding of the nature of the parameters in the problem-formulation process.

It should be observed here that, in most real-world situations, the possible values of these parameters are often only imprecisely or ambiguously known to the experts. With this observation in mind, it would be more appropriate to interpret the experts' understanding of the parameters as fuzzy numerical data which can be represented by means of fuzzy sets over the real line known as fuzzy numbers. The resulting mathematical programming problem with fuzzy parameters would be viewed as a more realistic version than the conventional one (Sakawa, 1993; Sakawa and Yano, 1990).

From this viewpoint, we assume that parameters involved in the objective functions and the constraints of the two-level linear programming problem are characterized by fuzzy numbers. As a result, a problem with fuzzy parameters corresponding to problem (4.4) is formulated as:

$$\underset{\text{for DM1}}{\text{minimize}} \; z_1(\boldsymbol{x}_1,\boldsymbol{x}_2) = \tilde{\boldsymbol{c}}_{11}\boldsymbol{x}_1 + \tilde{\boldsymbol{c}}_{12}\boldsymbol{x}_2 \tag{4.36a}$$

$$\underset{\text{for DM2}}{\text{minimize}} \; z_2(\boldsymbol{x}_1,\boldsymbol{x}_2) = \tilde{\boldsymbol{c}}_{21}\boldsymbol{x}_1 + \tilde{\boldsymbol{c}}_{22}\boldsymbol{x}_2 \tag{4.36b}$$

$$\text{subject to} \; \tilde{A}_1\boldsymbol{x}_1 + \tilde{A}_2\boldsymbol{x}_2 \leqq \tilde{\boldsymbol{b}} \tag{4.36c}$$

$$\boldsymbol{x}_1 \geqq \boldsymbol{0},\; \boldsymbol{x}_2 \geqq \boldsymbol{0}. \tag{4.36d}$$

where $\boldsymbol{x}_i$, $i = 1,2$ is an n_i-dimensional decision variable column vector of DMi, and $z_i(\boldsymbol{x}_1,\boldsymbol{x}_2)$, $i = 1,2$ is the objective function of DMi. In particular, $\tilde{\boldsymbol{c}}_{ij}$, $i,j = 1,2$ is an n_j-dimensional row vector of fuzzy parameters; $\tilde{A}_j$, $j = 1,2$ is an $m \times n_j$ matrix of fuzzy parameters; and $\tilde{\boldsymbol{b}}$ is an m-dimensional column vector of fuzzy parameters. For the sake of simplicity, we use the following notations: $\tilde{\boldsymbol{c}}_1 = (\tilde{\boldsymbol{c}}_{11},\tilde{\boldsymbol{c}}_{12}), \tilde{\boldsymbol{c}}_2 = (\tilde{\boldsymbol{c}}_{21},\tilde{\boldsymbol{c}}_{22})$, and $\tilde{A} = [\tilde{A}_1\ \tilde{A}_2]$.

Assume that the fuzzy parameters $\tilde{\boldsymbol{c}}_{11}, \tilde{\boldsymbol{c}}_{12}, \tilde{\boldsymbol{c}}_{12}, \tilde{\boldsymbol{c}}_{22}, \tilde{A}_1, \tilde{A}_2$ and $\tilde{\boldsymbol{b}}$ are characterized by fuzzy numbers, and let the corresponding membership functions be denoted by $(\mu_{\tilde{c}_{11,1}}(c_{11,1}),\ldots,\mu_{\tilde{c}_{11,n_1}}(c_{11,n_1}))$, $(\mu_{\tilde{c}_{12,1}}(c_{12,1}),\ldots,\mu_{\tilde{c}_{12,n_2}}(c_{12,n_2}))$, $(\mu_{\tilde{c}_{21,1}}(c_{21,1}),\ldots,$ $\mu_{\tilde{c}_{21,n_1}}(c_{21,n_1}))$, $(\mu_{\tilde{c}_{22,1}}(c_{22,1}),\ldots,\mu_{\tilde{c}_{22,n_2}}(c_{22,n_2}))$, $\mu_{\tilde{a}_{1,kj}}(a_{1,kj})$, $k = 1,\ldots,m$, $j = 1,\ldots,$ n_1, $\mu_{\tilde{a}_{2,kj}}(a_{2,kj})$, $k = 1,\ldots,m$, $j = 1,\ldots,n_2$, and $(\mu_{\tilde{b}_1}(b_1),\ldots,\mu_{\tilde{b}_m}(b_m))$.

We introduce an α-level set of the fuzzy numbers $\tilde{\boldsymbol{c}}$, $\tilde{\boldsymbol{b}}$ and $\tilde{A}$ defined as the following ordinary set $(\tilde{\boldsymbol{c}},\tilde{\boldsymbol{b}},\tilde{A})_\alpha$ in which the degree of their membership functions exceeds a given level α:

$$\begin{aligned}(\tilde{\boldsymbol{c}},\tilde{\boldsymbol{b}},\tilde{A})_\alpha = \{\, (\boldsymbol{c},\boldsymbol{b},A) \mid\ & \mu_{\tilde{c}_{11,j}}(c_{11,j}) \geqq \alpha,\ j = 1,\ldots,n_1,\\ & \mu_{\tilde{c}_{12,j}}(c_{12,j}) \geqq \alpha,\ j = 1,\ldots,n_2,\\ & \mu_{\tilde{c}_{21,j}}(c_{21,j}) \geqq \alpha,\ j = 1,\ldots,n_1,\\ & \mu_{\tilde{c}_{22,j}}(c_{22,j}) \geqq \alpha,\ j = 1,\ldots,n_2,\\ & \mu_{\tilde{b}_k}(b_k) \geqq \alpha,\ k = 1,\ldots,m,\\ & \mu_{\tilde{a}_{1,kj}}(a_{1,kj}) \geqq \alpha,\ k = 1,\ldots,m,\ j = 1,\ldots,n_1,\\ & \mu_{\tilde{a}_{2,kj}}(a_{2,kj}) \geqq \alpha,\ k = 1,\ldots,m,\ j = 1,\ldots,n_2\}. \end{aligned} \tag{4.37}$$

Now suppose that DM1 considers that the degree of all of the membership functions of the fuzzy numbers involved in the two-level linear programming problem should be greater than or equal to some value α. Then, for such a membership degree α, problem (4.36) can be interpreted as the following nonfuzzy two-level linear programming problem which depends on a coefficient vector $(\boldsymbol{c},\boldsymbol{b},A) \in (\tilde{\boldsymbol{c}},\tilde{\boldsymbol{b}},\tilde{A})_\alpha$ (Sakawa, 1993; Sakawa and Yano, 1990):

$$\begin{aligned} \underset{\text{for DM1}}{\text{minimize}} \quad & z_1(\boldsymbol{x}_1,\boldsymbol{x}_2) = \boldsymbol{c}_{11}\boldsymbol{x}_1 + \boldsymbol{c}_{12}\boldsymbol{x}_2 && (4.38a) \\ \underset{\text{for DM2}}{\text{minimize}} \quad & z_2(\boldsymbol{x}_1,\boldsymbol{x}_2) = \boldsymbol{c}_{21}\boldsymbol{x}_1 + \boldsymbol{c}_{22}\boldsymbol{x}_2 && (4.38b) \\ \text{subject to} \quad & A_1\boldsymbol{x}_1 + A_2\boldsymbol{x}_2 \leqq \boldsymbol{b} && (4.38c) \\ & \boldsymbol{x}_1 \geqq \boldsymbol{0},\ \boldsymbol{x}_2 \geqq \boldsymbol{0}. && (4.38d) \end{aligned}$$

Observe that there exist an infinite number of such problems (4.38) depending on the coefficient vector $(\boldsymbol{c},\boldsymbol{b},A) \in (\tilde{\boldsymbol{c}},\tilde{\boldsymbol{b}},\tilde{A})_\alpha$ and the values of $(\boldsymbol{c},\boldsymbol{b},A)$ are arbitrarily chosen from the α-level set $(\tilde{\boldsymbol{c}},\tilde{\boldsymbol{b}},\tilde{A})_\alpha$ in the sense that the degree of all of the membership functions for the fuzzy numbers in problem (4.38) exceeds the level α. However, if possible, it would be desirable for each decision maker to choose $(\boldsymbol{c},\boldsymbol{b},A) \in (\tilde{\boldsymbol{c}},\tilde{\boldsymbol{b}},\tilde{A})_\alpha$ in problem (4.38) so as to minimize the objective function under the constraints. Assuming that DM1 chooses a degree of the α-level, from such a point of view, it seems to be quite natural for DM1 to have understood the two-level linear programming problem with fuzzy parameters as the following nonfuzzy α-two-level linear programming problem (Sakawa, 1993; Sakawa and Yano, 1990):

$$\begin{aligned} \underset{\text{for DM1}}{\text{minimize}} \quad & z_1(\boldsymbol{x}_1,\boldsymbol{x}_2;\boldsymbol{c}_{11},\boldsymbol{c}_{12}) = \boldsymbol{c}_{11}\boldsymbol{x}_1 + \boldsymbol{c}_{12}\boldsymbol{x}_2 && (4.39a) \\ \underset{\text{for DM2}}{\text{minimize}} \quad & z_2(\boldsymbol{x}_1,\boldsymbol{x}_2;\boldsymbol{c}_{21},\boldsymbol{c}_{22}) = \boldsymbol{c}_{21}\boldsymbol{x}_1 + \boldsymbol{c}_{22}\boldsymbol{x}_2 && (4.39b) \\ \text{subject to} \quad & A_1\boldsymbol{x}_1 + A_2\boldsymbol{x}_2 \leqq \boldsymbol{b} && (4.39c) \\ & \boldsymbol{x}_1 \geqq \boldsymbol{0},\ \boldsymbol{x}_2 \geqq \boldsymbol{0} && (4.39d) \\ & (\boldsymbol{c},\boldsymbol{b},A) \in (\tilde{\boldsymbol{c}},\tilde{\boldsymbol{b}},\tilde{A})_\alpha. && (4.39e) \end{aligned}$$

It should be noted that the parameters $(\boldsymbol{c},\boldsymbol{b},A)$ are treated as decision variables rather than constants.

In a way similar to the two-level linear programming problems considered in the previous section, it is natural that the decision makers have fuzzy goals for their objective functions when they take fuzziness of human judgments into consideration. To identify membership functions of the fuzzy goals for the objective functions, we set $\alpha = 0$, and solve the following individual single-objective problems for $i = 1,2$:

$$\begin{aligned} \text{minimize} \quad & z_i(\boldsymbol{x};\boldsymbol{c}) = \boldsymbol{c}_i\boldsymbol{x} && (4.40a) \\ \text{subject to} \quad & A\boldsymbol{x} \leqq \boldsymbol{b} && (4.40b) \\ & \boldsymbol{x} \geqq \boldsymbol{0} && (4.40c) \\ & (\boldsymbol{c},\boldsymbol{b},A) \in (\tilde{\boldsymbol{c}},\tilde{\boldsymbol{b}},\tilde{A})_\alpha. && (4.40d) \end{aligned}$$

Although problem (4.40) is not a linear programming problem, it can be solved by linear programming techniques as we will show the solution method for the maximin problem (4.41).

After identifying the membership functions $\mu_i(z_i(\boldsymbol{x};\boldsymbol{c}_i))$, $i = 1,2$, to derive an overall satisfactory solution to the two-level linear programming problem with fuzzy parameters (4.39), we first solve the following maximin problem for obtaining a solution which maximizes the smaller degree between the satisfactory degrees of the two decision makers:

$$\begin{aligned} &\text{maximize} \;\; \min\{\mu_1(z_1(\boldsymbol{x};\boldsymbol{c}_1)),\mu_2(z_2(\boldsymbol{x};\boldsymbol{c}_2))\} && (4.41a)\\ &\text{subject to} \;\; A\boldsymbol{x} \leqq \boldsymbol{b} && (4.41b)\\ &\qquad \boldsymbol{x} \geqq \boldsymbol{0} && (4.41c)\\ &\qquad (\boldsymbol{c},\boldsymbol{b},A) \in (\tilde{\boldsymbol{c}},\tilde{\boldsymbol{b}},\tilde{A})_\alpha. && (4.41d) \end{aligned}$$

By introducing an auxiliary variable λ, this problem can be transformed into the following equivalent maximization problem:

$$\begin{aligned} &\text{maximize} \;\; \lambda && (4.42a)\\ &\text{subject to} \;\; \mu_1(z_1(\boldsymbol{x};\boldsymbol{c}_1)) \geqq \lambda && (4.42b)\\ &\qquad \mu_2(z_2(\boldsymbol{x};\boldsymbol{c}_2)) \geqq \lambda && (4.42c)\\ &\qquad A\boldsymbol{x} \leqq \boldsymbol{b} && (4.42d)\\ &\qquad \boldsymbol{x} \geqq \boldsymbol{0} && (4.42e)\\ &\qquad (\boldsymbol{c},\boldsymbol{b},A) \in (\tilde{\boldsymbol{c}},\tilde{\boldsymbol{b}},\tilde{A})_\alpha. && (4.42f) \end{aligned}$$

Unfortunately, problem (4.42) is not a linear programming problem even if all the membership functions $\mu_i(z_i(\boldsymbol{x};\boldsymbol{c}_i))$, $i = 1,2$ are linear. To solve problem (4.42) by using the linear programming technique, we introduce the set-valued functions:

$$\begin{aligned} &S_i(\boldsymbol{c}_i) = \{(\boldsymbol{x},\lambda) \mid \mu_i(z_i(\boldsymbol{x},\boldsymbol{c}_i)) \geqq \lambda\},\; i = 1,2 && (4.43a)\\ &T_j(A_{(j\cdot)},b_j) = \{\boldsymbol{x} \mid A_{(j\cdot)}\boldsymbol{x} \leqq b_j\},\; j = 1,\ldots,m, && (4.43b) \end{aligned}$$

where $A_{(j\cdot)}$ is a row vector corresponding to the jth row of the $m \times (n_1+n_2)$ matrix A. Then, it can be easily verified that the following relations hold for $S_i(\boldsymbol{c}_i)$ and $T_j(A_{(j\cdot)},b_j)$ when $\boldsymbol{x} \geqq \boldsymbol{0}$ (Sakawa and Yano, 1990; Sakawa, 1993).

Proposition 4.1. *For given coefficient vectors $\boldsymbol{c}_i^1$ and $\boldsymbol{c}_i^2$ of the objective functions, coefficient vectors $A_{(j\cdot)}^1$ and $A_{(j\cdot)}^2$ of the constraints, and constants b_j^1 and b_j^2 of the constraints, the following relations hold when $\boldsymbol{x} \geqq \boldsymbol{0}$.*

(i) *If $\boldsymbol{c}_i^1 \leqq \boldsymbol{c}_i^2$, then $S_i(\boldsymbol{c}_i^1) \supseteq S_i(\boldsymbol{c}_i^2)$.*
(ii) *If $A_{(j\cdot)}^1 \leqq A_{(j\cdot)}^2$, then $T_j(A_{(j\cdot)}^1,\cdot) \supseteq T_j(A_{(j\cdot)}^2,\cdot)$.*
(iii) *If $b_j^1 \leqq b_j^2$, then $T_j(\cdot,b_j^1) \subseteq T_j(\cdot,b_j^2)$.*

From the properties of the α-level sets for the vectors of fuzzy numbers $\tilde{\boldsymbol{c}}_1$, $\tilde{\boldsymbol{c}}_2$, $\tilde{\boldsymbol{b}}$ and the matrix of fuzzy numbers $\tilde{A}$, it should be noted that the individual feasible regions for $\boldsymbol{c}_1$, $\boldsymbol{c}_2$, b_j and $A_{(j\cdot)}$ in the α-level sets can be denoted by the closed intervals $[\boldsymbol{c}_1^L, \boldsymbol{c}_1^R]$, $[\boldsymbol{c}_2^L, \boldsymbol{c}_2^R]$, $[b_j^L, b_j^R]$ and $[A_{(j\cdot)}^L, A_{(j\cdot)}^R]$, respectively.

Therefore, from Proposition 4.1 and the properties of the α-level sets, we can obtain an optimal solution to problem (4.42) by solving the following linear programming problem:

$$\begin{aligned} &\text{maximize } \lambda && (4.44\text{a}) \\ &\text{subject to } \mu_1(z_1(\boldsymbol{x}; \boldsymbol{c}_1^L)) \geqq \lambda && (4.44\text{b}) \\ &\qquad \mu_2(z_2(\boldsymbol{x}; \boldsymbol{c}_2^L)) \geqq \lambda && (4.44\text{c}) \\ &\qquad A^L \boldsymbol{x} \leqq \boldsymbol{b}^R && (4.44\text{d}) \\ &\qquad \boldsymbol{x} \geqq \boldsymbol{0}. && (4.44\text{e}) \end{aligned}$$

For the problem which maximizes DM2's membership function under the condition that DM1's membership function $\mu_1(z_1(\boldsymbol{x}; \boldsymbol{c}_1))$ is larger than or equal to the minimal satisfactory level $\hat{\delta}$ specified by DM1, we can also formulate the following problem with fuzzy parameters:

$$\begin{aligned} &\text{maximize } \mu_2(z_2(\boldsymbol{x}; \boldsymbol{c}_2)) && (4.45\text{a}) \\ &\text{subject to } \mu_1(z_1(\boldsymbol{x}; \boldsymbol{c}_1)) \geqq \hat{\delta} && (4.45\text{b}) \\ &\qquad A\boldsymbol{x} \leqq \boldsymbol{b} && (4.45\text{c}) \\ &\qquad \boldsymbol{x} \geqq \boldsymbol{0} && (4.45\text{d}) \\ &\qquad (\boldsymbol{c}, \boldsymbol{b}, A) \in (\tilde{\boldsymbol{c}}, \tilde{\boldsymbol{b}}, \tilde{A})_\alpha. && (4.45\text{e}) \end{aligned}$$

From Proposition 4.1 and the properties of the α-level sets, Problem (4.45) can be also transformed into the following equivalent problem:

$$\begin{aligned} &\text{maximize } \mu_2(z_2(\boldsymbol{x}; \boldsymbol{c}_2^L)) && (4.46\text{a}) \\ &\text{subject to } \mu_1(z_1(\boldsymbol{x}; \boldsymbol{c}_1^L)) \geqq \hat{\delta} && (4.46\text{b}) \\ &\qquad A^L \boldsymbol{x} \leqq \boldsymbol{b}^R && (4.46\text{c}) \\ &\qquad \boldsymbol{x} \geqq \boldsymbol{0}. && (4.46\text{d}) \end{aligned}$$

For the two-level linear programming problem with fuzzy parameters (4.36), we can provide the termination conditions of the interactive process and the procedure for updating the minimal satisfactory level $\hat{\delta}$ which are the same as those of the interactive fuzzy programming for the two-level linear programming problem (4.4), and give a similar algorithm with problems (4.44) and (4.46) for deriving satisfactory solutions.

4.3.2 Numerical example

In this subsection, we provide an illustrative numerical example for a two-level linear programming problem with fuzzy parameters to demonstrate the feasibility of the method described in the previous subsection.

Consider the following two-level linear programming problem with fuzzy parameters:

$$\underset{\text{for DM1}}{\text{minimize}} \ \tilde{\boldsymbol{c}}_{11}\boldsymbol{x}_1 + \tilde{\boldsymbol{c}}_{12}\boldsymbol{x}_2 \tag{4.47a}$$

$$\underset{\text{for DM2}}{\text{minimize}} \ \tilde{\boldsymbol{c}}_{21}\boldsymbol{x}_1 + \tilde{\boldsymbol{c}}_{22}\boldsymbol{x}_2 \tag{4.47b}$$

$$\text{subject to} \ \tilde{A}_1\boldsymbol{x}_1 + \tilde{A}_2\boldsymbol{x}_2 \leqq \tilde{\boldsymbol{b}} \tag{4.47c}$$

$$\boldsymbol{x}_1 \geqq \boldsymbol{0},\ \boldsymbol{x}_2 \geqq \boldsymbol{0}. \tag{4.47d}$$

where $\boldsymbol{x}_1 = (x_{11}, \ldots, x_{110})^T$, $\boldsymbol{x}_2 = (x_{21}, \ldots, x_{210})^T$; each entry of the 25×10 coefficient matrices A_1 and A_2 is a randomly selected number from the interval $[-50, 50]$; each entry of the right-hand side constant column vector $\boldsymbol{b}$ is a sum of entries of the corresponding row vector of A_1 and A_2 multiplied by 0.6; there are 8 fuzzy parameters in each objective function and 30 fuzzy parameters in the constraints. Coefficients including fuzzy parameters are shown in Tables 4.14, 4.15 and 4.16 where $\tilde{c}_{ij,k}$, $\tilde{a}_{1,ij}$, $\tilde{a}_{2,ij}$ and $\tilde{b}_i$ denote fuzzy parameters; the content rate of the fuzzy parameters in the coefficients is seven percent. For simplicity, it is assumed that all of the membership functions for the fuzzy numbers involved in this example are triangular fuzzy numbers. In Table 4.16, the left side points, the means and the right side points of each triangular fuzzy number are shown.

To identify membership functions of the fuzzy goals for the objective functions, we first solve two individual minimization problems of the two decision makers. The individual minima and the corresponding optimal solutions are shown in Table 4.17.

Suppose that the decision makers employ the linear membership function (4.7) whose parameters are determined by the Zimmermann method (1978). Then, the maximin problem (4.44) for this numerical example can be formulated as

$$\text{maximize} \ \lambda \tag{4.48a}$$

$$\text{subject to} \ (z_1(\boldsymbol{x}; \boldsymbol{c}_1^L) + 384.32)/(-783.98 + 384.32) \geqq \lambda \tag{4.48b}$$

$$(z_2(\boldsymbol{x}; \boldsymbol{c}_2^L) - 84.39)/(-127.09 - 84.29) \geqq \lambda \tag{4.48c}$$

$$A^L\boldsymbol{x} \leqq \boldsymbol{b}^R \tag{4.48d}$$

$$\boldsymbol{x} \geqq \boldsymbol{0}. \tag{4.48e}$$

The result of the first iteration including an optimal solution to problem (4.48) is shown in Table 4.18.

Suppose that DM1 is not satisfied with the solution obtained in Iteration 1, and then he specifies the minimal satisfactory level at $\hat{\delta} = 0.75$ and the bounds of the

Table 4.14 Coefficients in problem (4.47).

$\boldsymbol{c}_{11}$	$\tilde{c}_{11,1}$	−15	−15	−39	−44	−12	−27	−22	−42	$\tilde{c}_{11,10}$
$\boldsymbol{c}_{21}$	−23	−22	$\tilde{c}_{21,3}$	−38	−29	−34	$\tilde{c}_{21,7}$	−1	−13	−1
A_1	$\tilde{a}_{1,11}$	50	29	16	−21	9	9	−47	−18	$\tilde{a}_{1,110}$
	44	16	−27	−20	1	47	−21	31	8	−30
	−20	−15	21	19	13	−35	28	−38	−28	42
	6	$\tilde{a}_{1,42}$	−25	−23	9	36	−10	−4	$\tilde{a}_{1,49}$	−41
	−38	−21	27	−15	42	2	−17	−7	−37	20
	−38	−48	−37	$\tilde{a}_{1,64}$	−9	42	$\tilde{a}_{1.67}$	−13	−5	41
	42	−21	−13	32	28	13	−35	49	11	−21
	30	18	22	3	−3	−2	−10	37	32	−15
	−13	−32	−25	−9	0	−32	−26	−48	−39	−27
	1	−24	43	31	$\tilde{a}_{1,105}$	−7	$\tilde{a}_{1,107}$	−4	−8	16
	31	1	−33	−33	47	5	−37	−15	−8	40
	32	−32	10	40	50	−35	28	−4	33	−24
	9	13	15	−38	44	5	7	34	29	17
	1	29	29	5	−38	24	−39	−49	−47	8
	12	−32	35	−35	29	−46	5	−21	−29	44
	−1	20	−13	−48	$\tilde{a}_{1,165}$	$\tilde{a}_{1,166}$	5	10	47	18
	−26	22	18	7	−24	40	30	−18	31	−37
	11	−46	8	−16	−42	−23	−17	−20	−24	−36
	44	−39	−31	−33	−18	−14	27	16	−23	−41
	−47	−15	14	7	−15	−3	34	$\tilde{a}_{1,208}$	47	$\tilde{a}_{1,2010}$
	−36	13	−5	42	24	−48	−18	−16	−48	44
	11	−17	−45	−1	2	−21	21	−35	41	16
	3	−31	39	−46	−2	25	−39	−26	−47	12
	−41	−29	7	28	−21	14	−2	40	−44	45
	−44	22	−24	25	−1	15	26	−5	−28	−8

Table 4.15 Coefficients in problem (4.47) (continued).

$\boldsymbol{c}_{12}$	−34	$\tilde{c}_{12,1}$	−19	−23	−15	−42	−29	−12	$\tilde{c}_{12,9}$	−40		
$\boldsymbol{c}_{22}$	34	21	$\tilde{c}_{22,3}$	23	15	42	29	$\tilde{c}_{22,8}$	10	40		
A_2	−1	35	$\tilde{a}_{2,13}$	−42	−13	21	17	$\tilde{a}_{2,18}$	18	41	$\boldsymbol{b}$	$\tilde{b}_1$
	−24	−39	−25	−18	49	−12	−47	45	−43	17		−28
	−6	−7	−45	45	−46	8	−34	−32	33	33		−38
	43	25	$\tilde{a}_{2,43}$	$\tilde{a}_{2,44}$	32	49	41	−39	3	23		$\tilde{b}_4$
	46	31	37	−36	−30	−47	18	26	22	36		35
	20	$\tilde{a}_{2,62}$	−25	45	50	$\tilde{a}_{2,66}$	39	−24	33	19		$\tilde{b}_6$
	−18	−11	−17	−29	49	5	−28	−32	15	27		27
	26	−28	49	3	−41	−14	−15	−31	−17	29		43
	8	15	−5	43	43	−10	−15	15	46	4		−64
	−31	35	−3	$\tilde{a}_{2,104}$	−18	$\tilde{a}_{2,106}$	−17	14	37	42		$\tilde{b}_{10}$
	−13	−49	−1	−13	48	44	−45	−6	28	−5		−8
	−46	34	47	39	−19	48	10	1	41	−12		144
	2	−7	34	−17	−34	31	−30	11	45	−43		76
	23	4	−1	42	−25	−15	−22	−30	32	−43		−67
	−31	−30	−18	34	−9	−4	−31	−43	−49	25		−116
	41	$\tilde{a}_{2,162}$	−49	$\tilde{a}_{2,164}$	−23	36	46	−1	17	−12		$\tilde{b}_{16}$
	45	−15	14	−11	−39	−13	47	36	−9	26		74
	−16	48	48	14	−42	−25	8	−3	39	−42		−105
	8	−12	50	26	−48	4	41	−33	−44	−49		−101
	$\tilde{a}_{2,201}$	49	8	−24	−13	34	47	−49	50	$\tilde{a}_{2,2010}$		$\tilde{b}_{20}$
	−45	−17	−2	45	−33	26	14	−44	−17	−16		−82
	−38	39	−11	−5	−28	36	−20	−30	−45	−45		−104
	−17	−28	−21	−32	−25	34	8	−19	−3	−22		−142
	−42	38	−49	43	49	25	−4	−21	23	27		51
	13	28	1	−7	−34	30	−48	−7	25	−13		−20

Table 4.16 Coefficients represented by fuzzy numbers in problem (4.47).

	left	mean	right		left	mean	right
$\tilde{c}_{11,1}$	−50	−45	−40	$\tilde{c}_{11,10}$	−16	−11	−6
$\tilde{c}_{12,1}$	−25	−20	−15	$\tilde{c}_{21,9}$	−14	−9	−4
$\tilde{c}_{21,3}$	−37	−32	−27	$\tilde{c}_{21,7}$	−38	−33	−28
$\tilde{c}_{22,3}$	14	19	24	$\tilde{c}_{22,8}$	7	12	17
$\tilde{a}_{1,11}$	21	26	31	$\tilde{a}_{1,110}$	14	19	24
$\tilde{a}_{2,13}$	7	12	17	$\tilde{a}_{2,18}$	−27	−22	−17
$\tilde{a}_{1,42}$	−38	−33	−28	$\tilde{a}_{1,49}$	0	5	10
$\tilde{a}_{2,43}$	43	48	53	$\tilde{a}_{2,44}$	6	11	16
$\tilde{a}_{1,64}$	8	13	18	$\tilde{a}_{1,67}$	−6	−1	4
$\tilde{a}_{2,62}$	−44	−39	−34	$\tilde{a}_{2,66}$	−7	−2	3
$\tilde{a}_{1,105}$	−23	−18	−13	$\tilde{a}_{1,107}$	−34	−29	−24
$\tilde{a}_{2,104}$	27	32	37	$\tilde{a}_{2,106}$	−11	−6	−1
$\tilde{a}_{1,165}$	20	25	30	$\tilde{a}_{1,166}$	20	25	30
$\tilde{a}_{2,162}$	−45	−40	−35	$\tilde{a}_{2,164}$	−30	−25	−20
$\tilde{a}_{1,208}$	−46	−41	−36	$\tilde{a}_{1,2010}$	10	15	20
$\tilde{a}_{2,201}$	32	37	42	$\tilde{a}_{2,2010}$	−42	−37	−32
$\tilde{b}_1$	78	83	88	$\tilde{b}_4$	87	92	97
$\tilde{b}_6$	30	35	40	$\tilde{b}_{10}$	50	55	60
$\tilde{b}_{16}$	40	45	50	$\tilde{b}_{20}$	52	57	62

Table 4.17 Optimal solutions to the individual problems.

$z_1^{\min}$	−783.988									
$\boldsymbol{x}_1$	3.358	1.406	0.071	1.555	0.648	0.000	3.026	1.463	0.856	2.895
$\boldsymbol{x}_2$	4.636	2.794	0.196	2.386	2.003	0.000	0.035	2.309	0.000	0.000
$z_2^{\min}$	−127.097									
$\boldsymbol{x}_1$	0.582	0.881	1.431	2.020	0.720	0.000	1.653	0.959	0.929	1.488
$\boldsymbol{x}_2$	0.888	0.533	0.620	0.000	1.250	0.000	0.116	1.214	0.000	0.000

ratio at the interval $[\Delta_{\min}, \Delta_{\max}] = [0.6, 1.0]$, taking account of the result of the first

Table 4.18 Iteration 1.

$\boldsymbol{x}_1^1$	1.141	2.715	1.382	1.794	2.334	0.000	2.145	0.930	0.834	1.442
$\boldsymbol{x}_2^1$	3.110	1.315	0.336	1.945	2.865	0.000	0.305	1.006	0.000	0.000
(z_1^1, z_2^1)	(−665.638, −64.468)									
$(\mu_1(z_1^1), \mu_2(z_2^1))$	(0.704, 0.704)									
λ^1	0.704									
Δ^1	1.000									

iteration. Then, the problem with the minimal satisfactory level (4.46) is formulated as

$$\begin{aligned}
&\text{maximize} \quad \mu_2(z_2(\boldsymbol{x}, \boldsymbol{c}_2^L)) && (4.49a)\\
&\text{subject to} \quad (z_1(\boldsymbol{x}; \boldsymbol{c}_1^L) + 384.32)/(-783.98 + 384.32) \geqq 0.75 && (4.49b)\\
&\qquad A^L \boldsymbol{x} \leqq \boldsymbol{b}^R && (4.49c)\\
&\qquad \boldsymbol{x} \geqq \boldsymbol{0}. && (4.49d)
\end{aligned}$$

The result of the second iteration including an optimal solution to problem (4.49) is shown in Table 4.19.

Table 4.19 Iteration 2.

$\boldsymbol{x}_1^2$	1.471	2.524	1.166	1.784	2.064	0	2.221	0.997	0.877	1.695
$\boldsymbol{x}_2^2$	3.338	1.555	0.310	2.022	2.712	0	0.254	1.234	0.000	0.000
(z_1^2, z_2^2)	$(-684.072, -41.473)$									
$(\mu_1(z_1^2), \mu_2(z_2^2))$	$(0.750, 0.595)$									
λ^2	0.595									
Δ^2	0.794									

At the second iteration, the satisfactory degree $\mu_1(z_1^2) = 0.75$ of DM1 becomes equal to the minimal satisfactory level $\hat{\delta} = 0.75$, and the ratio $\Delta^2 = 0.794$ of satisfactory degrees is in the valid interval $[0.6, 1.0]$ of the ratio. Therefore, this solution satisfies the termination conditions of the interactive process and becomes a satisfactory solution for both decision makers if DM1 accepts the solution.

4.4 Fuzzy two-level linear fractional programming

When we consider hierarchical decision problems in firms, it seems to be appropriate for decision makers or analysts not only to take the cooperative relationship between the decision makers into consideration but also to employ linear fractional objectives rather than linear ones. Objectives or criteria represented as fractional functions (Kornbluth and Steuer, 1981a; Steuer, 1986) are often observed in the following situations: for finance or corporate planning, debt-to-equity ratio, return on investment, current ratio, risk-assets to capital, actual capital to required capital, foreign loans to total loans, residential mortgages to total mortgages, etc; for production planning, inventory to sales, actual cost to standard cost, output per employee, and so forth.

For instance, by adopting a criterion with respect to finance or corporate planning as an objective function of the decision maker at the upper level and employing a criterion regarding production planning as an objective function of the decision maker at the lower level, a two-level linear fractional programming problem is formulated for a hierarchical decision problem in a firm.

An usual linear fractional programming problem is a special case of a nonlinear programming problem, but it can be transformed into a linear programming problem by using the variable transformation method by Charnes and Cooper (1962), or it can be solved by adopting the updated objective function method by Bitran and Novaes (1973). In their method, an optimal solution is derived by repeating two steps: computing the local gradient of the fractional objective function, and solving the resulting linear programming problem. Concerning multiobjective linear fractional programming, Kornbluth and Steuer (1981a,b) present two different approaches to multiobjective linear fractional programming based on the weighted Tchebyscheff norm. Luhandjula (1984) proposes a linguistic approach to multiobjective linear fractional programming by introducing linguistic variables to represent linguistic aspirations of the decision maker. Employing the fuzzy decision by Bellman and Zadeh (1970), Sakawa and Yumine (1983) present a fuzzy programming approach for solving multiobjective linear fractional programming problems by the combined use of the bisection method and the first phase in the simplex method of linear programming. As a generalization of the result of Sakawa and Yumine (1983), Sakawa and Yano (1988) propose a linear programming-based interactive fuzzy satisficing method for multiobjective linear fractional programming to derive the satisficing solution efficiently from a set of Pareto optimal solutions by updating the reference membership values of the decision maker.

In this section, we deal with two-level linear fractional programming problems with cooperative decision makers, and give the corresponding interactive fuzzy programming. In the interactive method, satisfactory solutions are derived in a procedure similar to that of the two-level linear programming problems, and optimal solutions to the formulated programming problems are obtained by the variable transformation by Charnes and Cooper and the combined use of the bisection method and the first phase of the simplex method.

4.4.1 Interactive fuzzy programming

In this section, we deal with two-level linear fractional programming problems in which the decision makers are essentially cooperative, and then, the problem can be expressed by the following formulation:

$$\underset{\text{for DM1}}{\text{minimize}}\ z_1(\boldsymbol{x}_1,\boldsymbol{x}_2) = \frac{p_1(\boldsymbol{x}_1,\boldsymbol{x}_2)}{q_1(\boldsymbol{x}_1,\boldsymbol{x}_2)} = \frac{\boldsymbol{c}_{11}\boldsymbol{x}_1+\boldsymbol{c}_{12}\boldsymbol{x}_2+c_{13}}{\boldsymbol{d}_{11}\boldsymbol{x}_1+\boldsymbol{d}_{12}\boldsymbol{x}_2+d_{13}} \tag{4.50a}$$

$$\underset{\text{for DM2}}{\text{minimize}}\ z_2(\boldsymbol{x}_1,\boldsymbol{x}_2) = \frac{p_2(\boldsymbol{x}_1,\boldsymbol{x}_2)}{q_2(\boldsymbol{x}_1,\boldsymbol{x}_2)} = \frac{\boldsymbol{c}_{21}\boldsymbol{x}_1+\boldsymbol{c}_{22}\boldsymbol{x}_2+c_{23}}{\boldsymbol{d}_{21}\boldsymbol{x}_1+\boldsymbol{d}_{22}\boldsymbol{x}_2+d_{23}} \tag{4.50b}$$

$$\text{subject to}\ A_1\boldsymbol{x}_1+A_2\boldsymbol{x}_2 \leqq \boldsymbol{b} \tag{4.50c}$$

$$\boldsymbol{x}_1 \geqq \boldsymbol{0},\ \boldsymbol{x}_2 \geqq \boldsymbol{0}, \tag{4.50d}$$

where $\boldsymbol{x}_i$ and $z_i(\boldsymbol{x}_1,\boldsymbol{x}_2)$, $i = 1,2$ are the decision variable vector and the objective function of DMi; $\boldsymbol{c}_{i1}$ and $\boldsymbol{d}_{i1}$, $i = 1,2$ are n_1-dimensional coefficient row vectors; $\boldsymbol{c}_{i2}$ and $\boldsymbol{d}_{i2}$, $i = 1,2$ are n_2-dimensional coefficient row vectors; c_{i3} and d_{i3}, $i = 1,2$ are constants; $\boldsymbol{b}$ is an m-dimensional constant column vector; A_i, $i = 1,2$ is an $m \times n_i$ coefficient matrix; and it is assumed that the denominators are positive, i.e., $q_i(\boldsymbol{x}_1,\boldsymbol{x}_2) > 0$, $i = 1,2$. For the sake of simplicity, we use the following notation: $\boldsymbol{x} = (\boldsymbol{x}_1^T,\boldsymbol{x}_2^T)^T \in \mathbb{R}^{n_1+n_2}$, $\boldsymbol{c}_i = [\boldsymbol{c}_{i1}\ \boldsymbol{c}_{i2}\ c_{i3}]$, $i = 1,2$, $\boldsymbol{d}_i = [\boldsymbol{d}_{i1}\ \boldsymbol{d}_{i2}\ d_{i3}]$, $i = 1,2$.

As we considered in the two-level linear programming problems in the previous section, it is natural for each of the decision makers to have a fuzzy goal for the objective function when they take fuzziness of human judgments into consideration.

Although the membership function does not need to be a linear function throughout this section, we assume that it is a continuous and strictly monotone decreasing function. In this connection, Sakawa and Yumine (1983) propose several types of nonlinear membership function: exponential, hyperbolic, hyperbolic inverse and piecewise linear membership functions.

To identify membership functions of the fuzzy goals for the objective functions, we solve the following two individual single-objective linear fractional programming problems for $i = 1,2$:

$$\text{minimize} \quad \frac{\boldsymbol{c}_{i1}\boldsymbol{x}_1 + \boldsymbol{c}_{i2}\boldsymbol{x}_2 + c_{i3}}{\boldsymbol{d}_{i1}\boldsymbol{x}_1 + \boldsymbol{d}_{i2}\boldsymbol{x}_2 + d_{i3}} \tag{4.51a}$$

$$\text{subject to} \quad A_1\boldsymbol{x}_1 + A_2\boldsymbol{x}_2 \leqq \boldsymbol{b} \tag{4.51b}$$

$$\boldsymbol{x}_1 \geqq \boldsymbol{0},\ \boldsymbol{x}_2 \geqq \boldsymbol{0}. \tag{4.51c}$$

Problem (4.51) is a conventional linear fractional programming problem, and it can be solved by using the Charnes and Cooper method (1962) or the Bitran and Novaes method (1973).

After identifying the membership functions $\mu_i(z_i(\boldsymbol{x}))$, $i = 1,2$, to derive an overall satisfactory solution to the two-level linear fractional programming problem (4.50), we first solve the following maximin problem for obtaining a solution which maximizes the smaller degree between the satisfactory degrees of the two decision makers:

$$\text{minimize} \quad \min\{\mu_1(z_1(\boldsymbol{x})), \mu_2(z_2(\boldsymbol{x}))\} \tag{4.52a}$$

$$\text{subject to} \quad \boldsymbol{x} \in S, \tag{4.52b}$$

where S denotes the feasible region of problem (4.50).

By introducing an auxiliary variable λ, this problem can be transformed into the following equivalent maximization problem:

$$\text{maximize} \quad \lambda \tag{4.53a}$$

$$\text{subject to} \quad \mu_1(z_1(\boldsymbol{x})) \geqq \lambda \tag{4.53b}$$

$$\mu_2(z_2(\boldsymbol{x})) \geqq \lambda \tag{4.53c}$$

$$\boldsymbol{x} \in S. \tag{4.53d}$$

Let $\mu_i^{-1}(\cdot)$, $i=1,2$ be an inverse function of the continuous and strictly monotone decreasing membership function $\mu_i(\cdot)$, $i=1,2$, and then, problem (4.53) can be transformed into the following equivalent problem:

$$\begin{aligned}
&\text{maximize } \lambda && (4.54a)\\
&\text{subject to } p_1(\boldsymbol{x}) \leqq \mu_1^{-1}(\lambda) q_1(\boldsymbol{x}) && (4.54b)\\
&\qquad p_2(\boldsymbol{x}) \leqq \mu_2^{-1}(\lambda) q_2(\boldsymbol{x}) && (4.54c)\\
&\qquad \boldsymbol{x} \in S. && (4.54d)
\end{aligned}$$

By solving problem (4.54), we can obtain a solution maximizing a smaller satisfactory degree between the two decision makers. In problem (4.54), however, even if the membership function $\mu_i(\cdot)$ is linear, problem (4.54) is not a linear programming problem because $\mu_i^{-1}(\lambda)q_i(\boldsymbol{x})$, $i=1,2$ is nonlinear. Thus, we cannot directly apply the linear programming techniques, but from the following facts, we can solve problem (4.54) by using the bisection method and solving the phase one problem in the simplex method (Sakawa and Yumine, 1983; Sakawa, 1993). In problem (4.54), if the value of λ, which ought to be in the interval $[0,1]$, is fixed at a certain value, the constraints of problem (4.54) can be reduced to a set of linear inequalities. Finding the optimal value λ^* to the above problem is equivalent to determining the maximal value of λ so that there exists a feasible solution satisfying the constraints of problem (4.54). After the value of λ is fixed at $\lambda^1 = 0.5$, to check out whether or not there exists a feasible solution satisfying the corresponding constraints, we solve the phase one problem of the linear programming problem with the fixed value of λ^1 in the simplex method, and update the value of λ using the bisection method as follows:

$$\lambda^{n+1} = \begin{cases} \lambda^n + 1/2^{n+1} & \text{if there exists a feasible solution,} \\ \lambda^n - 1/2^{n+1} & \text{if there does not exist any feasible solution.} \end{cases} \tag{4.55}$$

By executing the above mentioned procedure iteratively, we obtain the maximal value of λ such that there exists a feasible solution satisfying the constraints of problem (4.54). It should be noted that the above mentioned procedure can be applicable not only to linear membership functions but also to nonlinear membership functions if they are continuous and strictly monotone decreasing.

After λ^* has been found, to obtain the corresponding solution $\boldsymbol{x}^*$, we solve the following problem in which, the objective function $z_2(\boldsymbol{x})$ of DM2 is minimized under the condition (4.54b) with $\lambda = \lambda^*$:

$$\begin{aligned}
&\text{minimize } z_2(\boldsymbol{x}) = \frac{\boldsymbol{c}_{21}\boldsymbol{x}_1 + \boldsymbol{c}_{22}\boldsymbol{x}_2 + c_{23}}{\boldsymbol{d}_{21}\boldsymbol{x}_1 + \boldsymbol{d}_{22}\boldsymbol{x}_2 + d_{23}} && (4.56a)\\
&\text{subject to } p_1(\boldsymbol{x}) \leqq \mu_1^{-1}(\lambda^*) q_1(\boldsymbol{x}) && (4.56b)\\
&\qquad \boldsymbol{x} \in S. && (4.56c)
\end{aligned}$$

In problem (4.56), because DM1 specifies the minimal satisfactory level $\hat{\delta}$ later if DM1 is not satisfied with an optimal solution to problem (4.56), the objective function $z_2(\boldsymbol{x})$ of DM2 is minimized and the constraint for the fuzzy goal of DM2 is eliminated.

In order to solve the linear fractional programming problem (4.56), by using the following variable transformation by Charnes and Cooper (1962)

$$t = \frac{1}{q_2(\boldsymbol{x})}, \quad \begin{pmatrix} \boldsymbol{y}_1 \\ \boldsymbol{y}_2 \\ t \end{pmatrix} = \begin{pmatrix} \boldsymbol{x}_1 \\ \boldsymbol{x}_2 \\ 1 \end{pmatrix} t, \tag{4.57}$$

problem (4.56) can be equivalently transformed into:

$$\begin{aligned} &\text{minimize } \boldsymbol{c}_2\boldsymbol{y} && (4.58\text{a})\\ &\text{subject to } [A_1\ A_2\ -\boldsymbol{b}]\boldsymbol{y} \leqq \boldsymbol{0} && (4.58\text{b})\\ &\qquad \boldsymbol{c}_1\boldsymbol{y} \leqq \mu_1^{-1}(\lambda^*)\boldsymbol{d}_1\boldsymbol{y} && (4.58\text{c})\\ &\qquad \boldsymbol{d}_2\boldsymbol{y} = 1 && (4.58\text{d})\\ &\qquad \boldsymbol{y} \geqq \boldsymbol{0}, && (4.58\text{e}) \end{aligned}$$

where, for simplicity, we use the notation $\boldsymbol{y} = (\boldsymbol{y}_1^T, \boldsymbol{y}_2^T, t)^T \in \mathbb{R}^{n_1+n_2+1}$.

The problem with the minimal satisfactory level $\hat{\delta}$ specified by DM1 is formulated as follows:

$$\begin{aligned} &\text{maximize } \mu_2(z_2(\boldsymbol{x})) && (4.59\text{a})\\ &\text{subject to } \mu_1(z_1(\boldsymbol{x})) \geqq \hat{\delta} && (4.59\text{b})\\ &\qquad \boldsymbol{x} \in S. && (4.59\text{c}) \end{aligned}$$

Because the membership function $\mu_2(\cdot)$ is strictly monotone decreasing, problem (4.59) is equivalent to the following linear fractional problem:

$$\begin{aligned} &\text{minimize } z_2(\boldsymbol{x}) = \frac{\boldsymbol{c}_{21}\boldsymbol{x}_1 + \boldsymbol{c}_{22}\boldsymbol{x}_2 + c_{23}}{\boldsymbol{d}_{21}\boldsymbol{x}_1 + \boldsymbol{d}_{22}\boldsymbol{x}_2 + d_{23}} && (4.60\text{a})\\ &\text{subject to } p_1(\boldsymbol{x}) \leqq \mu_1^{-1}(\hat{\delta})q_1(\boldsymbol{x}) && (4.60\text{b})\\ &\qquad \boldsymbol{x} \in S. && (4.60\text{c}) \end{aligned}$$

Using the variable transformation again, we can formulate the linear programming problem:

$$\begin{aligned} &\text{minimize } \boldsymbol{c}_2\boldsymbol{y} && (4.61\text{a})\\ &\text{subject to } [A_1\ A_2\ -\boldsymbol{b}]\boldsymbol{y} \leqq \boldsymbol{0} && (4.61\text{b})\\ &\qquad \boldsymbol{c}_1\boldsymbol{y} \leqq \mu_1^{-1}(\hat{\delta})\boldsymbol{d}_1\boldsymbol{y} && (4.61\text{c})\\ &\qquad \boldsymbol{d}_2\boldsymbol{y} = 1 && (4.61\text{d})\\ &\qquad \boldsymbol{y} \geqq \boldsymbol{0}. && (4.61\text{e}) \end{aligned}$$

For the two-level linear fractional programming problem (4.50), we can provide the termination conditions of the interactive process and the procedure for updating the minimal satisfactory level $\hat{\delta}$ which are the same as those of interactive fuzzy programming for the two-level linear programming problem (4.4). We also give a similar algorithm for deriving satisfactory solutions; the phase one problem in the simplex method is repeatedly solved through the bisection method in order to find maximin value, and then problems (4.58) and (4.61) transformed by the Charnes and Cooper method are solved.

4.4.2 Numerical example

To illustrate the interactive fuzzy programming method described in the previous subsection, consider the following two-level linear fractional programming problem:

$$\underset{\text{for DM1}}{\text{minimize}}\ z_1(\boldsymbol{x}) = \frac{\boldsymbol{c}_1\boldsymbol{x}_1 + \boldsymbol{c}_2\boldsymbol{x}_2 + 1}{\boldsymbol{d}_1\boldsymbol{x}_1 + \boldsymbol{d}_2\boldsymbol{x}_2 + 1} \tag{4.62a}$$

$$\underset{\text{for DM2}}{\text{minimize}}\ z_2(\boldsymbol{x}) = \frac{\boldsymbol{c}_3\boldsymbol{x}_1 + \boldsymbol{c}_4\boldsymbol{x}_2 + 1}{\boldsymbol{d}_3\boldsymbol{x}_1 + \boldsymbol{d}_4\boldsymbol{x}_2 + 1} \tag{4.62b}$$

$$\text{subject to}\ A_1\boldsymbol{x}_1 + A_2\boldsymbol{x}_2 \leqq \boldsymbol{b} \tag{4.62c}$$

$$x_{ij} \geqq 0,\ i = 1,2,\ j = 1,\dots,5, \tag{4.62d}$$

where $\boldsymbol{x}_1 = (x_{11},\dots,x_{15})^T$, $\boldsymbol{x}_2 = (x_{21},\dots,x_{25})^T$; each entry of 5-dimensional coefficient vectors $\boldsymbol{c}_i$, $\boldsymbol{d}_i$, $i = 1,2,3,4$, and 11×5 coefficient matrices A_1 and A_2 is a randomly selected number from the interval $[-50, 50]$; each entry of the right-hand side constant column vector $\boldsymbol{b}$ is a sum of entries of the corresponding row vector of A_1 and A_2 multiplied by 0.6. The coefficients are shown in Table 4.20.

To identify membership functions of the fuzzy goals for the objective functions, we first solve two individual minimization problems of both levels. The individual minima and the corresponding optimal solutions are shown in Table 4.21.

Suppose that the decision makers employ the linear membership function (4.7) whose parameters are determined by the Zimmermann method (1978). Then, we have $z_1^0 = z_1^{\mathrm{m}} = 0.559$ and $z_2^0 = z_2^{\mathrm{m}} = 0.525$, and the maximin problem (4.54) for this numerical example can be formulated as

$$\text{maximize}\ \lambda \tag{4.63a}$$

$$\text{subject to}\ (z_1(\boldsymbol{x}) + 0.559)/(-1.436 + 0.559) \geqq \lambda \tag{4.63b}$$

$$(z_2(\boldsymbol{x}) - 0.525)/(-0.497 - 0.525) \geqq \lambda \tag{4.63c}$$

$$\boldsymbol{x} \in S, \tag{4.63d}$$

where S denotes the feasible region of problem (4.62). The result of the first iteration including an optimal solution to problem (4.63) is shown in Table 4.22.

Table 4.20 Coefficients in problem (4.62).

c_1	−34	−7	−44	−22	−10	c_2	−13	−25	−1	−5	−27		
d_1	22	36	4	31	49	d_2	1	48	33	37	4		
c_3	−11	−33	−31	−14	−7	c_4	13	25	1	5	27		
d_3	38	5	42	10	8	d_4	1	48	33	37	4		
A_1	−6	38	−30	24	−24	A_2	−10	−41	42	20	2	b	8
	−29	−22	16	43	−15		15	3	−9	11	24		22
	−28	−22	43	−45	−37		30	−45	−18	−20	−27		−101
	5	44	46	11	−49		−5	−11	49	6	16		67
	36	47	−39	20	−23		−45	34	−43	10	−11		−8
	21	40	−17	36	50		−4	11	−28	36	16		96
	32	18	−7	−15	−23		−41	−44	−31	42	17		−31
	3	−41	−14	15	−8		−39	35	−15	−20	22		−37
	−31	−36	27	−17	17		−8	22	44	−4	45		35
	3	−6	5	−23	−22		−31	−42	31	4	−21		−61
	13	−43	−10	49	−2		34	−13	0	27	−19		21

Table 4.21 Optimal solutions to the individual problems.

$z_1^{\min}$	−1.436				
x_1	1.883	0.668	0.018	0.000	0.000
x_2	2.362	0.861	0.242	0.000	2.299
$z_2^{\min}$	−0.497				
x_1	0.000	0.921	0.859	1.019	0.982
x_2	0.223	0.674	0.661	0.000	0.000

Table 4.22 Iteration 1.

x_1^1	2.784	0.241	1.410	0.000	1.101
x_2^1	1.059	0.827	0.701	0.000	0.725
(z_1^1, z_2^1)	(−1.131, −0.141)				
$(\mu_1(z_1^1), \mu_2(z_2^1))$	(0.652, 0.652)				
λ^1	0.652				
Δ^1	1.000				

Suppose that DM1 is not satisfied with the solution obtained in Iteration 1, and then he specifies the minimal satisfactory level at $\hat{\delta} = 0.7$ and the bounds of the ratio at the interval $[\Delta_{\min}, \Delta_{\max}] = [0.6, 1.0]$, taking account of the result of Iteration 1. Then, the problem with the minimal satisfactory level (4.59) is formulated as

$$\text{maximize } \mu_2(z_2(\boldsymbol{x})) \quad (4.64a)$$

$$\text{subject to } (z_1(\boldsymbol{x}) + 0.557)/(-1.436 + 0.559) \geqq 0.7 \quad (4.64b)$$

$$\boldsymbol{x} \in S. \quad (4.64c)$$

The result of Iteration 2 including an optimal solution to problem (4.64) is shown in Table 4.23.

Table 4.23 Iteration 2.

x_1^2	2.604	0.323	1.136	0.000	0.887
x_2^2	1.316	0.830	0.613	0.000	1.041
(z_1^2, z_2^2)	$(-1.173, -0.059)$				
$(\mu_1(z_1^2), \mu_2(z_2^2))$	$(0.700, 0.572)$				
λ^2	0.572				
Δ^2	0.817				

At Iteration 2, the satisfactory degree $\mu_1(z_1^2) = 0.700$ of DM1 becomes equal to the minimal satisfactory level $\hat{\delta} = 0.7$ and the ratio $\Delta^2 = 0.817$ of satisfactory degrees is in the valid interval $[0.6, 1.0]$ of the ratio. Therefore, this solution satisfies the conditions of termination of the interactive procedure, and it becomes a satisfactory solution for both decision makers if DM1 accepts the solution.

4.5 Fuzzy decentralized two-level linear programming

We consider decentralized two-level programming problems in which there are a single decision maker at the upper level and two or more decision makers at the lower level, and the objective functions of the decision makers and the constraint functions are linear functions. For such decentralized two-level linear programming problems, Simaan and Cruz (1973b) and Anandalingam (1988) assume that the decision makers at the lower level make decisions so as to equilibrate their objective function values for a decision of the decision maker at the upper level on condition that all the decision makers at the lower level do not have any motivation to cooperate mutually. Namely, they suppose that the decision maker at the upper level assumes that rational responses of the decision makers at the lower level with respect to the decision of himself are a Nash equilibrium solution, and the decision maker at the upper level selects a decision which optimizes the objective function of himself.

If the decision makers can coordinate their actions, the noncooperative solution concept described above is not always appropriate. From a viewpoint similar to those in the previous sections, in this section, assuming cooperative behavior of the decision makers, we present interactive fuzzy programming for the decentralized two-level linear programming problems with a single decision maker at the upper level and multiple decision makers at the lower level. The method described here consists of two phases. In the first phase, the decision makers at both levels identify membership functions of their fuzzy goals for the objective functions. Taking account of the overall satisfactory balance between the two levels, the decision maker at the upper level specifies the minimal satisfactory level and updates it if necessary, and then a tentative solution is obtained. In this phase, the decision makers at the lower level are treated impartially, and therefore, they can be regarded as a group. In the second phase, with a ratio of satisfaction degree of the decision maker at the upper level to that of each of the decision makers at the lower level in mind, the decision maker at the upper level specifies maximal satisfactory levels to some of the decision makers at the lower level and updates them if necessary. By coordinating the satisfaction degrees of the decision makers, the final satisfactory solution can be derived.

4.5.1 Interactive fuzzy programming

In this section, we extend interactive fuzzy programming for two-level linear programming problems to that for decentralized two-level linear programming problems. Suppose that there are a single decision maker at the upper level and k decision makers at the lower level in a decentralized organization, which is depicted in Figure 4.6. Let DM0 denote the decision maker at the upper level and DM1, ..., DMk denote the k decision makers at the lower level.

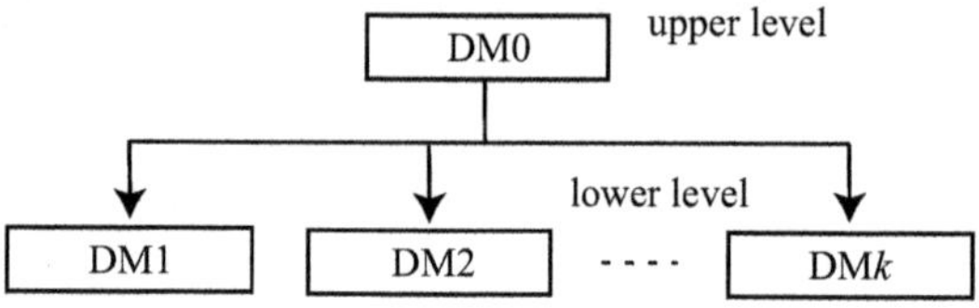

Fig. 4.6 Decentralized two-level organization.

A decentralized two-level linear programming problem is formally represented as:

$$\underset{\substack{\text{for DM0}\\ \text{(upper level)}}}{\text{minimize}}\ z_0(\boldsymbol{x}) = \boldsymbol{c}_{00}\boldsymbol{x}_0 + \boldsymbol{c}_{01}\boldsymbol{x}_1 + \cdots + \boldsymbol{c}_{0k}\boldsymbol{x}_k \tag{4.65a}$$

$$\underset{\substack{\text{for DM1}\\ \text{(lower level)}}}{\text{minimize}}\ z_1(\boldsymbol{x}) = \boldsymbol{c}_{10}\boldsymbol{x}_0 + \boldsymbol{c}_{11}\boldsymbol{x}_1 + \cdots + \boldsymbol{c}_{1k}\boldsymbol{x}_k \tag{4.65b}$$

$$\cdots\cdots\cdots\cdots\cdots\cdots$$

$$\underset{\substack{\text{for DM}k\\ \text{(lower level)}}}{\text{minimize}}\ z_k(\boldsymbol{x}) = \boldsymbol{c}_{k0}\boldsymbol{x}_0 + \boldsymbol{c}_{k1}\boldsymbol{x}_1 + \cdots + \boldsymbol{c}_{kk}\boldsymbol{x}_k \tag{4.65c}$$

$$\text{subject to}\ A_0\boldsymbol{x}_0 + A_1\boldsymbol{x}_1 + \cdots + A_k\boldsymbol{x}_k \leqq \boldsymbol{b} \tag{4.65d}$$

$$\boldsymbol{x}_i \geqq \boldsymbol{0},\ i = 0, 1, \ldots, k, \tag{4.65e}$$

where $\boldsymbol{x}_i$, $i = 0, 1, \ldots, k$ is an n_i-dimensional decision variable column vector of DMi, and $\boldsymbol{x} = (\boldsymbol{x}_0^T, \boldsymbol{x}_1^T, \ldots, \boldsymbol{x}_k^T)^T$; $\boldsymbol{c}_{ij}$, $i, j = 0, 1 \ldots, k$ is an n_j-dimensional coefficient row vector; A_i, $i = 0, 1, \ldots, k$ is an $m \times n_i$ coefficient matrix; $\boldsymbol{b}$ is an m-dimensional constant column vector; $z_i(\boldsymbol{x})$, $i = 0, 1, \ldots, k$ is the objective function of DMi; for the sake of simplicity, let $\boldsymbol{A} = [A_0\ A_1\ \cdots\ A_k]$, $\boldsymbol{c}_i = (\boldsymbol{c}_{i1}, \ldots, \boldsymbol{c}_{ik})$ and $i = 0, 1, \ldots, k$.

In a way similar to the two-level linear programming problems considered in the previous section, it is natural that each of the decision makers has a fuzzy goal for the objective function when the decision maker takes fuzziness of human judgments into consideration. To identify membership functions of the fuzzy goals for the objective functions, we solve $(k+1)$ individual single-objective problems for $i = 0, 1, \ldots, k$:

$$\text{minimize} \;\; z_i(\boldsymbol{x}) = \boldsymbol{c}_i\boldsymbol{x} \tag{4.66a}$$
$$\text{subject to} \;\; A_0\boldsymbol{x}_0 + A_1\boldsymbol{x}_1 + \cdots + A_k\boldsymbol{x}_k \leqq \boldsymbol{b} \tag{4.66b}$$
$$\boldsymbol{x}_i \geqq \boldsymbol{0},\; i = 0, 1, \ldots, k. \tag{4.66c}$$

Suppose that, for the decentralized two-level programming problems, the decision makers employ the linear membership function (4.7) of the fuzzy goal whose parameters are determined by the Zimmermann method (1978). Let $\boldsymbol{x}^{io}$ denote an optimal solution to problem (4.66). The individual minimum is $z_i^{\min} = z_i(\boldsymbol{x}^{io})$, and let $z_i^{\rm m}$ be defined by

$$z_i^{\rm m} = \max_{j=0,1,\ldots,k,\; j \neq i} \{z_i(\boldsymbol{x}^{jo})\}. \tag{4.67}$$

Then, the parameters z_i^0 and z_i^1 of the linear membership function are determined by choosing $z_i^1 = z_i^{\min}$, $z_i^0 = z_i^{\rm m}$, $i = 0, 1, \ldots, k$.

The interactive fuzzy programming method consists of the two phases, and they are described as follows.

First phase

After identifying the membership functions $\mu_i(z_i(\boldsymbol{x}))$, $i = 0, 1, \ldots, k$, to derive an overall satisfactory solution to the formulated problem (4.65), the first phase of interactive fuzzy programming starts to solve the following problem yielding a solution which maximizes the smallest degree among satisfactory degrees of all the decision makers:

$$\text{maximize} \;\; \min\left\{\mu_0(z_0(\boldsymbol{x})),\; \min_{i=1,\ldots,k} \mu_i(z_i(\boldsymbol{x}))\right\} \tag{4.68a}$$
$$\text{subject to} \;\; \boldsymbol{x} \in S, \tag{4.68b}$$

where S denotes the feasible region of problem (4.65). By introducing an auxiliary variable λ, this problem can be transformed into the following equivalent maximization problem:

$$\begin{aligned} &\text{maximize } \lambda && (4.69a)\\ &\text{subject to } \mu_0(z_0(\boldsymbol{x})) \geqq \lambda && (4.69b)\\ &\qquad \mu_1(z_1(\boldsymbol{x})) \geqq \lambda && (4.69c)\\ &\qquad \cdots\cdots \\ &\qquad \mu_k(z_k(\boldsymbol{x})) \geqq \lambda && (4.69d)\\ &\qquad \boldsymbol{x} \in S. && (4.69e) \end{aligned}$$

Let $\boldsymbol{x}^*$ denote an optimal solution to problem (4.69). The satisfactory degree of both levels is defined as

$$\lambda^* = \min\left\{\mu_0(z_0(\boldsymbol{x}^*)), \min_{i=1,\ldots,k} \mu_i(z_i(\boldsymbol{x}^*))\right\}. \qquad (4.70)$$

If DM0 is satisfied with the optimal solution $\boldsymbol{x}^*$ to problem (4.69), it follows that the optimal solution $\boldsymbol{x}^*$ becomes a satisfactory solution. Otherwise, DM0 specifies the minimal satisfactory level $\hat{\delta}$, taking account of the satisfactory degree of both levels (4.70). Then, the following problem with the minimal satisfactory level $\hat{\delta}$ is formulated:

$$\begin{aligned} &\text{maximize } \min_{i=1,\ldots,k} \mu_i(z_i(\boldsymbol{x})) && (4.71a)\\ &\text{subject to } \mu_0(z_0(\boldsymbol{x})) \geqq \hat{\delta} && (4.71b)\\ &\qquad \boldsymbol{x} \in S. && (4.71c) \end{aligned}$$

By introducing an auxiliary variable λ, this problem can be transformed into the following equivalent maximization problem:

$$\begin{aligned} &\text{maximize } \lambda && (4.72a)\\ &\text{subject to } \mu_0(z_0(\boldsymbol{x})) \geqq \hat{\delta} && (4.72b)\\ &\qquad \mu_1(z_1(\boldsymbol{x})) \geqq \lambda && (4.72c)\\ &\qquad \cdots\cdots \\ &\qquad \mu_k(z_k(\boldsymbol{x})) \geqq \lambda && (4.72d)\\ &\qquad \boldsymbol{x} \in S. && (4.72e) \end{aligned}$$

If there exists an optimal solution to problem (4.72), it follows that DM0 obtains a solution with a satisfactory degree larger than or equal to the minimal satisfactory level specified by DM0. However, the larger the minimal satisfactory level is assessed, the smaller the whole satisfactory degree of the decision makers at the lower level, DM1, ..., DMk, becomes. Consequently, a relative difference between the satisfactory degree of DM0 and the whole satisfactory degree of DM1, ..., DMk might be large, and as a result, the satisfactory degrees of both levels would become ill-balanced.

To take account of the overall satisfactory balance between both levels, DM0 sometimes needs to compromise with DM1, ..., DMk on DM0's own minimal satisfactory level. Because DM0 treats DM1, ..., DMk as a group in the first phase, the following ratio of satisfactory degree of the upper level to that of the lower level is useful:

$$\Delta = \frac{\min\{\mu_1(z_1(\boldsymbol{x})), \ldots, \mu_k(z_k(\boldsymbol{x}))\}}{\mu_0(z_0(\boldsymbol{x}))}. \tag{4.73}$$

For the decentralized two-level linear programming problem (4.65), we can provide the termination conditions of the first phase and the procedure for updating the minimal satisfactory level $\hat{\delta}$ which are the same as that of interactive fuzzy programming for the two-level linear programming problem (4.4).

Second phase

There is a possibility that the satisfactory degrees of some decision makers at the lower level have unexpected high values. According to circumstances, it may be necessary that DM0 coordinates the satisfactory degrees of the decision makers at both levels. In such cases, we start out the second phase of interactive fuzzy programming; to coordinate satisfactory degrees of DM0 at the upper level and each of the decision makers at the lower level, the following ratios of satisfactory degrees of DM0 and each of DM1, ..., DMk are calculated:

$$\Delta_i = \frac{\mu_i(z_i(\boldsymbol{x}))}{\mu_0(z_0(\boldsymbol{x}))}, \quad i = 1, \ldots, k. \tag{4.74}$$

In the second phase, for some decision maker, say DMi, DM0 may specify the lower bound $\Delta^i_{\min}$ and the upper bound $\Delta^i_{\max}$ for the ratio of satisfactory degrees. Let $[\Delta^i_{\min}, \Delta^i_{\max}]$ be the interval between the lower and the upper bounds for the ratio of satisfactory degree of DM0 at the upper level to that of DMi at the lower level.

For DMi at the lower level whose ratio of satisfactory degrees is larger than the upper bound $\Delta^i_{\max}$ or is smaller than the lower bound $\Delta^i_{\min}$, DM0 specifies the permissible maximal level $\bar{\delta}^i_{\max}$ or the permissible minimal level $\bar{\delta}^i_{\min}$ to a satisfactory degree of DMi, and the following problem is formulated:

$$\begin{aligned}
&\text{maximize} \quad \min_{i=1,\ldots,k} \{\mu_i(z_i(\boldsymbol{x}))\} && (4.75\text{a})\\
&\text{subject to} \quad \mu_0(z_0(\boldsymbol{x})) \geqq \hat{\delta} && (4.75\text{b})\\
&\qquad \bar{\delta}^i_{\min} \leqq \mu_i(z_i(\boldsymbol{x})) \leqq \bar{\delta}^i_{\max}, \ \forall i \in V && (4.75\text{c})\\
&\qquad \boldsymbol{x} \in S, && (4.75\text{d})
\end{aligned}$$

where V is the index set of decision makers whose ratio Δ_i of satisfactory degrees is larger than the upper bound $\Delta^i_{\max}$ or is smaller than the lower bound $\Delta^i_{\min}$. The value $\Delta^i_{\max}\hat{\delta}$ is considered to be one of appropriate values for the permissible maximal level $\bar{\delta}^i_{\max}$ because $\mu_i(z_i(\boldsymbol{x}))/\mu_0(z_0(\boldsymbol{x})) \leqq \Delta^i_{\max}$ if $\bar{\delta}^i_{\max} \leqq \Delta^i_{\max}\hat{\delta}$. The value $\Delta^i_{\min}\hat{\delta}$ also seems to be appropriate as a candidate for the permissible minimal level $\bar{\delta}^i_{\min}$.

Problem (4.75) can be rewritten as an equivalent maximization problem by introducing an auxiliary variable λ:

$$\text{maximize } \lambda \tag{4.76a}$$

$$\text{subject to } \mu_0(z_0(\boldsymbol{x})) \geqq \hat{\delta} \tag{4.76b}$$

$$\bar{\delta}^i_{\text{min}} \leqq \mu_i(z_i(\boldsymbol{x})) \leqq \bar{\delta}^i_{\text{max}},\ \forall i \in V \tag{4.76c}$$

$$\mu_1(z_1(\boldsymbol{x})) \geqq \lambda \tag{4.76d}$$

$$\cdots\cdots$$

$$\mu_k(z_k(\boldsymbol{x})) \geqq \lambda \tag{4.76e}$$

$$\boldsymbol{x} \in S. \tag{4.76f}$$

Suppose that there exists an optimal solution to problem (4.76). Then, individual ratios of satisfactory degrees Δ_i, $i = 1,\ldots,k$ are calculated again, and it is verified if all of Δ_i, $i = 1,\ldots,k$ are in the interval $[\Delta^i_{\text{min}}, \Delta^i_{\text{max}}]$. Unless there exists an optimal solution to problem (4.76) which satisfies this condition and DM0 accepts the obtained solution, DM0 needs to update the permissible maximal level $\bar{\delta}^i_{\text{max}}$ or the permissible minimal level $\bar{\delta}^i_{\text{min}}$ for DMi.

[Procedure for updating the permissible maximal level $\bar{\delta}^i_{\text{max}}$ for DMi]

Case 1 If there does not exit any feasible solution of problem (4.76), then DM0 increases some of the permissible maximal levels $\bar{\delta}^i_{\text{max}}$, $i \in V$.

Case 2 If $\Delta_i > \Delta^i_{\text{max}}$ for $i \in V$, then DM0 decreases the permissible maximal level $\bar{\delta}^i_{\text{max}}$ for DMi.

Case 3 If there exists i such that $\Delta_i > \Delta^i_{\text{max}}$ and i does not belong to the index set V, then DM0 newly specifies $\bar{\delta}^i_{\text{max}}$.

Case 4 Although Δ_i is in the interval $[\Delta^i_{\text{min}}, \Delta^i_{\text{max}}]$, if DM0 is not satisfied with the obtained solution and judges that it is desirable to increase the satisfactory degree of DM0 at the expense of the satisfactory degree of DMi, then DM0 decreases the permissible maximal level $\bar{\delta}^i_{\text{max}}$. Conversely, if DM0 judges that it is desirable to increase the satisfactory degree of DMi at the expense of the satisfactory degree of DM0, then DM0 increases the permissible maximal level $\bar{\delta}^i_{\text{max}}$.

The procedure for updating the permissible minimal level $\bar{\delta}^i_{\text{min}}$ for DMi can be given in a similar way. The algorithm for decentralized two-level linear programming problems is summarized in the following, and it is also illustrated with a flowchart in Figure 4.7.

[Algorithm of interactive fuzzy programming for decentralized two-level linear programming problems]

First phase:

Step 1-0 Solve the individual problems (4.66) for the $(k+1)$ decision makers.

Step 1-1 DM0 at the upper level identifies the membership function $\mu_0(z_0)$ of the fuzzy goal, and DM1, ..., DMk at the lower level also identify their membership functions $\mu_1(z_1)$, ..., $\mu_k(z_k)$.

Step 1-2 Set $l := 1$ and solve problem (4.69). If DM0 is satisfied with the solution, stop the algorithm and the solution becomes a satisfactory solution. Otherwise, with the satisfactory degree of both levels λ^l and the related information in mind, DM0 specifies the minimal satisfactory level $\hat{\delta}$ and the lower and the upper bounds $[\Delta_{\min}, \Delta_{\max}]$ of the ratio of satisfactory degrees.

Step 1-3 Set $l := l+1$, and solve problem (4.72). The obtained solution $\boldsymbol{x}^l$ is proposed to DM0 together with $(z_0^l, z_1^l, \dots, z_k^l)$, λ^l, $\mu_0(z_0^l)$, $\mu_1(z_1^l)$, $\dots, \mu_k(z_k^l)$ and Δ^l.

Step 1-4 If the proposed solution meets the termination conditions of the first phase, the interactive process of the first phase terminates. Moreover, if DM0 is satisfied with the solution, stop the algorithm and the solution becomes a satisfactory solution. If it is necessary to coordinate satisfactory degrees of the decision makers at the lower level individually, go to the second phase: Step 2-1. Otherwise, DM0 updates the minimal satisfactory level $\hat{\delta}$ in accordance with the procedure of updating the minimal satisfactory level and then return to Step 1-3.

Second phase:

Step 2-1 Calculate individual ratios of satisfactory degrees Δ_i, $i = 1, \dots, k$ with respect to the solution obtained in the first phase. DM0 specifies the lower and the upper bounds $[\Delta^i_{\min}, \Delta^i_{\max}]$ of the ratio of satisfactory degree of DM0 to that of DMi for all $i = 1, \dots, k$. Let V be the set of indices of decision makers at the lower level such that the ratio Δ_i of satisfactory degrees is not in the interval $[\Delta^i_{\min}, \Delta^i_{\max}]$. DM0 specifies the permissible minimal or the maximal levels, $\bar{\delta}^i_{\min}$ and $\bar{\delta}^i_{\max}$, for DMi, $i \in V$.

Step 2-2 Solve problem (4.76). The obtained solution $\boldsymbol{x}^l$ is proposed to DM0 together with $(z_0^l, z_1^l, \dots, z_k^l)$, λ^l, $\mu_0(z_0^l)$, $\mu_1(z_1^l)$, $\dots, \mu_k(z_k^l)$ and $(\Delta_1^l, \dots, \Delta_k^l)$.

Step 2-3 If each of Δ_i, $i = 1, \dots, k$ is in the corresponding interval $[\Delta^i_{\min}, \Delta^i_{\max}]$ and DM0 accepts the solution, stop the algorithm and the solution becomes a satisfactory solution. Otherwise, after DM0 updates the permissible minimal or maximal levels, $\bar{\delta}^i_{\min}$ and $\bar{\delta}^i_{\max}$, in accordance with the procedure of updating them, set $l := l+1$ and return to Step 2-3.

4.5.2 Numerical example

To illustrate the interactive fuzzy programming method for decentralized two-level linear programming problems, we consider the following problem in which there are a single decision maker (DM0) at the upper level and three decision makers (DM1, DM2 and DM3) at the lower level:

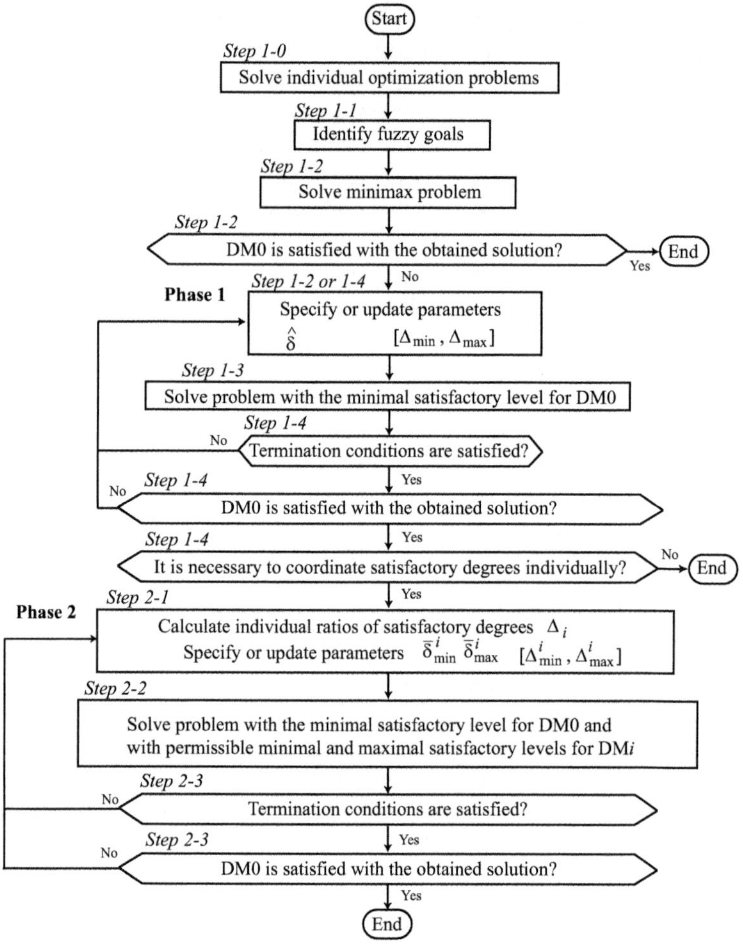

Fig. 4.7 Flowchart of interactive fuzzy programming.

$$\underset{\substack{\text{for DM0}\\ \text{(upper level)}}}{\text{minimize}} \quad z_0(\boldsymbol{x}) = \boldsymbol{c}_{00}\boldsymbol{x}_0 + \boldsymbol{c}_{01}\boldsymbol{x}_1 + \boldsymbol{c}_{02}\boldsymbol{x}_2 + \boldsymbol{c}_{03}\boldsymbol{x}_3 \tag{4.77a}$$

$$\underset{\substack{\text{for DM1}\\ \text{(lower level)}}}{\text{minimize}} \quad z_1(\boldsymbol{x}) = \boldsymbol{c}_{10}\boldsymbol{x}_0 + \boldsymbol{c}_{11}\boldsymbol{x}_1 + \boldsymbol{c}_{12}\boldsymbol{x}_2 + \boldsymbol{c}_{13}\boldsymbol{x}_3 \tag{4.77b}$$

$$\underset{\substack{\text{for DM2}\\ \text{(lower level)}}}{\text{minimize}} \quad z_2(\boldsymbol{x}) = \boldsymbol{c}_{20}\boldsymbol{x}_0 + \boldsymbol{c}_{21}\boldsymbol{x}_1 + \boldsymbol{c}_{22}\boldsymbol{x}_2 + \boldsymbol{c}_{23}\boldsymbol{x}_3 \tag{4.77c}$$

$$\underset{\substack{\text{for DM3}\\ \text{(lower level)}}}{\text{minimize}} \quad z_3(\boldsymbol{x}) = \boldsymbol{c}_{30}\boldsymbol{x}_0 + \boldsymbol{c}_{31}\boldsymbol{x}_1 + \boldsymbol{c}_{32}\boldsymbol{x}_2 + \boldsymbol{c}_{33}\boldsymbol{x}_3 \tag{4.77d}$$

$$\text{subject to} \quad A_0\boldsymbol{x}_0 + A_1\boldsymbol{x}_1 + A_2\boldsymbol{x}_2 + A_3\boldsymbol{x}_3 \leqq \boldsymbol{b} \tag{4.77e}$$

$$\boldsymbol{x}_i \geqq \boldsymbol{0}, \; i = 0, 1, 2, 3, \tag{4.77f}$$

where $\boldsymbol{x}_i = (x_{i1}, x_{i2})^T$, $i = 0,1,2,3$ is a two-dimensional decision variable vector of DMi, $i = 0,1,2,3$; each entry of coefficient vectors $\boldsymbol{c}_{ij}$, $i, j = 0,1,2,3$ and coefficient matrices A_i, $i = 0,1,2,3$ is a randomly selected number from the interval $[0,100]$; each entry of the right-hand side constant column vector $\boldsymbol{b}$ is a sum of entries of the corresponding row vector of A_i, $i = 0,1,2,3$ multiplied by 0.6. These coefficients are shown in Table 4.24.

Table 4.24 Coefficients in problem (4.77).

$(\boldsymbol{c}_{00}\,\boldsymbol{c}_{01}\,\boldsymbol{c}_{02}\,\boldsymbol{c}_{03})$	18	−10	−18	11	−23	−40	−8	32		
$(\boldsymbol{c}_{10}\,\boldsymbol{c}_{11}\,\boldsymbol{c}_{12}\,\boldsymbol{c}_{13})$	17	−40	45	−31	−49	15	31	4		
$(\boldsymbol{c}_{20}\,\boldsymbol{c}_{21}\,\boldsymbol{c}_{22}\,\boldsymbol{c}_{23})$	35	9	−20	44	−10	−7	−46	24		
$(\boldsymbol{c}_{30}\,\boldsymbol{c}_{31}\,\boldsymbol{c}_{32}\,\boldsymbol{c}_{33})$	19	−8	30	−3	40	−36	-40	1		
$[A_0\,A_1\,A_2\,A_3]$	−9	−18	12	13	37	−11	−13	19	$\boldsymbol{b}$	18
	47	−14	−1	4	1	−49	21	16		12
	−23	2	45	−35	12	41	−17	13		22
	6	−19	−1	−2	−49	−11	3	31		−23
	−31	−8	2	17	47	−25	5	−2		5
	46	3	−28	17	−36	−3	41	2		24
	−45	34	−44	44	16	−2	42	39		50
	29	−13	38	19	−2	7	−10	−29		20
	13	10	27	−29	−49	−38	−12	23		−33

To identify membership functions of the fuzzy goals for the objective functions, we first solve the individual minimization problems of all the decision makers. The individual minima and the corresponding optimal solutions are shown in Table 4.25.

Table 4.25 Optimal solutions to the individual problems.

$z_0^{\min}$	−31.220			
$(\boldsymbol{x}_0, \boldsymbol{x}_1)$	0.247	0.851	0.094	0.852
$(\boldsymbol{x}_2, \boldsymbol{x}_3)$	0.000	0.593	0.000	0.000
$z_1^{\min}$	−25.680			
$(\boldsymbol{x}_0, \boldsymbol{x}_1)$	0.492	0.462	0.000	0.758
$(\boldsymbol{x}_2, \boldsymbol{x}_3)$	0.348	0.202	0.000	0.116
$z_2^{\min}$	−223.418			
$(\boldsymbol{x}_0, \boldsymbol{x}_1)$	1.843	3.589	0.343	0.000
$(\boldsymbol{x}_2, \boldsymbol{x}_3)$	1.678	0.527	0.000	0.005
$z_3^{\min}$	−30.407			
$(\boldsymbol{x}_0, \boldsymbol{x}_1)$	0.660	0.000	0.000	0.000
$(\boldsymbol{x}_2, \boldsymbol{x}_3)$	0.760	0.409	0.000	0.017

Suppose that the decision makers employ the linear membership function (4.7) whose parameters are determined by the Zimmermann method (1978); the parameters of the linear membership function are determined as: $(z_0^0, z_0^1) = (z_0^{\rm m}, z_0^{\min}) =$

$(66.538, -31.220)$, $(z_1^0, z_1^1) = (z_1^{\rm m}, z_1^{\rm min}) = (11.326, -25.680)$, $(z_2^0, z_2^1) = (z_2^{\rm m}, z_2^{\rm min}) = (-26.249, -223.418)$ and $(z_3^0, z_3^1) = (z_3^{\rm m}, z_3^{\rm min}) = (46.200, -30.407)$. After the membership functions of the decision makers are identified, the first phase of the algorithm starts in formulating the following linear programming problem which maximizes the minimum among the satisfactory degrees of the four decision makers.

$$
\begin{aligned}
&\text{maximize } \lambda && \text{(4.78a)}\\
&\text{subject to } (z_0(\boldsymbol{x}) - 66.538)/(-31.220 - 66.538) \geqq \lambda && \text{(4.78b)}\\
&\qquad (z_1(\boldsymbol{x}) - 11.326)/(-25.680 - 11.326) \geqq \lambda && \text{(4.78c)}\\
&\qquad (z_2(\boldsymbol{x}) + 26.249)/(-223.418 + 26.249) \geqq \lambda && \text{(4.78d)}\\
&\qquad (z_3(\boldsymbol{x}) - 46.200)/(-30.407 - 46.200) \geqq \lambda && \text{(4.78e)}\\
&\qquad \boldsymbol{x} \in S, && \text{(4.78f)}
\end{aligned}
$$

where S denotes the feasible region of problem (4.77). An optimal solution to problem (4.78) and the related information are shown in Table 4.26.

Table 4.26 Iteration 1.

$(\boldsymbol{x}_0, \boldsymbol{x}_1)$	1.129	1.262	0.184	0.613
$(\boldsymbol{x}_2, \boldsymbol{x}_3)$	1.000	0.749	0.000	0.628
$(z_0^1, z_1^1, z_2^1, z_3^1)$	$(23.004, -6.125, -114.053, 12.085)$			
$(\mu_0(z_0^1), \mu_1(z_1^1), \mu_2(z_2^1), \mu_3(z_3^1))$	$(0.445, 0.472, 0.445, 0.445)$			
λ^1	0.445			
Δ^1	1.000			

Suppose that DM0 at the upper level is not satisfied with the obtained solution, and specifies the minimal satisfactory level at $\hat{\delta} = 0.55$ for the membership function $\mu_0(z_0(\boldsymbol{x}))$ by taking account of the satisfactory degree of both levels $\lambda^1 = 0.445$ and the related information. Moreover, suppose that DM0 sets the lower and the upper bounds of ratio Δ of the satisfactory degree of DM0 to the whole satisfactory degree of the decision makers at the lower level at 0.6 and 0.75, respectively, i.e., $[\Delta_{\min}, \Delta_{\max}] = [0.6, 0.75]$.

Then, a problem corresponding to problem (4.72), which maximizes the minimum among the satisfactory degrees of DM1, DM2 and DM3 under the given constraints and the condition with respect to DM0's satisfaction, is formulated as

$$\begin{aligned}
&\text{maximize } \lambda && (4.79a)\\
&\text{subject to } (z_0(\boldsymbol{x})-66.538)/(-31.220-66.538) \geqq 0.55 && (4.79b)\\
&\qquad (z_1(\boldsymbol{x})-11.326)/(-25.680-11.326) \geqq \lambda && (4.79c)\\
&\qquad (z_2(\boldsymbol{x})+26.249)/(-223.418+26.249) \geqq \lambda && (4.79d)\\
&\qquad (z_3(\boldsymbol{x})-46.200)/(-30.407-46.200) \geqq \lambda && (4.79e)\\
&\qquad \boldsymbol{x} \in S. && (4.79f)
\end{aligned}$$

An optimal solution to problem (4.79) and the related information are shown in Table 4.27.

Table 4.27 Iteration 2.

$(\boldsymbol{x}_0, \boldsymbol{x}_1)$	0.988	1.099	0.224	0.680
$(\boldsymbol{x}_2, \boldsymbol{x}_3)$	0.815	0.669	0.000	0.650
$(z_0^2, z_1^2, z_2^2, z_3^2)$	$(12.771, -5.528, -103.684, 16.114)$			
$(\mu_0(z_0^2), \mu_1(z_1^2), \mu_2(z_2^2), \mu_3(z_3^2))$	$(0.550, 0.455, 0.393, 0.393)$			
λ^2	0.393			
Δ^2	0.714			

At Iteration 2, the satisfactory degree $\mu_0(z_0^2) = 0.550$ of DM0 becomes equal to the minimal satisfactory level $\hat{\delta} = 0.55$ and the ratio $\Delta^2 = 0.714$ of satisfactory degrees is in the specified interval $[0.6, 0.75]$. Therefore, this solution satisfies both of the termination conditions of the interactive procedure. Suppose that DM0 accepts the solution, and then the first phase of the algorithm is completed.

At the second phase, the individual ratios between the satisfactory degree of DM0 and those of DM1, DM2 and DM3 are calculated as follows:

$$\Delta_1^2 = 0.828, \ \ \Delta_2^2 = 0.714, \ \ \Delta_3^2 = 0.714.$$

Suppose that DM0 specifies the lower and the upper bounds $\Delta_{\min}^i$, $\Delta_{\max}^i$ of the individual ratio between the satisfactory degrees of DM0 and DMi for all $i = 1, 2, 3$ at the same value as that of the first phase, i.e., $[\Delta_{\min}^i, \Delta_{\max}^i] = [0.6, 0.75]$, $i = 1, 2, 3$.

The ratios between DM0 and each of DM2 and DM3 are in the specified interval, but the ratio between DM0 and DM1 is not in the interval. Suppose that DM0 specifies the permissible maximal level at $\bar{\delta}_{\max}^1 = 0.55 \cdot 0.75 = 0.4125$ because the ratio between the satisfactory degrees of DM0 and DM1 is over the upper bound $\Delta_{\max}^1 = 0.75$. The problem with the permissible maximal level for DM1 is formulated as

$$
\begin{aligned}
&\text{maximize } \lambda && \text{(4.80a)}\\
&\text{subject to } (z_0(\boldsymbol{x}) - 66.538)/(-31.220 - 66.538) \geqq 0.55 && \text{(4.80b)}\\
&\quad (z_1(\boldsymbol{x}) - 11.326)/(-25.680 - 11.326) \leqq 0.4125 && \text{(4.80c)}\\
&\quad (z_1(\boldsymbol{x}) - 11.326)/(-25.680 - 11.326) \geqq \lambda && \text{(4.80d)}\\
&\quad (z_2(\boldsymbol{x}) + 26.249)/(-223.418 + 26.249) \geqq \lambda && \text{(4.80e)}\\
&\quad (z_3(\boldsymbol{x}) - 46.200)/(-30.407 - 46.200) \geqq \lambda && \text{(4.80f)}\\
&\quad \boldsymbol{x} \in S. && \text{(4.80g)}
\end{aligned}
$$

An optimal solution to problem (4.80) and the related information are shown in Table 4.28.

Table 4.28 Iteration 3.

$(\boldsymbol{x}_0, \boldsymbol{x}_1)$	0.822	0.916	0.331	0.699
$(\boldsymbol{x}_2, \boldsymbol{x}_3)$	0.780	0.601	0.000	0.660
$(z_0^3, z_1^3, z_2^3, z_3^3)$	$(10.854, -4.256, -97.578, 17.773)$			
$(\mu_0(z_0^3), \mu_1(z_1^3), \mu_2(z_2^3), \mu_3(z_3^3))$	$(0.583, 0.413, 0.380, 0.380)$			
λ^3	0.380			
Δ^3	0.652			
$(\Delta_1^3, \Delta_3^3, \Delta_3^3)$	$(0.708, 0.652, 0.652)$			

At the third iteration, the ratio $\Delta_1^3 = 0.708$ of satisfactory degrees is in the specified interval $[0.6, 0.75]$. Therefore, this solution satisfies all the termination conditions of the interactive procedure. If DM0 accepts the solution, then it becomes the satisfactory solution and the algorithm stops. By the interactive fuzzy programming method, we efficiently obtain the solution such that DM0's degree of satisfaction exceeds 0.55 and the degrees of satisfaction of DM1, DM2 and DM3 at the lower level are between 60% and 75% of DM0's degree of satisfaction.

4.6 Fuzzy two-level linear 0-1 programming

In interactive fuzzy programming shown in the previous sections, the linear programming techniques are utilized to solve the formulated mathematical programming problems. In this section, because we deal with two-level 0-1 programming problems, the linear programming techniques cannot be utilized obviously, and therefore efficient computational methods are required for solving 0-1 programming problems.

For solving 0-1 programming problems, the branch-and-bound scheme is well known to be the most practical solution method. However, it is feared that, computational time for searching an optimal solution exceedingly increases in proportion to the size of the problem, and therefore some efficient approximate solution methods are required to solve 0-1 programming problems formulated in interactive fuzzy programming. For this end, genetic algorithms which have attracted attention in various research fields are considered as promising optimization techniques for solving discrete programming problems and other hard optimization problems such as nonconvex nonlinear programming problems.

In this section, we consider a two-level 0-1 programming problem with cooperative decision makers and show interactive fuzzy programming for the problem. In the interactive method, satisfactory solutions are derived in a procedure similar to that of the two-level linear programming problems, and optimal solutions to the formulated programming problems are obtained through a genetic algorithm in which 0-1 bit strings are employed to represent individuals in an artificial genetic system.

4.6.1 Interactive fuzzy programming

A two-level 0-1 programming problem with cooperative decision makers is formally represented as:

$$\underset{\text{for DM1}}{\text{minimize}} \quad z_1(\boldsymbol{x}_1, \boldsymbol{x}_2) = \boldsymbol{c}_{11}\boldsymbol{x}_1 + \boldsymbol{c}_{12}\boldsymbol{x}_2 \tag{4.81a}$$

$$\underset{\text{for DM2}}{\text{minimize}} \quad z_2(\boldsymbol{x}_1, \boldsymbol{x}_2) = \boldsymbol{c}_{21}\boldsymbol{x}_1 + \boldsymbol{c}_{22}\boldsymbol{x}_2 \tag{4.81b}$$

$$\text{subject to} \quad A_1\boldsymbol{x}_1 + A_2\boldsymbol{x}_2 \leqq \boldsymbol{b} \tag{4.81c}$$

$$\boldsymbol{x}_1 \in \{0,1\}^{n_1}, \ \boldsymbol{x}_2 \in \{0,1\}^{n_2}, \tag{4.81d}$$

where $\boldsymbol{x}_i$, $i = 1,2$ is an n_i-dimensional 0-1 decision variable vector of DMi; $\boldsymbol{c}_{i1}$, $i = 1,2$ is an n_1-dimensional coefficient row vector; $\boldsymbol{c}_{i2}$, $i = 1,2$ is an n_2-dimensional coefficient row vector; $\boldsymbol{b}$ is an m-dimensional constant column vector; A_i, $i = 1,2$ is an $m \times n_i$ coefficient matrix; $z_1(\boldsymbol{x}_1,\boldsymbol{x}_2)$ and $z_2(\boldsymbol{x}_1,\boldsymbol{x}_2)$ represent the objective functions of DM1 and DM2, respectively. For the sake of simplicity, we use the following notation: $\boldsymbol{x} = (\boldsymbol{x}_1^T, \boldsymbol{x}_2^T)^T \in \{0,1\}^{n_1+n_2}$, $\boldsymbol{c}_i = [\boldsymbol{c}_{i1}\ \boldsymbol{c}_{i2}]$, $i = 1,2$.

In a way similar to the two-level linear programming problems considered in the previous sections, it is natural that each of the decision makers has a fuzzy goal for the objective function when the decision maker takes fuzziness of human judgments into consideration. The membership function does not need to be a linear function throughout this section, but we assume that it is a continuous and strictly monotone decreasing function. To identify membership functions of the fuzzy goals for the objective functions, we first solve two individual single-objective 0-1 programming problems for $i = 1,2$:

$$\text{minimize} \quad z_i(\boldsymbol{x}) = \boldsymbol{c}_{i1}\boldsymbol{x}_1 + \boldsymbol{c}_{i2}\boldsymbol{x}_2 \tag{4.82a}$$

$$\text{subject to} \quad A_1\boldsymbol{x}_1 + A_2\boldsymbol{x}_2 \leqq \boldsymbol{b} \tag{4.82b}$$

$$\boldsymbol{x}_1 \in \{0,1\}^{n_1}, \ \boldsymbol{x}_2 \in \{0,1\}^{n_2}. \tag{4.82c}$$

After identifying the membership functions $\mu_i(z_i(\boldsymbol{x}))$, $i = 1,2$, to derive an overall satisfactory solution to the two-level 0-1 programming problem (4.81), we solve the following maximin problem for obtaining a solution which maximizes the smaller degree between the satisfactory degrees of the two decision makers:

$$\begin{aligned} &\text{maximize} \quad \min\{\mu_1(z_1(\boldsymbol{x})), \mu_2(z_2(\boldsymbol{x}))\} && (4.83a) \\ &\text{subject to} \quad A_1\boldsymbol{x}_1 + A_2\boldsymbol{x}_2 \leqq \boldsymbol{b} && (4.83b) \\ &\qquad\qquad \boldsymbol{x}_1 \in \{0,1\}^{n_1},\ \boldsymbol{x}_2 \in \{0,1\}^{n_2}. && (4.83c) \end{aligned}$$

Let $\boldsymbol{x}^*$ denote an optimal solution to problem (4.83). Then, we define the satisfactory degree of both decision makers under the constraints as

$$\lambda^* = \min\{\mu_1(z_1(\boldsymbol{x}^*)), \mu_2(z_2(\boldsymbol{x}^*))\}. \tag{4.84}$$

If DM1 is satisfied with an optimal solution $\boldsymbol{x}^*$ to problem (4.83), it follows that the optimal solution $\boldsymbol{x}^*$ becomes a satisfactory solution; however, DM1 is not always satisfied with the solution $\boldsymbol{x}^*$. It is quite natural to assume that DM1 specifies a minimal satisfactory level $\hat{\delta} \in [0,1]$ for the membership function $\mu_1(z_1(\boldsymbol{x}))$ subjectively.

Consequently, if DM1 is not satisfied with the solution $\boldsymbol{x}^*$ to problem (4.83), the following problem is formulated:

$$\begin{aligned} &\text{minimize} \quad \mu_2(z_2(\boldsymbol{x})) && (4.85a) \\ &\text{subject to} \quad \mu_1(z_1(\boldsymbol{x})) \geqq \hat{\delta} && (4.85b) \\ &\qquad\qquad A_1\boldsymbol{x}_1 + A_2\boldsymbol{x}_2 \leqq \boldsymbol{b} && (4.85c) \\ &\qquad\qquad \boldsymbol{x}_1 \in \{0,1\}^{n_1},\ \boldsymbol{x}_2 \in \{0,1\}^{n_2}. && (4.85d) \end{aligned}$$

Problem (4.85) maximizes DM2's membership function under the condition that DM1's membership function $\mu_1(z_1(\boldsymbol{x}))$ is larger than or equal to the minimal satisfactory level $\hat{\delta}$ specified by DM1.

Problems (4.82), (4.83) and (4.85) are solved by the genetic algorithm with double strings which will be described below. As for the termination conditions of the interactive process and the procedure for updating the minimal satisfactory level $\hat{\delta}$, those of interactive fuzzy programming for the two-level linear programming problem (4.4) can be utilized for the two-level 0-1 programming problem (4.81). We also give a similar algorithm with problems (4.83) and (4.85) for deriving satisfactory solutions.

4.6.2 Genetic algorithm with double strings

The genetic algorithm with double strings proposed by Sakawa *et al.* (Sakawa and Shibano, 1996; Sakawa *et al.*, 1997; Sakawa, 2001) is applicable to solving the problems (4.83) and (4.85) if the constraints of the problems are linear and all the coefficients of the constraints are positive. This genetic algorithm with double strings is extended so as to deal with 0-1 programming problems not only with positive coefficients but also with negative coefficients and integer programming problems (Sakawa and Kato, 2000, 2003; Sakawa, 2001). Some techniques to apply the original genetic algorithm with double strings (Sakawa and Shibano, 1996; Sakawa *et al.*, 1997) to the two-level 0-1 programming problems with cooperative decision makers are summarized as follows.

Coding and decoding

In genetic algorithms, an individual is usually represented by a binary 0-1 string (Goldberg, 1989; Michalewicz, 1996). However, this representation may weaken the ability of genetic algorithms because a phenotype of an individual is not always feasible. In the genetic algorithm with double strings, only feasible solutions are generated if all the coefficients are positive; and a double string is represented as shown in Figure 4.8 for a solution to the two-level 0-1 programming problem.

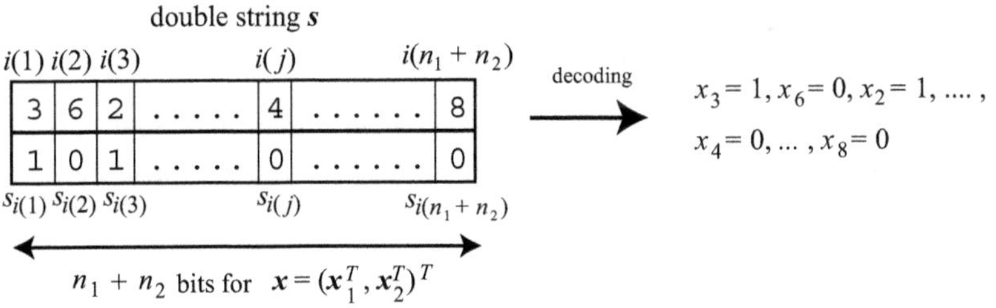

Fig. 4.8 Double string representation of individuals.

In the double string representation, $i(j)$ and $s_{i(j)}$ denote the index of an element in a decision variable vector and the value of the element, respectively, i.e., for a given index $i(j) \in \{1,\ldots,n_1+n_2\}$ of the element of the decision variable vector $\boldsymbol{x} = (\boldsymbol{x}_1^T, \boldsymbol{x}_2^T)^T$, the value of the element $x_{i(j)}$ is $s_{i(j)} \in \{1,0\}$, and for $j \neq j'$, $i(j) \neq i(j')$. A double string $\boldsymbol{s}$ can be transformed into a solution:

$$\boldsymbol{x} = (x_1,\ldots,x_{n_1+n_2})^T$$
$$x_{i(j)} := s_{i(j)},\ j = 1,\ldots,n_1+n_2.$$

Because only this simple mapping may generate infeasible solutions, the following decoding algorithm for eliminating infeasible solutions is employed. It should be noted that the following algorithm generates only feasible solutions for the problem (4.83), but it does not always generate only feasible solutions for the problem (4.85). In the algorithm, n_1+n_2, j, $i(j)$, $x_{i(j)}$ and $\boldsymbol{a}_{i(j)}$ denote the length of a string, a position in a string, an index of a variable, a 0-1 value of the variable with the index

$i(j)$ decoded from a string, and an $i(j)$th column vector of the coefficient matrix $[A_1 A_2]$, respectively. Let $\boldsymbol{sum}$ be an m-dimensional temporal vector.

Step 1 Set $j := 1$ and $\boldsymbol{sum} := \boldsymbol{0}$.
Step 2 If $s_{i(j)} = 1$, set $j := j+1$, and go to step 3. Otherwise, i.e., if $s_{i(j)} = 0$, set $j := j+1$, and go to step 4.
Step 3 If $\boldsymbol{sum} + \boldsymbol{a}_{i(j)} \leqq \boldsymbol{b}$, set $x_{i(j)} := 1$ and $\boldsymbol{sum} := \boldsymbol{sum} + \boldsymbol{a}_{i(j)}$, and go to step 4. Otherwise, set $x_{i(j)} := 0$, and go to step 4.
Step 4 If $j > n_1 + n_2$, stop the algorithm, and a solution $\boldsymbol{x} = (x_1, \ldots, x_{n_1+n_2})^T$ is obtained from the individuals represented by the double string. Otherwise, return to step 2.

Fitness and reproduction

Let $\boldsymbol{s}$ and $\boldsymbol{x}$ denote an individual expressed by the double string representation and the corresponding phenotype (solution), respectively. The decoding algorithm described above generates only feasible solutions for problem (4.83), and then the fitness of an individual in problem (4.83) is simply defined as follows:

$$f(\boldsymbol{s}) = \min\{\mu_1(z_1(\boldsymbol{x})), \mu_2(z_2(\boldsymbol{x}))\}. \tag{4.86}$$

In contrast, for problem (4.85), the decoding algorithm of the double string representation does not always generate feasible solutions due to the constraint (4.85b). Therefore, we adopt the following fitness function with a penalty in case of violating the constraint:

$$f(\boldsymbol{s}) = \min\{\mu_2(z_2(\boldsymbol{x})), \chi(\boldsymbol{x})\} \tag{4.87}$$

$$\chi(\boldsymbol{x}) = \begin{cases} 1 & \text{if } \mu_1(z_1(\boldsymbol{x})) \geqq \hat{\delta} \\ -1 & \text{otherwise.} \end{cases} \tag{4.88}$$

As a reproduction operator, the elitist roulette wheel selection is adopted. Let $\boldsymbol{x}^*(t)$ be the best individual, which has the maximal value of the fitness function, up to the generation t. If there does not exist the best individual $\boldsymbol{x}^*(t)$ in the population of the next generation $t+1$, $\boldsymbol{x}^*(t)$ is included in the population of the next generation $t+1$, and then the population size temporarily increases by one. The roulette wheel selection uses a roulette wheel with slots sized according to fitness values, and then individuals are reproduced with a probability corresponding to the size of slot given by $f(\boldsymbol{s}_i)/\sum_{i=1}^{pop\_size} f(\boldsymbol{s}_i)$.

Genetic operations

As for the crossover operation, because there is some possibility that multiple indices take the same number when a single-point or multi-point crossover operator is applied to individuals represented by double strings, the revised PMX for treating the double string structure, which is described in section 3.3 of chapter 3, is employed.

For the mutation operator playing a role of local random search in the genetic algorithm with double strings, the mutation of bit-reverse type is adopted for the lower string of a double string, together with the inversion operator.

4.6.3 Numerical example

In this subsection, to illustrate the interactive fuzzy programming, we consider the following two-level 0-1 programming problem:

$$\underset{\text{for DM1}}{\text{minimize}} \quad \boldsymbol{c}_{11}\boldsymbol{x}_1 + \boldsymbol{c}_{12}\boldsymbol{x}_2 \tag{4.89a}$$

$$\underset{\text{for DM2}}{\text{minimize}} \quad \boldsymbol{c}_{21}\boldsymbol{x}_1 + \boldsymbol{c}_{22}\boldsymbol{x}_2 \tag{4.89b}$$

$$\text{subject to} \quad A_1\boldsymbol{x}_1 + A_2\boldsymbol{x}_2 \leqq \boldsymbol{b} \tag{4.89c}$$

$$\boldsymbol{x}_1 \in \{0,1\}^{n_1},\ \boldsymbol{x}_2 \in \{0,1\}^{n_2}, \tag{4.89d}$$

where $\boldsymbol{x}_1 = (x_{11},\dots,x_{110})^T$ and $\boldsymbol{x}_2 = (x_{21},\dots,x_{210})^T$; each entry of 3×10 coefficient matrices A_1 and A_2 is a randomly selected number from the interval $[0,100]$; each entry of the right-hand side constant column vector $\boldsymbol{b}$ is a sum of entries of the corresponding row vector of A_1 and A_2 multiplied by 0.6. In Tables 4.29, coefficients of the problem are shown.

Table 4.29 Coefficients in problem (4.89).

$\boldsymbol{c}_{11}$	−44	−91	−57	−12	−57	−25	−50	−24	−48	−41		
$\boldsymbol{c}_{12}$	−87	−43	−36	−38	−5	−16	−52	−69	−10	−40		
$\boldsymbol{c}_{21}$	−77	−25	−34	−23	−30	−31	−88	−4	−65	−40		
$\boldsymbol{c}_{22}$	87	43	36	38	5	16	52	69	10	40		
A_1	51	18	31	53	94	18	70	23	49	13		
	82	16	62	32	35	91	52	40	61	78		
	42	46	97	13	22	95	74	41	78	76		
A_2	9	39	28	37	98	54	76	65	76	78	$\boldsymbol{b}$	588
	93	87	86	67	76	58	39	36	20	82		715
	95	3	32	75	25	59	5	95	32	6		606

To identify membership functions of the fuzzy goals for the objective functions, we first solve two individual minimization problems of both decision makers by using the genetic algorithm with double strings. Parameters of the genetic algorithms are specified as follows: the population size $pop\_size = 100$, the maximum generation number $max\_gen = 1000$, the probability of crossover $p_c = 0.7$, and the probability of mutation $p_m = 0.05$. The number of runs of the genetic algorithms is ten for each problem. The individual minima and the corresponding approximate optimal solutions are shown in Table 4.30.

Table 4.30 Optimal solutions to the individual problems.

$z_1^{\min}$	-654									
$\boldsymbol{x}_1$	0	1	1	0	1	0	1	1	1	0
$\boldsymbol{x}_2$	1	1	1	0	0	0	1	1	0	1
$z_2^{\min}$	-417									
$\boldsymbol{x}_1$	1	1	1	1	1	1	1	1	1	1
$\boldsymbol{x}_2$	0	0	0	0	0	0	0	0	0	0

Suppose that the decision makers employ the linear membership function (4.7) whose parameters are determined by the Zimmermann method (1978). From the results of the individual minimization problems, one finds that $(z_1^0, z_1^1) = (z_1^{\mathrm{m}}, z_1^{\min}) = (-449, -654)$ and $(z_2^0, z_2^1) = (z_2^{\mathrm{m}}, z_2^{\min}) = (81, -417)$, and obtain the membership functions:

$$\mu_1(z_1(\boldsymbol{x})) = (z_1(\boldsymbol{x}) + 449)/(-654 + 449)$$
$$\mu_2(z_2(\boldsymbol{x})) = (z_2(\boldsymbol{x}) - 81)/(-417 - 81).$$

Then, the maximin problem (4.83) for this numerical example can be formulated as

$$\text{maximize} \quad \min\{\mu_1(z_1(\boldsymbol{x})), \mu_2(z_2(\boldsymbol{x}))\} \qquad (4.90a)$$
$$\text{subject to} \quad \boldsymbol{x} \in S, \qquad (4.90b)$$

where S denotes the feasible region of problem (4.89). The result of the first iteration including an optimal solution to problem (4.90) is shown in Table 4.31.

Table 4.31 Iteration 1.

$\boldsymbol{x}_1^1$	1	1	1	1	1	0	1	1	1	1
$\boldsymbol{x}_2^1$	0	1	0	0	0	0	0	1	0	1
(z_1^1, z_2^1)	$(-576, -234)$									
$(\mu_1(z_1^1), \mu_2(z_2^1))$	$(0.620, 0.633)$									
λ^1	0.620									
Δ^1	1.021									

Suppose that DM1 is not satisfied with the solution obtained in Iteration 1, and then, he specifies the minimal satisfactory level at $\hat{\delta} = 0.8$ and the bounds of the ratio at the interval $[\Delta_{\min}, \Delta_{\max}] = [0.6, 1.0]$ with the result of Iteration 1 in mind. Then, the problem with the minimal satisfactory level (4.85) is formulated as

$$\begin{aligned} &\text{maximize} && \mu_2(z_2(\boldsymbol{x})) && (4.91\text{a})\\ &\text{subject to} && (z_1(\boldsymbol{x})+449)/(-654+449) \geqq 0.8 && (4.91\text{b})\\ & && \boldsymbol{x} \in S. && (4.91\text{c}) \end{aligned}$$

Because, at the second iteration, the ratio of satisfactory degrees is not in the valid interval $[0.6, 1.0]$ of the ratio, suppose that DM1 updates the minimal satisfactory level at $\hat{\delta} = 0.7$. Then, the problem with the revised minimal satisfactory level (4.85) is solved, and the result of the third iteration is shown in Table 4.32.

Table 4.32 Iteration 3.

$\boldsymbol{x}_1^3$	1	1	1	1	1	0	1	1	1	1
$\boldsymbol{x}_2^3$	1	0	0	0	0	0	1	0	0	1
(z_1^3, z_2^3)	$(-603, -207)$									
$(\mu_1(z_1^3), \mu_2(z_2^3))$	$(0.751, 0.578)$									
λ^3	0.578									
Δ^3	0.770									

In Iteration 3, the satisfactory degree $\mu_1(z_1^3) = 0.751$ of DM1 becomes larger than the minimal satisfactory level $\hat{\delta} = 0.7$, and the ratio $\Delta^3 = 0.770$ of satisfactory degrees is in the valid interval $[0.6, 1.0]$ of the ratio. Therefore, this solution satisfies the conditions of termination of the interactive procedure, and it becomes a satisfactory solution for both decision makers if DM1 accepts the solution.

4.7 Fuzzy two-level nonlinear programming

In this section, we deal with two-level nonconvex nonlinear programming problem. For a convex nonlinear programming problem such that a feasible region is a convex set and an objective function is a convex function, there exist several effective solution methods such as the generalized reduced gradient method, the multiplier method, the recursive quadratic programming method and so forth. However, these methods are not always effective for obtaining global optimal solutions to the nonconvex nonlinear programming problems. Attempting to find a promising solution, i.e., an approximate global optimal solution, we desire to efficiently search approximate global optimal solutions to the nonconvex nonlinear programming problems.

To find approximate global optimal solutions, it has been acknowledged that genetic algorithms are useful. While we have employed the genetic algorithm with double strings for solving two-level 0-1 programming problems in the previous section, we employ a different type of genetic algorithm with a vector of floating point numbers for the chromosomal representation to solve nonconvex nonlinear programming problems in this section.

For solving general nonlinear programming problems, in 1995, Michalewicz *et al.* (Michalewicz and Nazhiyath, 1995; Michalewicz and Schoenauer, 1996) proposed the GENOCOP (GEnetic algorithm for Numerical Optimization of COnstrained Problems) III. The GENOCOP III incorporates the original GENOCOP system for problems with only linear constraints (Michalewicz and Janikow, 1991; Michalewicz, 1995, 1996), and extends it by maintaining two separate populations, where a development in one population influences evaluations of individuals in the other population. The first population is a set of so-called search points which satisfy linear constraints of a nonlinear programming problem as in the original GENOCOP system. The second population consists of so-called reference points which satisfy all the constraints of the problem.

Unfortunately, however, in the GENOCOP III, because an initial reference point is generated randomly from individuals satisfying the condition of the lower and the upper bounds, it is quite difficult to generate an appropriate initial reference point in practice. Furthermore, because a new search point is randomly generated on the line segment between a search point and a reference point, the effectiveness and speed of search may be quite low.

To overcome such difficulties, Sakawa and Yauchi (1998; 1999; 2000) propose the coevolutionary genetic algorithm, called the revised GENOCOP III, which has a generating method of an initial reference point by minimizing the sum of squares of violated nonlinear constraints and the bisection method for generating a new feasible point on the line segment between a search point and a reference point efficiently.

In this section, for two-level nonconvex nonlinear programming problems, we deal with more complicated version of problems with the fuzzy parameters, which are assumed to be characterized as fuzzy numbers (Dubois and Prade, 1978, 1980; Sakawa, 1993), in order to take into account the experts' vague or fuzzy understanding of the nature of parameters in the problem-formulation process. We formulate two-level nonconvex nonlinear programming problems involving fuzzy parameters and present interactive fuzzy programming through the revised GENOCOP III.

Using the α-level sets of fuzzy numbers, we formulate the corresponding nonfuzzy α-two-level nonconvex nonlinear programming problem. The fuzzy goals for the nonconvex objective functions of the decision makers at both levels are quantified by identifying the corresponding membership functions. In the interactive fuzzy programming method, having specified the level sets of the fuzzy parameters involved in the two-level nonconvex nonlinear programming problems, a satisfactory solution can be derived efficiently by updating the satisfactory degrees of the decision maker at the upper level with considerations of overall satisfactory balance between both decision makers. The satisfactory solution well-balanced between both levels can be obtained by the interactive fuzzy programming method using the revised GENOCOP III (Sakawa and Nishizaki, 2002b).

4.7.1 Interactive fuzzy programming

Two-level nonconvex nonlinear programming problems are formulated as

$$\underset{\text{for DM1}}{\text{minimize}} \quad f_1(\boldsymbol{x}_1,\boldsymbol{x}_2) \tag{4.92a}$$

$$\underset{\text{for DM2}}{\text{minimize}} \quad f_2(\boldsymbol{x}_1,\boldsymbol{x}_2) \tag{4.92b}$$

$$\text{subject to} \quad g_j(\boldsymbol{x}_1,\boldsymbol{x}_2) \leqq 0,\ j=1,\ldots,q \tag{4.92c}$$

$$h_j(\boldsymbol{x}_1,\boldsymbol{x}_2) = 0,\ j=q+1,\ldots,m \tag{4.92d}$$

$$l_{1k_1} \leqq x_{1k_1} \leqq u_{1k_1},\ k_1=1,\ldots,n_1 \tag{4.92e}$$

$$l_{2k_2} \leqq x_{2k_2} \leqq u_{2k_2},\ k_2=1,\ldots,n_2, \tag{4.92f}$$

where $\boldsymbol{x}_1$ and $\boldsymbol{x}_2$ are n_1- and n_2-dimensional vectors of the decision variables of DM1 and DM2, respectively; $f_1(\boldsymbol{x}_1,\boldsymbol{x}_2)$ and $f_2(\boldsymbol{x}_1,\boldsymbol{x}_2)$ represent the objective functions of DM1 and DM2, respectively; $g_j(\boldsymbol{x}_1,\boldsymbol{x}_2)$, $j=1,\ldots,q$, and $h_j(\boldsymbol{x}_1,\boldsymbol{x}_2)$, $j=q+1,\ldots,m$ are constraint functions; l_{ik_i} and u_{ik_i}, $i=1,2$ are the lower and the upper bounds of the decision variables x_{ik_i}, respectively. Obviously, in problem (4.92), the convexity conditions of the objective functions and/or the feasible region are not satisfied.

We have already considered the two-level linear programming problems involving fuzzy parameters from the viewpoint of experts' imprecise or fuzzy understanding of the nature of parameters in a problem-formulation process. For the two-level nonconvex nonlinear programming problems, in this section, we also deal with the problems with fuzzy parameters which are formulated as follows:

$$\underset{\text{for DM1}}{\text{minimize}} \quad f_1(\boldsymbol{x}_1,\boldsymbol{x}_2,\tilde{\boldsymbol{a}}_1) \tag{4.93a}$$

$$\underset{\text{for DM2}}{\text{minimize}} \quad f_2(\boldsymbol{x}_1,\boldsymbol{x}_2,\tilde{\boldsymbol{a}}_2) \tag{4.93b}$$

$$\text{subject to} \quad g_j(\boldsymbol{x}_1,\boldsymbol{x}_2,\tilde{\boldsymbol{b}}_j) \leqq 0,\ j=1,\ldots,q \tag{4.93c}$$

$$h_j(\boldsymbol{x}_1,\boldsymbol{x}_2,\tilde{\boldsymbol{b}}_j) = 0,\ j=q+1,\ldots,m \tag{4.93d}$$

$$\tilde{l}_{1k_1} \leqq x_{1k_1} \leqq \tilde{u}_{1k_1},\ k_1=1,\ldots,n_1 \tag{4.93e}$$

$$\tilde{l}_{2k_2} \leqq x_{2k_2} \leqq \tilde{u}_{2k_2},\ k_2=1,\ldots,n_2, \tag{4.93f}$$

where $\tilde{\boldsymbol{a}}_i = (\tilde{a}_{i1},\ldots,\tilde{a}_{ip_i}), i=1,2$, denote vectors of fuzzy parameters of the objective function; $\tilde{\boldsymbol{b}}_j = (\tilde{b}_{j1},\ldots,\tilde{b}_{jm_j}), j=1,\ldots,m$ denote vectors of fuzzy parameters of the constraints; $\tilde{l}_{ik_i}$ and $\tilde{u}_{ik_i}$, $i=1,2$, $k_i=1,\ldots,n_i$ denote fuzzy parameters representing the lower and the upper bounds for x_{ik_i}, respectively. For notational convenience, let $\tilde{\boldsymbol{a}} = (\tilde{\boldsymbol{a}}_1,\tilde{\boldsymbol{a}}_2)$, $\tilde{\boldsymbol{b}} = (\tilde{\boldsymbol{b}}_1,\ldots,\tilde{\boldsymbol{b}}_m)$, $\tilde{\boldsymbol{l}}_i = (\tilde{l}_{i1},\ldots,\tilde{l}_{in_i})$, $\tilde{\boldsymbol{l}} = (\tilde{\boldsymbol{l}}_1,\tilde{\boldsymbol{l}}_2)$, $\tilde{\boldsymbol{u}}_i = (\tilde{u}_{i1},\ldots,\tilde{u}_{in_i})$, and $\tilde{\boldsymbol{u}} = (\tilde{\boldsymbol{u}}_1,\tilde{\boldsymbol{u}}_2)$. These fuzzy parameters are assumed to be characterized as fuzzy numbers introduced by Dubois and Prade (1978, 1980).

Observing that problem (4.93) involves fuzzy numbers both in the objective functions and in the constraints, we first introduce the α-level set of the fuzzy numbers $\tilde{\boldsymbol{a}}$, $\tilde{\boldsymbol{b}}$, $\tilde{\boldsymbol{l}}$ and $\tilde{\boldsymbol{u}}$ defined as the ordinary set $(\tilde{\boldsymbol{a}},\tilde{\boldsymbol{b}},\tilde{\boldsymbol{l}},\tilde{\boldsymbol{u}})_\alpha$ in which the degrees of the membership functions exceeds the level α:

$$
\begin{aligned}
(\tilde{\boldsymbol{a}},\tilde{\boldsymbol{b}},\tilde{\boldsymbol{l}},\tilde{\boldsymbol{u}})_\alpha = \{(\boldsymbol{a},\boldsymbol{b},\boldsymbol{l},\boldsymbol{u}) \mid\ & \mu_{\tilde{a}_{1r}}(a_{1r}) \geqq \alpha,\ r=1,\ldots,p_1,\\
& \mu_{\tilde{a}_{2r}}(a_{2r}) \geqq \alpha,\ r=1,\ldots,p_2,\\
& \mu_{\tilde{b}_{js}}(b_{js}) \geqq \alpha,\ j=1,\ldots,m,\ s=1,\ldots,m_j,\\
& \mu_{\tilde{l}_{1k}}(l_{1k}) \geqq \alpha,\ k=1,\ldots,n_1,\\
& \mu_{\tilde{l}_{2k}}(l_{2k}) \geqq \alpha,\ k=1,\ldots,n_2,\\
& \mu_{\tilde{u}_{1k}}(u_{1k}) \geqq \alpha,\ k=1,\ldots,n_1,\\
& \mu_{\tilde{u}_{2k}}(u_{2k}) \geqq \alpha,\ k=1,\ldots,n_2\}.
\end{aligned} \tag{4.94}
$$

where $\boldsymbol{a}_i = (a_{i1},\ldots,a_{ip_i})$, $\boldsymbol{b}_j = (b_{j1},\ldots,b_{jm_j})$, $\boldsymbol{a} = (\boldsymbol{a}_1,\boldsymbol{a}_2)$, $\boldsymbol{b} = (\boldsymbol{b}_1,\ldots,\boldsymbol{b}_m)$, $\boldsymbol{l}_i = (l_{i1},\ldots,l_{in_i})$, $\boldsymbol{l} = (\boldsymbol{l}_1,\boldsymbol{l}_2)$, $\boldsymbol{u}_i = (u_{i1},\ldots,u_{in_i})$, $\boldsymbol{u} = (\boldsymbol{u}_1,\boldsymbol{u}_2)$; and $\mu_{\tilde{a}_{ir}}(\cdot)$, $\mu_{\tilde{b}_{js}}(\cdot)$, $\mu_{\tilde{l}_{ik_i}}(\cdot)$ and $\mu_{\tilde{u}_{ik_i}}(\cdot)$ denote the membership functions of $\tilde{a}_{ir}$, $\tilde{b}_{js}$, $\tilde{l}_{ik_i}$ and $\tilde{u}_{ik_i}$, respectively.

Now suppose that DM1 considers that the degrees of all of the membership functions of the fuzzy numbers involved in problem (4.93) should be larger than or equal to a certain value α. Then, for such a degree α, problem (4.93) can be interpreted as a nonfuzzy two-level nonconvex nonlinear programming problem which depends on the coefficient vector $(\boldsymbol{a},\boldsymbol{b},\boldsymbol{l},\boldsymbol{u}) \in (\tilde{\boldsymbol{a}},\tilde{\boldsymbol{b}},\tilde{\boldsymbol{l}},\tilde{\boldsymbol{u}})_\alpha$. Observe that there exist an infinite number of such problems depending on the coefficient vector $(\boldsymbol{a},\boldsymbol{b},\boldsymbol{l},\boldsymbol{u}) \in (\tilde{\boldsymbol{a}},\tilde{\boldsymbol{b}},\tilde{\boldsymbol{l}},\tilde{\boldsymbol{u}})_\alpha$, and the values of $(\boldsymbol{a},\boldsymbol{b},\boldsymbol{l},\boldsymbol{u})$ are arbitrary as long as $(\boldsymbol{a},\boldsymbol{b},\boldsymbol{l},\boldsymbol{u}) \in (\tilde{\boldsymbol{a}},\tilde{\boldsymbol{b}},\tilde{\boldsymbol{l}},\tilde{\boldsymbol{u}})_\alpha$ in the sense that the degrees of all of the membership functions for the fuzzy numbers in problem (4.93) exceed the level α. However, if possible, it would be desirable for the decision makers to choose $(\boldsymbol{a},\boldsymbol{b},\boldsymbol{l},\boldsymbol{u}) \in (\tilde{\boldsymbol{a}},\tilde{\boldsymbol{b}},\tilde{\boldsymbol{l}},\tilde{\boldsymbol{u}})_\alpha$ so as to minimize the objective functions under the constraints. From such a point of view (Sakawa, 1993), for a certain degree α, it seems to be quite natural to have problem (4.93) as the following nonfuzzy α-two-level nonconvex nonlinear programming problem:

$$
\begin{aligned}
&\underset{\text{for DM1}}{\text{minimize}} && f_1(\boldsymbol{x},\boldsymbol{a}_1) && (4.95\text{a})\\
&\underset{\text{for DM2}}{\text{minimize}} && f_2(\boldsymbol{x},\boldsymbol{a}_2) && (4.95\text{b})\\
&\text{subject to} && g_j(\boldsymbol{x},\boldsymbol{b}_j) \leqq 0,\ j=1,\ldots,q && (4.95\text{c})\\
& && h_j(\boldsymbol{x},\boldsymbol{b}_j) = 0,\ j=q+1,\ldots,m && (4.95\text{d})\\
& && l_{1k_1} \leqq x_{1k_1} \leqq u_{1k_1},\ k_1=1,\ldots,n_1 && (4.95\text{e})\\
& && l_{2k_2} \leqq x_{2k_2} \leqq u_{2k_2},\ k_2=1,\ldots,n_2 && (4.95\text{f})\\
& && (\boldsymbol{a},\boldsymbol{b},\boldsymbol{l},\boldsymbol{u}) \in (\tilde{\boldsymbol{a}},\tilde{\boldsymbol{b}},\tilde{\boldsymbol{l}},\tilde{\boldsymbol{u}})_\alpha, && (4.95\text{g})
\end{aligned}
$$

where $\boldsymbol{x} = (\boldsymbol{x}_1^T,\boldsymbol{x}_2^T)^T \in \mathbb{R}^{n_1+n_2}$ is used for compact notation.

It should be emphasized here that, in problem (4.95), $(\boldsymbol{a},\boldsymbol{b},\boldsymbol{l},\boldsymbol{u})$ are treated as decision variables rather than constants. For compact notation, the feasible region satisfying the constraints with respect to $\boldsymbol{x}$ is denoted by $S(\boldsymbol{b},\boldsymbol{l},\boldsymbol{u})$.

For problem (4.95), considering the vague or fuzzy nature of human judgements, it is quite natural to assume that the decision makers at both levels have fuzzy goals for the objective functions $f_i(\boldsymbol{x},\boldsymbol{a}_i)$ similarly to in the previous sections. To identify a membership function $\mu_i(f_i(\boldsymbol{x},\boldsymbol{a}_i))$ for each of the objective functions $f_i(\boldsymbol{x},\boldsymbol{a}_i)$, $i = 1,2$, in problem (4.95), we first calculate the individual minimum of each objective function under the given constraints for $\alpha = 0$ by solving the following problems for $i = 1,2$:

$$\begin{aligned} &\text{minimize} \;\; f_i(\boldsymbol{x},\boldsymbol{a}_i) && (4.96\text{a})\\ &\text{subject to} \;\; \boldsymbol{x} \in S(\boldsymbol{b},\boldsymbol{l},\boldsymbol{u}) && (4.96\text{b})\\ &\qquad (\boldsymbol{a},\boldsymbol{b},\boldsymbol{l},\boldsymbol{u}) \in (\tilde{\boldsymbol{a}},\tilde{\boldsymbol{b}},\tilde{\boldsymbol{l}},\tilde{\boldsymbol{u}})_\alpha. && (4.96\text{c}) \end{aligned}$$

After identifying the membership functions $\mu_i(f_i(\boldsymbol{x},\boldsymbol{a}_i))$, $i = 1,2$, to derive an overall satisfactory solution to the non-fuzzy α-two-level nonconvex nonlinear programming problem (4.95), we first solve the following maximin problem for obtaining a solution which maximizes the smaller degree between the satisfactory degrees of the two decision makers:

$$\begin{aligned} &\text{maximize} \;\; \min\{\mu_1(f_1(\boldsymbol{x},\boldsymbol{a}_1)),\mu_2(f_2(\boldsymbol{x},\boldsymbol{a}_2))\} && (4.97\text{a})\\ &\text{subject to} \;\; \boldsymbol{x} \in S(\boldsymbol{b},\boldsymbol{l},\boldsymbol{u}) && (4.97\text{b})\\ &\qquad (\boldsymbol{a},\boldsymbol{b},\boldsymbol{l},\boldsymbol{u}) \in (\tilde{\boldsymbol{a}},\tilde{\boldsymbol{b}},\tilde{\boldsymbol{l}},\tilde{\boldsymbol{u}})_\alpha. && (4.97\text{c}) \end{aligned}$$

Using the properties of the α-level sets for the vectors of the fuzzy numbers $\tilde{\boldsymbol{a}}$, $\tilde{\boldsymbol{b}}$, $\tilde{\boldsymbol{l}}$ and $\tilde{\boldsymbol{u}}$, the feasible regions for $\boldsymbol{a}$, $\boldsymbol{b}$, $\boldsymbol{l}$ and $\boldsymbol{u}$ can be denoted by the closed intervals $[\boldsymbol{a}^L,\boldsymbol{a}^R]$, $[\boldsymbol{b}^L,\boldsymbol{b}^R]$, $[\boldsymbol{l}^L,\boldsymbol{l}^R]$ and $[\boldsymbol{u}^L,\boldsymbol{u}^R]$, respectively, and therefore, we can obtain an optimal solution to (4.97) by solving the following problem:

$$\begin{aligned} &\text{maximize} \;\; \min\{\mu_1(f_1(\boldsymbol{x},\boldsymbol{a}_1)),\mu_2(f_2(\boldsymbol{x},\boldsymbol{a}_2))\} && (4.98\text{a})\\ &\text{subject to} \;\; \boldsymbol{x} \in S(\boldsymbol{b},\boldsymbol{l}^L,\boldsymbol{u}^R) && (4.98\text{b})\\ &\qquad (\boldsymbol{a},\boldsymbol{b}) \in (\tilde{\boldsymbol{a}},\tilde{\boldsymbol{b}})_\alpha. && (4.98\text{c}) \end{aligned}$$

By solving problem (4.98), we can obtain a solution which maximizes the smaller degree between the satisfactory degrees of both decision makers. Let $(\boldsymbol{x}^*,\boldsymbol{a}_1^*,\boldsymbol{a}_2^*,\boldsymbol{b}^*)$ denote an optimal solution to problem (4.98). Then, we define the satisfactory degree of both decision makers under the constraints as

$$\lambda^* = \min\{\mu_1(f_1(\boldsymbol{x}^*,\boldsymbol{a}_1^*)),\mu_2(f_2(\boldsymbol{x}^*,\boldsymbol{a}_2^*))\}. \qquad (4.99)$$

If DM1 is satisfied with the optimal solution $\boldsymbol{x}^*$ to problem (4.98), it follows that the optimal solution $\boldsymbol{x}^*$ becomes a satisfactory solution; however, DM1 is not always satisfied with the solution $\boldsymbol{x}^*$, and it is assumed that DM1 specifies a minimal satisfactory level $\hat{\delta} \in [0,1]$ for the membership function $\mu_1(f_1(\boldsymbol{x},\boldsymbol{a}_1))$ subjectively.

Consequently, if DM1 is not satisfied with the solution $\boldsymbol{x}^*$ to problem (4.98), the following problem is formulated:

$$\begin{aligned} \text{maximize} \quad & \mu_2(f_2(\boldsymbol{x},\boldsymbol{a}_2)) && (4.100\text{a}) \\ \text{subject to} \quad & \mu_1(f_1(\boldsymbol{x},\boldsymbol{a}_1)) \geqq \hat{\delta} && (4.100\text{b}) \\ & \boldsymbol{x} \in S(\boldsymbol{b},\boldsymbol{l}^L,\boldsymbol{u}^R) && (4.100\text{c}) \\ & (\boldsymbol{a},\boldsymbol{b}) \in (\tilde{\boldsymbol{a}},\tilde{\boldsymbol{b}})_\alpha, && (4.100\text{d}) \end{aligned}$$

where DM2's membership function is maximized under the condition that DM1's membership function $\mu_1(f_1(\boldsymbol{x}))$ is larger than or equal to the minimal satisfactory level $\hat{\delta}$ specified by DM1.

At iteration l, let $\mu_1(f_1^l)$, $\mu_2(f_2^l)$, λ^l and $\Delta^l = \mu_2(f_2^l)/\mu_1(f_1^l)$ denote DM1's and DM2's satisfactory degrees, a satisfactory degree of both levels and the ratio between the satisfactory degrees of both decision makers, respectively; let the corresponding optimal solution be $\boldsymbol{x}^l$. The interactive process terminates if the two conditions which are similar to those of the two-level linear programming problems shown in the previous section are satisfied and DM1 concludes the solution as an overall satisfactory solution. Furthermore, the procedure for updating the minimal satisfactory level $\hat{\delta}$ is also given in a way similar to that of the two-level linear programming problems.

However, because we deal with nonconvex nonlinear programming problems, for a certain bounds $[\Delta_{\min}, \Delta_{\max}]$ specified by DM1, there would be a gap of the ratio of satisfactory degrees with respect to continuous changes of the minimal satisfactory level $\hat{\delta}$, and we could not find any satisfactory solution satisfying Condition 2, i.e., $\Delta \in [\Delta_{\min}, \Delta_{\max}]$. However, because, for such a situation, all of the feasible solutions satisfying the two conditions are not Pareto optimal solutions, it seems appropriate to us that DM1 attempts to find a Pareto optimal solution by enlarging the bounds.

Suppose that DM1 specifies the upper and the lower bounds $[\Delta_{\min}, \Delta_{\max}]$ of the ratio Δ between the satisfactory degrees of both decision makers like in Figure 4.9. On these occasions , DM1 cannot obtain any solution satisfying Condition 2, and then DM1 should update the upper and the lower bounds $[\Delta_{\min}, \Delta_{\max}]$. In the example shown in Figure 4.9, by increasing the upper bound, a candidate $\boldsymbol{x}^2$ of the satisfactory solution can be obtained.

We provide a procedure of updating the parameters $\Delta_{\min}$ and $\Delta_{\max}$ as follows: First, DM1 specifies two parameters N and ε which represent the maximal iteration number for Case 2 in updating the minimal satisfactory level $\hat{\delta}$ given in section 4.2 and the minimal gap of the ratio of satisfactory degrees, respectively. If Case 2 of the procedure of updating the minimal satisfactory level $\hat{\delta}$ is repeated N times, we verify whether or not there exists a gap of the ratio of satisfactory degrees by the following procedure.

[Procedure for updating the parameters $\Delta_{\min}$ and $\Delta_{\max}$]

Step 1 If the ratio Δ exceeds the upper bound, revise the minimal satisfactory level $\hat{\delta}$ to $\hat{\delta}+\varepsilon$, and if the ratio Δ is below the lower bound, revise the minimal satisfactory level $\hat{\delta}$ to $\hat{\delta}-\varepsilon$. After revising the minimal satisfactory level $\hat{\delta}$ to $\hat{\delta}\pm\varepsilon$, solve the following problem:

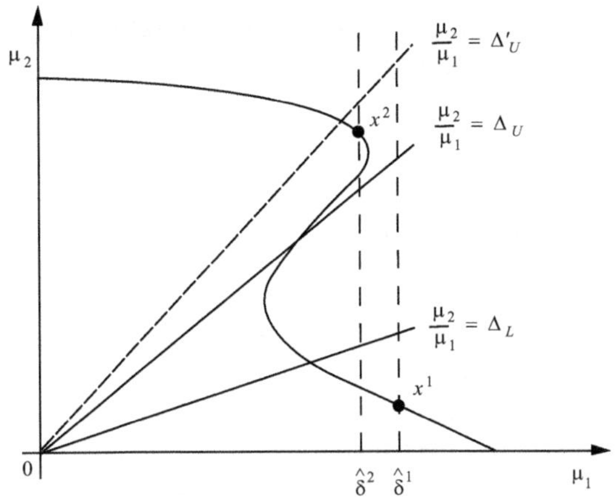

Fig. 4.9 Gap of the ratio of satisfactory degrees.

$$
\begin{aligned}
\text{maximize} \quad & \mu_2(f_2(\boldsymbol{x},\boldsymbol{a}_2)) \\
\text{subject to} \quad & \boldsymbol{x} \in X(\boldsymbol{b},\boldsymbol{l}^L,\boldsymbol{u}^R) \\
& \mu_1(f_1(\boldsymbol{x},\boldsymbol{a}_1)) \geqq \hat{\delta} \pm \varepsilon \\
& (\boldsymbol{a},\boldsymbol{b}) \in (\tilde{\boldsymbol{a}},\tilde{\boldsymbol{b}})_\alpha.
\end{aligned}
$$

If the ratio Δ is in the bounds $[\Delta_{\min},\Delta_{\max}]$, DM1 does not need to update the bounds.

Step 2 If the ratio Δ corresponding to the problem with $\hat{\delta}$ exceeds the upper bound $\Delta_{\max}$ and the new ratio Δ corresponding to the problem with $\hat{\delta}+\varepsilon$ is below the lower bound $\Delta_{\min}$, or the ratio Δ corresponding to the problem with $\hat{\delta}$ is below the lower bound $\Delta_{\min}$ and the new ratio Δ corresponding to the problem with $\hat{\delta}-\varepsilon$ exceeds the upper bound $\Delta_{\max}$, then we determine that there exists a gap of the ratio of satisfactory degrees. Otherwise, go to Step 1.

Step 3 Let the maximal ratio Δ obtained so far such that $\Delta < \Delta_{\min}$ denote $\bar{\Delta}_{\min}$, and let the minimal ratio Δ obtained so far such that $\Delta > \Delta_{\max}$ denote $\underline{\Delta}_{\max}$. Ask DM1 to change the bounds $[\Delta_{\min},\Delta_{\max}]$ of the ratio to new bounds $[\Delta_{\min}^{\text{new}},\Delta_{\max}]$ or $[\Delta_{\min},\Delta_{\max}^{\text{new}}]$ in accordance with preference of DM1, where $\Delta_{\min}^{\text{new}} \leqq \bar{\Delta}_{\min}$ and $\underline{\Delta}_{\max} \leqq \Delta_{\max}^{\text{new}}$.

For the two-level nonlinear programming problem with fuzzy parameters (4.93), we can provide an algorithm of the interactive fuzzy programming similar to that for two-level linear programming problem together with the termination conditions of the interactive process and the procedure for updating the minimal satisfactory level $\hat{\delta}$. In the interactive fuzzy programming, it is required to solve the nonconvex nonlinear programming problems (4.98) and (4.100), and for that purpose, we apply the revised GENOCOP III (Sakawa and Yauchi, 1998, 1999, 2000).

4.7.2 Genetic algorithm for nonlinear programming problems: Revised GENOCOP III

The GENOCOP III proposed by Michalewicz *et al.* (Michalewicz and Nazhiyath, 1995; Michalewicz and Schoenauer, 1996) for solving general nonlinear programming problems. However, in the GENOCOP III, because an initial reference point is generated randomly from individuals satisfying the condition of the lower and the upper bounds, it is difficult to generate an appropriate initial reference point efficiently. Furthermore, to evolve the two population jointly, a new search point is generated from a pair of a search point and a reference point, but because a new search point is randomly generated on the line segment between the search point and the reference point, the effectiveness and speed of search may be quite low.

To overcome such difficulties, Sakawa and Yauchi (1998, 1999, 2000) propose the coevolutionary genetic algorithm, called the revised GENOCOP III by employing a method generating initial reference points and new search points efficiently. We can solve the nonconvex nonlinear programming problems (4.98) and (4.100) in the interactive fuzzy two-level nonlinear programming method by using the revised GENOCOP III, and the algorithm of the revised GENOCOP III can be summarized as follows:

[Algorithm of the revised GENOCOP III]

Step 1 Generate two separate initial populations. An initial population of search points is created randomly from individuals satisfying the lower and the upper bounds determined by both the linear constraints and the original lower and the upper bounds of decision variables. An initial reference point is generated by minimizing the sum of squares of violated nonlinear constraints. Then, an initial population of reference points is created via multiple copies of the initial reference point obtained in this way.

Step 2 Apply the crossover and the mutation operators to the population of search points.

Step 3 Find a new feasible point on a segment between a search point and a reference point based on the bisection method. If the evaluation of a newly found feasible point is better than that of the reference point, replace the reference point by the new point. Moreover, with a given probability of replacement, replace the search point by the new point.

Step 4 After evaluating the individuals, apply the selection operator for generating individuals of the next generation.

Step 5 If the termination conditions are met, stop the algorithm. Otherwise, return to Step 2.

Representation of individuals and initial population

In the revised GENOCOP III for obtaining an satisfactory solution to the two-level nonlinear programming problem with fuzzy parameters, the decision variable $\boldsymbol{x}$ and the fuzzy parameters $\boldsymbol{a}$ and $\boldsymbol{b}$ of problems (4.98) and (4.100) are represented as real number strings of the floating-point representation. Thus, one string consists of $n_1 + n_2 + p_1 + p_2 + m$ floating-point numbers, and an example of the string is shown in Figure 4.10.

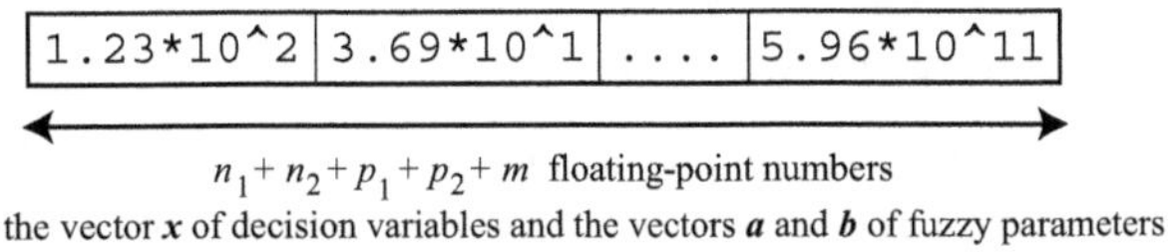

Fig. 4.10 Real number string.

The revised GENOCOP III has the two initial separate populations of search points and reference points, respectively. The initial population of search points is created randomly from individuals satisfying the lower and upper bounds determined by both the linear constraints and the original lower and upper bounds of decision variables. An initial reference point is generated by minimizing the sum of squares of violated nonlinear constraints. Namely, the following problem that minimizes the sum of squares of violated nonlinear constraints is formulated:

$$\underset{\boldsymbol{x} \in LC}{\text{minimize}} \; \sum_{j \in I_g} (g_j(\boldsymbol{x}))^2 + \sum_{j \in I_h} (h_j(\boldsymbol{x}))^2, \tag{4.101}$$

where LS is the set of points satisfying only the linear constraints; I_g and I_h are the index sets violating the nonlinear inequality and equality constraints, respectively. To find one initial reference point, problem (4.101) is solved by the original GENOCOP system (Michalewicz and Janikow, 1991; Michalewicz, 1995, 1996), and then the initial population of reference points is created via multiple copies of the obtained initial reference point.

Reproduction

The exponential ranking method, which is one version of ranking selection, is used as a reproduction operator, where the population is sorted from the best to the worst and the selection probability of each individual is assigned according to the ranking. To be more precise, the selection probability p_i for the individual of rank i is determined by

$$p_i = c(1-c)^{i-1}, \tag{4.102}$$

where $c \in (0,1)$ represents the selection probability that an individual of rank 1 is selected. It follows that a larger value of c implies stronger selective pressure.

Genetic operations

As genetic operators for crossover and mutation, the simple crossover, the single arithmetic crossover, the whole arithmetic crossover, the heuristic crossover, the uniform and the nonuniform mutations, the boundary mutation, and the whole nonuniform mutation (Michalewicz, 1996) are adopted, and in practice one operation of them is randomly selected to be executed.

Simple crossover For two parent strings $\boldsymbol{s}^1$ and $\boldsymbol{s}^2$, choose one crossover point in the strings at random. Let the crossover point be between the ith and the $i+1$th numbers. From the two parents $\boldsymbol{s}^1 = (s_1^1,\dots,s_n^1)$ and $\boldsymbol{s}^2 = (s_1^2,\dots,s_n^2)$, two offspring $\boldsymbol{s}^{1\prime}$ and $\boldsymbol{s}^{2\prime}$ are generated as follows:

$$\boldsymbol{s}^{1\prime} = (s_1^1,\dots,s_i^1, as_{i+1}^2 + (1-a)s_{i+1}^1,\dots,as_n^2 + (1-a)s_n^1) \tag{4.103a}$$

$$\boldsymbol{s}^{2\prime} = (s_1^2,\dots,s_i^2, as_{i+1}^1 + (1-a)s_{i+1}^2,\dots,as_n^1 + (1-a)s_n^2), \tag{4.103b}$$

where $a \in [0,1]$ and $n = n_1 + n_2 + p_1 + p_2 + m$. The operation of the simple crossover is illustrated in Figure 4.11.

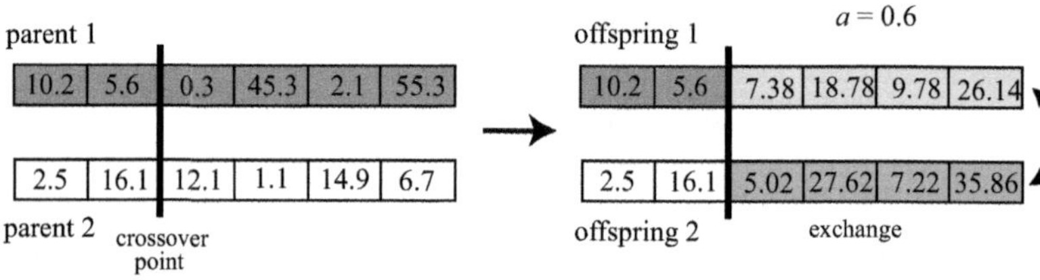

Fig. 4.11 Simple crossover.

Single arithmetic crossover For two parents $\boldsymbol{s}^1$ and $\boldsymbol{s}^2$, if they are crossed at the ith position, the resulting offspring are

$$\boldsymbol{s}^{1\prime} = (s_1^1,\dots,s_{i-1}^1,\ as_i^2 + (1-a)s_i^1,\ s_{i+1}^1,\dots,s_n^1) \tag{4.104a}$$

$$\boldsymbol{s}^{2\prime} = (s_1^2,\dots,s_{i-1}^2,\ as_i^1 + (1-a)s_i^2,\ s_{i+1}^2,\dots,s_n^2), \tag{4.104b}$$

where a is a parameter. If the feasible region is convex, when the remaining components are fixed, the two offspring $\boldsymbol{s}^{1\prime}$ and $\boldsymbol{s}^{2\prime}$ are feasible by selecting a as

$$a \in \begin{cases} [\max(\alpha,\beta),\min(\gamma,\delta)] & \text{if } s_i^1 > s_i^2 \\ [0,0] & \text{if } s_i^1 = s_i^2 \\ [\max(\gamma,\delta),\min(\alpha,\beta)] & \text{if } s_i^1 < s_i^2, \end{cases} \tag{4.105}$$

where

$$\alpha = (l(s_i^2) - s_i^2)/(s_i^1 - s_i^2), \qquad \beta = (u(s_i^1) - s_i^1)/(s_i^2 - s_i^1),$$
$$\gamma = (l(s_i^1) - s_i^1)/(s_i^2 - s_i^1), \qquad \delta = (u(s_i^2) - s_i^2)/(s_i^1 - s_i^2),$$

and $l(s_i^j)$ and $u(s_i^j)$ denotes the lower and the upper bounds of s_i^j, $j = 1,2$, respectively.

Whole arithmetic crossover From two parents $\boldsymbol{s}^1$ and $\boldsymbol{s}^2$, the following two offspring $\boldsymbol{s}^{1\prime}$ and $\boldsymbol{s}^{2\prime}$ are generated:

$$\boldsymbol{s}^{1\prime} = a\boldsymbol{s}^2 + (1-a)\boldsymbol{s}^1 \tag{4.106a}$$
$$\boldsymbol{s}^{2\prime} = a\boldsymbol{s}^1 + (1-a)\boldsymbol{s}^2, \tag{4.106b}$$

where a is randomly chosen in the interval $[0,1]$. The resulting offsprings are convex combinations of their parents, and they always become feasible when the feasible region is convex.

Heuristic crossover The heuristic crossover uses values of the objective function in determining search direction. For a given two parents $\boldsymbol{s}^1$ and $\boldsymbol{s}^2$, assume that the parent $\boldsymbol{s}^2$ is not worse than $\boldsymbol{s}^1$, i.e., $f(\boldsymbol{s}^2) \leqq f(\boldsymbol{s}^1)$, for minimization problems. Then, this operator generates only one offspring $\boldsymbol{s}'$ from the two parents $\boldsymbol{s}^1$ and $\boldsymbol{s}^2$, according to the following rule:

$$\boldsymbol{s}' = r(\boldsymbol{s}^2 - \boldsymbol{s}^1) + \boldsymbol{s}^2, \tag{4.107}$$

where r is randomly chosen in the interval $[0,1]$.

Uniform and boundary mutation A ceratin element, say i, is chosen from the index set $\{1,\ldots,n\}$. Then, in the uniform mutation, the randomly selected string $\boldsymbol{s} = (s_1,\ldots,s_i,\ldots,s_n)$ is changed to $\boldsymbol{s}' = (s_1,\ldots, s_i',\ldots,s_n)$; s_i' is a random value from the interval $(u(s_i), l(s_i))$, where $u(s_i)$ and $l(s_i)$ denote the upper bound and the lower bound of s_i, respectively. Note that s_i' is a random variable of the uniform distribution. The boundary mutation differs from the uniform mutation by the choice of s_i' which is either $u(s_i)$ or $l(s_i)$.

Nonuniform mutation Assume that the ith element s_i is selected for mutation, and let x be a random variable which is either 0 or 1 with the equal probability. For a given string $\boldsymbol{s}$, the resulting offspring is $\boldsymbol{s}' = (s_1,\ldots,s_i',\ldots,s_n)$, where

$$s_i' = \begin{cases} v_i + \Delta(t, u(v_i) - v_i) & \text{if } x = 0 \\ v_i - \Delta(t, v_i - l(v_i)), & \text{if } x = 1. \end{cases} \tag{4.108}$$

The value of the function $\Delta(t,y)$ becomes close to 0 as the generation number t increases; In Michalewicz, Logan and Swaminathan (1994), the function

$$\Delta(t,y) = yr\left(1 - \frac{t}{max\_gen}\right)^b \tag{4.109}$$

is used, where r is a random number in the interval $[0,1]$, max_gen is the maximal generation number, and b is a parameter determining the degree of nonuniformity.

Whole nonuniform mutation While in the nonuniform mutation described above, the operation is applied to only the selected one element, it is applied to the whole of the string in the whole nonuniform mutation.

4.7.3 Numerical example

To illustrate the interactive fuzzy programming method described in the previous section, consider the following two-level nonconvex programming problem involving fuzzy numbers:

$$\begin{aligned}
\underset{\text{for DM1}}{\text{minimize}}\ & f_1(\boldsymbol{x},\tilde{\boldsymbol{a}}_1) = 7x_{11}^2 - x_{12}^2 + x_{11}x_{12} + \tilde{a}_{11}x_{11} + \tilde{a}_{12}x_{12} + 8(x_{13}-10)^2 \\
& + \tilde{a}_{13}(x_{14}-5)^2 + (x_{15}-3)^2 + 2(x_{21}-1)^2 + \tilde{a}_{14}x_{22}^2 + \tilde{a}_{15}(x_{23}-11)^2 \\
& + 2(x_{24}-10)^2 + x_{25}^2 + 45 && (4.110\text{a}) \\
\underset{\text{for DM2}}{\text{minimize}}\ & f_2(\boldsymbol{x},\tilde{\boldsymbol{a}}_2) = (x_{11}-5)^2 + \tilde{a}_{21}(x_{12}-12)^2 + 0.5x_{13}^4 + \tilde{a}_{22}(x_{14}-11)^2 \\
& + 0.2x_{15}^5 + \tilde{a}_{23}x_{21}^2 + 0.1x_{22}^4 + \tilde{a}_{24}x_{21}x_{22} + \tilde{a}_{25}x_{21} - 8x_{22} + x_{23}^2 \\
& + 3(x_{24}-5)^2 + (x_{25}-5)^2 && (4.110\text{b}) \\
\text{subject to}\ & \tilde{b}_{11}(x_{11}-2)^2 + 4(x_{12}-3)^2 + 2x_{13}^2 - 7x_{14} + \tilde{b}_{12}x_{15}x_{21}x_{23} - 120 \leqq 0 \\
& && (4.110\text{c}) \\
& 5x_{11}^2 + \tilde{b}_{21}x_{12} + (x_{13}-6)^2 + \tilde{b}_{22}x_{14} - 40 \leqq 0 && (4.110\text{d}) \\
& x_{11}^2 + 2(x_{12}-2)^2 + \tilde{b}_{31}x_{11}x_{12} + \tilde{b}_{32}x_{15} + 6x_{15}x_{21} \leqq 0 && (4.110\text{e}) \\
& \tilde{b}_{41}(x_{11}-8)^2 + 2(x_{12}-4)^2 + 3x_{15}^2 + \tilde{b}_{42}x_{15}x_{23} - 30 \leqq 0 && (4.110\text{f}) \\
& -3x_{11} + 6x_{12} + \tilde{b}_{51}(x_{24}-8)^2 + \tilde{b}_{52}x_{25} \leqq 0 && (4.110\text{g}) \\
& \tilde{b}_{61}x_{11} + 5x_{12} + \tilde{b}_{62}x_{22} + 9x_{23} - 105 \leqq 0 && (4.110\text{h}) \\
& 10x_{11} + \tilde{b}_{71}x_{12} + \tilde{b}_{72}x_{22} + 2x_{23} \leqq 0 && (4.110\text{i}) \\
& -8x_{11} + 2x_{12} + 5x_{24} + \tilde{b}_{81}x_{25} \leqq 0 && (4.110\text{j})
\end{aligned}$$

$$x_{12}x_{13} + \tilde{b}_{91}x_{14}x_{15} + \tilde{b}_{92}x_{21}x_{22} - 5x_{23}x_{24} = 0 \quad (4.110k)$$
$$\tilde{b}_{10,1}x_{11}^3 + \tilde{b}_{10,2}x_{21}^3 = 0 \quad (4.110l)$$
$$-5.0 \leqq x_{1k_1} \leqq 10.0,\ k_1 = 3,4,5 \quad (4.110m)$$
$$\tilde{l}_{11} \leqq x_{11} \leqq 10.0 \quad (4.110n)$$
$$\tilde{l}_{12} \leqq x_{12} \leqq 10.0 \quad (4.110o)$$
$$-5.0 \leqq x_{2k_2} \leqq 10.0,\ k_2 = 2,\ldots,5 \quad (4.110p)$$
$$\tilde{l}_{21} \leqq x_{21} \leqq \tilde{u}_{21}, \quad (4.110q)$$

where $\boldsymbol{x}_1 = (x_{11},\ldots,x_{15})^T$, $\boldsymbol{x}_2 = (x_{21},\ldots,x_{25})^T$ and $\boldsymbol{x} = (\boldsymbol{x}_1^T, \boldsymbol{x}_2^T)^T$.

For simplicity, it is assumed that all of the membership functions for the fuzzy numbers involved in this problem are triangular ones; the left, the mean and the right points of each triangular fuzzy number are shown in Table 4.33.

Table 4.33 Fuzzy numbers in problem (4.110).

$\tilde{a}_{11}$	−14.5	−14	−13.5	$\tilde{b}_{41}$	0.2	0.5	0.9
$\tilde{a}_{12}$	−16.5	−16	−15.5	$\tilde{b}_{42}$	−1.2	−1	−0.8
$\tilde{a}_{13}$	3.7	4	4.4	$\tilde{b}_{51}$	11.2	12	12.8
$\tilde{a}_{14}$	4.5	5	5.5	$\tilde{b}_{52}$	−7.5	−7	−6.5
$\tilde{a}_{15}$	6.4	7	7.5	$\tilde{b}_{61}$	3.6	4	4.4
$\tilde{a}_{21}$	4.5	5	5.5	$\tilde{b}_{62}$	−3.4	−3	−2.6
$\tilde{a}_{22}$	2.7	3	3.3	$\tilde{b}_{71}$	−8.4	−8	−7.6
$\tilde{a}_{23}$	6.5	7	7.5	$\tilde{b}_{72}$	−18	−17	−16
$\tilde{a}_{24}$	−4.5	−4	−3.6	$\tilde{b}_{81}$	−2.2	−2	−1.8
$\tilde{a}_{25}$	−10.7	−10	−9.3	$\tilde{b}_{91}$	−5.5	−5	−4.5
$\tilde{b}_{11}$	2.6	3	3.3	$\tilde{b}_{92}$	0.8	1	1.2
$\tilde{b}_{12}$	1.8	2	2.2	$\tilde{b}_{10,1}$	0.5	1	1.3
$\tilde{b}_{21}$	7.4	8	8.6	$\tilde{b}_{10,2}$	0.7	1	1.5
$\tilde{b}_{22}$	−2.2	−2	−1.8	$\tilde{l}_{11}$	−6	−5.0	−4
$\tilde{b}_{31}$	−2.4	−2	−1.7	$\tilde{l}_{12}$	−5.5	−5.0	−4.5
$\tilde{b}_{32}$	13.2	14	14.8	$\tilde{l}_{21}$	−5.7	−5.0	−4.3
				$\tilde{u}_{21}$	9.0	10.0	10.5

For this problem, the parameter values of the revised GENOCOP III are specified as follows: the population sizes $pop\_size\_search = pop\_size\_reference = 70$, the maximum generation number $max\_gen = 5000$, the probability of replacement $p_r = 0.2$, and the parameter of the exponential ranking selection $c = 0.1$.

Suppose that the decision makers employ the linear membership function

$$\mu_i(f_i(\boldsymbol{x})) = \begin{cases} 0 & \text{if } f_i(\boldsymbol{x}) > f_i^0 \\ \dfrac{f_i(\boldsymbol{x}) - f_i^0}{f_i^1 - f_i^0} & \text{if } f_i^1 < f_i(\boldsymbol{x}) \leqq f_i^0 \\ 1 & \text{if } f_i(\boldsymbol{x}) \leqq f_i^1, \end{cases}$$

where parameters are determined by the Zimmermann method (1978). The individual minimization problems with $\alpha = 0$ are solved by using the revised GENOCOP III, and we have the parameters $(f_1^0, f_1^1) = (f_1^{\mathrm{m}}, f_1^{\mathrm{min}}) = (1690.796, 797.917)$ and $(f_2^0, f_2^1) = (f_2^{\mathrm{m}}, f_2^{\mathrm{min}}) = (3479.330, 473.590)$ of the membership functions; the membership functions are expressed as

$$\mu_1(f_1(\boldsymbol{x},\boldsymbol{a}_1)) = (f_1(\boldsymbol{x},\boldsymbol{a}_1) - 1690.796)/(797.917 - 1690.796) \tag{4.111}$$
$$\mu_2(f_2(\boldsymbol{x},\boldsymbol{a}_2)) = (f_2(\boldsymbol{x},\boldsymbol{a}_2) - 3479.330)/(473.590 - 3479.330). \tag{4.112}$$

Suppose that DM1 specifies the membership degree of the α-level set at $\alpha = 0.9$. Then, the maximin problem (4.98) for this numerical example, which maximizes the smaller degree between the of satisfactory degrees of the two decision makers, is formulated as

$$\text{maximize} \quad \min\{\mu_1(f_1(\boldsymbol{x},\boldsymbol{a}_1)), \mu_2(f_2(\boldsymbol{x},\boldsymbol{a}_2))\} \tag{4.113a}$$
$$\text{subject to} \quad \boldsymbol{x} \in S(\boldsymbol{b}, \boldsymbol{l}^L, \boldsymbol{u}^R) \tag{4.113b}$$
$$(\boldsymbol{a},\boldsymbol{b}) \in (\tilde{\boldsymbol{a}}, \tilde{\boldsymbol{b}})_\alpha, \tag{4.113c}$$

where $S(\boldsymbol{b}, \boldsymbol{l}^L, \boldsymbol{u}^R)$ denotes the feasible region of $\boldsymbol{x}$ for $(\boldsymbol{b}, \boldsymbol{l}^L, \boldsymbol{u}^R)$. Solving this problem through the revised GENOCOP III, we obtain the result of Iteration 1 which is shown in Table 4.34.

Table 4.34 Iteration 1.

λ^1	f_1^1	$\mu_1(f_1^1)$	f_2^1	$\mu_2(f_2^1)$	Δ^1
0.844251	936.982	0.844251	941.732	0.844251	1.000000

Suppose that DM1 is not satisfied with the current approximate optimal solution, and then he specifies the minimal satisfactory level at $\delta = 0.90$ together with the lower and the upper bounds of Δ at $[0.8, 1.0]$. Then, the problem with the minimal satisfactory level (4.100) for this numerical example is formulated as:

$$\text{maximize} \quad \mu_2(f_2(\boldsymbol{x},\boldsymbol{a}_2)) \tag{4.114a}$$
$$\text{subject to} \quad \mu_1(f_1(\boldsymbol{x},\boldsymbol{a}_1)) \geqq 0.9 \tag{4.114b}$$
$$\boldsymbol{x} \in S(\boldsymbol{b}, \boldsymbol{l}^L, \boldsymbol{u}^R) \tag{4.114c}$$
$$(\boldsymbol{a},\boldsymbol{b}) \in (\tilde{\boldsymbol{a}}, \tilde{\boldsymbol{b}})_\alpha. \tag{4.114d}$$

Solving this problem by using the revised GENOCOP III, we obtain the result of Iteration 2 which is shown Table 4.35.

Table 4.35 Iteration 2.

λ^2	f_1^2	$\mu_1(f_1^2)$	f_2^2	$\mu_2(f_2^2)$	Δ^2
0.727285	887.166	0.900044	1293.300	0.727285	0.808055

As can be seen from Table 4.35, the satisfactory degree $\mu_1(f_1^2) = 0.900044$ of DM1 exceeds the minimal satisfactory level $\hat{\delta} = 0.90$, and the ratio $\Delta^2 = 0.808055$ of satisfactory degrees lies in the valid interval $[0.8, 1.0]$. The obtained solution satisfies both the termination conditions of the interactive procedure, and it becomes an overall satisfactory solution for both decision makers if DM1 accepts the solution. Otherwise, a similar procedure continues in this manner until an overall satisfactory solution is derived.

4.8 Fuzzy multiobjective two-level linear programming

In real-world problems, the diversity of evaluation has been a matter of great importance to us, and therefore it is natural for decision makers to want to achieve several goals simultaneously. Namely, they should have multiple objectives and evaluate alternatives, considering trade-offs among the multiple objectives. From this viewpoint, it is desirable to develop an interactive fuzzy programming method to be able to handle multiple objectives.

In this section, assuming that two decision makers at the upper level and at the lower level have own fuzzy goals with respect to their multiple objective functions and also have partial information on their preferences among the multiple objectives, we deal with two-level linear programming problems in multiobjective environments, and give an interactive fuzzy programming method for the multiobjective problems (Sakawa, Nishizaki and Oka, 2000). In this method, after determining the fuzzy goals of the two decision makers, the decision maker at the upper level subjectively specifies minimal satisfactory levels for all of the fuzzy goals, and the decision maker at the lower level also specifies aspiration levels for all the fuzzy goals. In an interactive process of this method, tentative solutions are obtained and evaluated by using the partial information on preferences of the decision makers. Taking into consideration overall satisfactory balance between the two decision makers, they update some of the minimal satisfactory levels and the aspiration levels, if necessary, in order to derive a satisfactory solution.

The algorithm begins with solving a mathematical programming problem treating all the fuzzy goals for the objective functions of the decision makers equally, i.e., maximizing the minimum among the satisfactory degrees of all the fuzzy goals. The decision maker at the upper level specifies minimal satisfactory levels for the fuzzy goals with an optimal solution to the above mentioned maximin problem in mind, and the decision maker at the lower level also specifies aspiration levels for the fuzzy goals. Subsequently, we formulate and solve the second mathematical programming problem yielding a tentative solution which gives satisfactory degrees larger than the minimal satisfactory levels for the fuzzy goals of the decision maker at the upper level and is closest to the vector of the aspiration levels of the decision maker at the lower level.

If the decision maker at the upper level is satisfied with the tentative solution, it becomes the final satisfactory solution. Otherwise, the decision maker at the upper level updates the minimal satisfactory levels. By using the partial information on preference of the decision makers, the aggregated satisfactory degrees of the decision makers are evaluated as fuzzy numbers. The ratio of the satisfactory degree of the decision maker at the upper level to that of the decision maker at the lower level and the related information obtained by solving the second problem are helpful to the decision maker at the upper level in updating the minimal satisfactory levels. The above interactive procedure is repeated until a satisfactory solution is obtained.

4.8.1 Interactive fuzzy programming

In this section, we deal with a two-level linear programming problem in which each decision maker has multiple objective functions to be minimized and the two decision makers can determine their decisions cooperatively. Let DM1 and DM2 denote the decision makers at the upper level and the lower level, respectively. Such a multiobjective two-level linear programming problem is formulated as

$$\underset{\text{for DM1}}{\text{minimize}} \quad z_1^1(\boldsymbol{x}_1,\boldsymbol{x}_2) = \boldsymbol{c}_{11}^1\boldsymbol{x}_1 + \boldsymbol{c}_{12}^1\boldsymbol{x}_2 \tag{4.115a}$$

$$\cdots\cdots\cdots\cdots\cdots\cdots$$

$$\underset{\text{for DM1}}{\text{minimize}} \quad z_1^{k_1}(\boldsymbol{x}_1,\boldsymbol{x}_2) = \boldsymbol{c}_{21}^{k_2}\boldsymbol{x}_1 + \boldsymbol{c}_{22}^{k_2}\boldsymbol{x}_2 \tag{4.115b}$$

$$\underset{\text{for DM2}}{\text{minimize}} \quad z_2^1(\boldsymbol{x}_1,\boldsymbol{x}_2) = \boldsymbol{c}_{21}^1\boldsymbol{x}_1 + \boldsymbol{c}_{22}^1\boldsymbol{x}_2 \tag{4.115c}$$

$$\cdots\cdots\cdots\cdots\cdots\cdots$$

$$\underset{\text{for DM2}}{\text{minimize}} \quad z_2^{k_2}(\boldsymbol{x}_1,\boldsymbol{x}_2) = \boldsymbol{c}_{21}^{k_2}\boldsymbol{x}_1 + \boldsymbol{c}_{22}^{k_2}\boldsymbol{x}_2 \tag{4.115d}$$

$$\text{subject to} \quad A_1\boldsymbol{x}_1 + A_2\boldsymbol{x}_2 \leqq \boldsymbol{b} \tag{4.115e}$$

$$\boldsymbol{x}_1 \geqq \boldsymbol{0},\ \boldsymbol{x}_2 \geqq \boldsymbol{0}, \tag{4.115f}$$

where $\boldsymbol{x}_1$ and $\boldsymbol{x}_2$ are n_1- and n_2-dimensional decision variable vectors of DM1 and DM2, respectively; $z_1^j(\boldsymbol{x}_1,\boldsymbol{x}_2)$, $j = 1,\dots,k_1$ and $z_2^j(\boldsymbol{x}_1,\boldsymbol{x}_2)$, $j = 1,\dots,k_2$ are the objective functions of DM1 and DM2, respectively; $\boldsymbol{c}_{i1}^j$ and $\boldsymbol{c}_{i2}^j$ are n_i-dimensional coefficient row vectors; A_i, $i = 1,2$ are $m \times n_i$ coefficient matrices; and $\boldsymbol{b}$ is an m-dimensional constant column vector. For the sake of simplicity, we use the following notations: $\boldsymbol{x}^T = (\boldsymbol{x}_1^T, \boldsymbol{x}_2^T) \in \mathbb{R}^{n_1+n_2}$, $\boldsymbol{c}_i^k = (\boldsymbol{c}_{i1}^k, \boldsymbol{c}_{i2}^k)$, $i = 1,2$, $k = 1,\ldots,k_i$, and $A = [A_1\ A_2]$.

Similarly to the two-level linear programming problems considered in the previous section, it is natural that a decision maker has a fuzzy goal for each of the multiple objective functions when the decision maker takes fuzziness of human judgments into consideration. To identify membership functions of the fuzzy goals for the objective functions, the following individual single-objective minimization problems are solved for $i = 1,2$ and $j = 1,\ldots,k_i$:

$$\begin{aligned} &\text{minimize} \ \ z_i^j(\boldsymbol{x}_1,\boldsymbol{x}_2) = \boldsymbol{c}_{i1}^j \boldsymbol{x}_1 + \boldsymbol{c}_{i2}^j \boldsymbol{x}_2 && (4.116\text{a}) \\ &\text{subject to} \ \ A_1\boldsymbol{x}_1 + A_2\boldsymbol{x}_2 \leqq \boldsymbol{b} && (4.116\text{b}) \\ &\qquad\qquad \boldsymbol{x}_1 \geqq \boldsymbol{0},\ \boldsymbol{x}_2 \geqq \boldsymbol{0}. && (4.116\text{c}) \end{aligned}$$

Suppose that, for the multiobjective two-level linear programming problems, the decision makers employ the linear membership function (4.7) of the fuzzy goal whose parameters are determined by the Zimmermann method (1978). The individual minimum is provided as

$$\begin{aligned} z_i^{j\min} &= z_i^j(\boldsymbol{x}^{ijo}) \\ &= \min\{z_i^j(\boldsymbol{x}) \mid A_1\boldsymbol{x}_1 + A_2\boldsymbol{x}_2 \leqq \boldsymbol{b}, \boldsymbol{x}_1 \geqq \boldsymbol{0},\ \boldsymbol{x}_2 \geqq \boldsymbol{0}\}, \end{aligned} \tag{4.117}$$

and $z_i^{j\mathrm{m}}$ is defined by

$$z_i^{j\mathrm{m}} = \max_{m=1,2,l=1,\ldots,k_i} \{z_i^j(\boldsymbol{x}^{mlo})\}. \tag{4.118}$$

Then, the parameters z_i^{j0} and z_i^{j1} of the linear membership function are determined by choosing $z_i^{j1} = z_i^{j\min}$, $z_i^{j0} = z_i^{j\mathrm{m}}$, $i = 1,2$, $j = 1,\ldots,k_i$.

We assume that each decision maker evaluates a solution $\boldsymbol{x}$ by aggregating the weighted membership functions additively. The aggregated membership function of DMi is represented as

$$\sum_{j=1}^{k_i} \lambda_j^i \mu_i^j(z_i^j(\boldsymbol{x})), \tag{4.119}$$

where $\boldsymbol{\lambda}^i = (\lambda_1^i,\ldots,\lambda_{k_i}^i)$ denotes a weighting coefficient vector satisfying

$$\boldsymbol{\lambda}^i \in \left\{ \boldsymbol{\lambda}^i \in \mathbb{R}^{k_i} \;\middle|\; \sum_{j=1}^{k_i} \lambda_j^i = 1,\ \lambda_j^i \geqq 0,\ j = 1,\ldots,k_i \right\}. \tag{4.120}$$

Moreover, we assume that the decision makers cannot identify the weighting coefficients precisely, but they have some partial information on their preferences (Malakooti, 1989; Marmol, Puerto and Fernandez, 1998). Suppose that such partial information can be represented by the following two inequalities for DMi:

$$LB^i_j \leqq \lambda^i_j \leqq UB^i_j, \tag{4.121a}$$

$$\lambda^i_p \geqq \lambda^i_q + \varepsilon,\ p \neq q, \tag{4.121b}$$

where ε is a small positive constant or a zero. In the condition (4.121a), the upper bound UB^i_j and the lower bound LB^i_j are specified for the weight λ^i_j to the membership function $\mu^j_i(z^j_i(\boldsymbol{x}))$ of the fuzzy goal for the jth objective function of DMi. The condition (4.121b) represents an order relation between the pth and the qth fuzzy goals. Let Λ^i denote a set of weighting coefficient vectors $\boldsymbol{\lambda}^i = (\lambda^i_1, \ldots, \lambda^i_{k_i})$ of DMi satisfying the conditions (4.121) as well as the condition (4.120). For example, when DMi has two objectives, suppose that DMi thinks μ^i_1 is more important than μ^i_2, but there does not exist a great difference between them. Then, DMi could specify the partial information of preference like $\lambda^i_2 \geqq 0.4$ and $\lambda^i_1 \geqq \lambda^i_2$. As a result, λ^i_1 and λ^i_2 are restricted as $0.5 \leqq \lambda^i_1 \leqq 0.6$ and $0.4 \leqq \lambda^i_2 \leqq 0.5$, respectively.

To derive a satisfactory solution to problem (4.115), we first solve the following maximin problem for obtaining a solution which maximizes the smallest degree among the satisfactory degrees of all the fuzzy goals of the two decision makers:

$$\text{maximize } \min\{\mu^1_1(z^1_1(\boldsymbol{x})), \ldots, \mu^{k_1}_1(z^{k_1}_1(\boldsymbol{x})), \mu^2_2(z^1_2(\boldsymbol{x})), \ldots, \mu^{k_2}_2(z^{k_2}_2(\boldsymbol{x}))\} \tag{4.122a}$$

$$\text{subject to } \boldsymbol{x} \in S, \tag{4.122b}$$

where S denotes the feasible region of problem (4.115).

By introducing an auxiliary variable η, problem (4.122) can be transformed into the following equivalent maximization problem:

$$\text{maximize } \eta \tag{4.123a}$$

$$\text{subject to } \mu^j_1(z^j_1(\boldsymbol{x})) \geqq \eta,\ j = 1, \ldots, k_1 \tag{4.123b}$$

$$\mu^j_2(z^j_2(\boldsymbol{x})) \geqq \eta,\ j = 1, \ldots, k_2 \tag{4.123c}$$

$$\boldsymbol{x} \in S, \tag{4.123d}$$

If DM1 is satisfied with an optimal solution $\boldsymbol{x}^*$ to problem (4.123), it follows that the optimal solution $\boldsymbol{x}^*$ becomes a satisfactory solution; however, DM1 is not always satisfied with the solution $\boldsymbol{x}^*$. Then, it is quite natural to assume that DM1 expects the satisfactory degrees for the membership functions μ^j_1, $j = 1, \ldots, k_1$ to be larger than certain minimal satisfactory levels $\hat{\delta}^1_j \in [0, 1]$, $j = 1, \ldots, k_1$, and DM2 also holds certain aspiration levels $\bar{\mu}^2_j$, $j = 1, \ldots, k_2$ to values of the membership functions μ^2_j, $j = 1, \ldots, k_2$. To specify the minimal satisfactory levels $\hat{\delta}^1_j$, $j = 1, \ldots, k_1$ and the aspiration levels $\bar{\mu}^2_j$, $j = 1, \ldots, k_2$, it seems reasonable for DM1 and DM2 to take account of the optimal solution to the maximin problem (4.123) and the related information.

Consequently, if DM1 is not satisfied with the solution $\boldsymbol{x}^*$ to problem (4.123), the following problem is formulated:

$$\begin{aligned}
&\text{minimize } \eta && (4.124a)\\
&\text{subject to } \mu_1^j(z_1^j(\boldsymbol{x})) \geqq \hat{\delta}_j^1,\ j=1,\ldots,k_1 && (4.124b)\\
&\qquad \bar{\mu}_j^2 - \mu_2^j(z_2^j(\boldsymbol{x})) \leqq \eta,\ j=1,\ldots,k_2 && (4.124c)\\
&\qquad \boldsymbol{x} \in S, && (4.124d)
\end{aligned}$$

where η is an auxiliary variable. In problem (4.124), the distance between a vector of the membership function values of DM2 and that of the aspiration levels is minimized under the conditions that the membership function values of DM1 are larger than or equal to the minimal satisfactory levels specified by DM1.

After obtaining an optimal solution to problem (4.124), with the preference of herself in mind, DM2 updates the membership values $\bar{\mu}_j^2$, $j=1,\ldots,k_2$ representing the aspiration levels and problem (4.124) with the updated aspiration levels is solved again if necessary.

If there exists an optimal solution to problem (4.124), it follows that DM1 obtains a solution whose satisfactory degrees are larger than or equal to the minimal satisfactory levels specified by himself. However, the larger the minimal satisfactory levels $\hat{\delta}_j^1$, $j=1,\ldots,k_1$ are assessed, the smaller the DM2's satisfactory degrees become. Consequently, a relative difference between the aggregated satisfactory degrees of DM1 and DM2 becomes larger, and it cannot be anticipated that the obtained solution becomes a satisfactory solution balancing the aggregated satisfactory degree of DM1 with that of DM2.

To obtain a satisfactory solution acceptable for both decision makers, a candidate for the satisfactory solution, that is, an optimal solution $\boldsymbol{x}^*$ to problem (4.124) should be evaluated. To do so, the aggregated membership functions with weighting coefficients is utilized. Because a possible weighting coefficient vector $\boldsymbol{\lambda}^i$ is in the set Λ^i, the minimum and the maximum of the aggregated membership function with weighting coefficients of DMi with respect to $\boldsymbol{x}^*$ can be represented by

$$S_{\min}^i = \min_{\boldsymbol{\lambda}^i \in \Lambda^i} \sum_{j=1}^{k_i} \lambda_j^i \mu_i^j(z_i^j(\boldsymbol{x}^*)),\ i=1,2, \tag{4.125a}$$

$$S_{\max}^i = \max_{\boldsymbol{\lambda}^i \in \Lambda^i} \sum_{j=1}^{k_i} \lambda_j^i \mu_i^j(z_i^j(\boldsymbol{x}^*)),\ i=1,2. \tag{4.125b}$$

Using the two values $S_{\min}^i$ and $S_{\max}^i$, we define an aggregated satisfactory degree of DMi with respect to $\boldsymbol{x}^*$ as the following L-L fuzzy number:

$$\tilde{S}^i = ((S_{\min}^i + S_{\max}^i)/2, (S_{\max}^i - S_{\min}^i)/2)_{LL}. \tag{4.126}$$

The L-L fuzzy number $\tilde{S}^i$ is prescribed by the following membership function:

$$\mu_{\tilde{S}^i}(p) = \begin{cases} L\left(\dfrac{S^i_{\min}+S^i_{\max}-2p}{S^i_{\max}-S^i_{\min}}\right) & \text{if } p \leqq (S^i_{\min}+S^i_{\max})/2 \\ L\left(\dfrac{2p-S^i_{\min}-S^i_{\max}}{S^i_{\max}-S^i_{\min}}\right) & \text{if } p > (S^i_{\min}+S^i_{\max})/2, \end{cases} \tag{4.127}$$

where $L(p) = \max(0, 1-|p|)$. The fuzzy number representing the satisfactory degree of DMi is depicted in Figure 4.12; $S^j_{\min}$ and $S^j_{\max}$ are values such that the satisfaction degrees are 0s, and $(S^j_{\max}+S^j_{\min})/2$ is a value such that the satisfaction degree is 1.

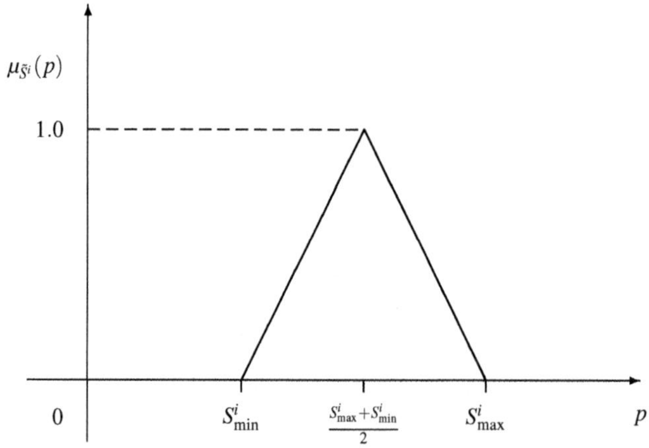

Fig. 4.12 Satisfactory degree $\tilde{S}^i$ of DMi.

In order to take account of the overall satisfactory balance between both levels, the ratio between the satisfactory degrees of the two decision makers is defined as a quotient of the two L-L fuzzy numbers $\tilde{S}^1 = (S^1, s^1)$ and $\tilde{S}^2 = (S^2, s^2)$ (Dubois and Prade, 1980):

$$\tilde{S}^2 \oslash \tilde{S}^1 \simeq \tilde{\Delta} = \left(\frac{S^2}{S^1}, \frac{s^2 S^1 + s^1 S^2}{(S^1)^2}\right)_{LL}. \tag{4.128}$$

Let $\mu_{\tilde{\Delta}}(p)$ denote a membership function for the ratio $\tilde{\Delta}$ between the satisfactory degrees of the two decision makers.

DM1 is guaranteed to have satisfactory degrees larger than or equal to the minimal satisfactory levels for all of the fuzzy goals because the corresponding constraints are involved in problem (4.124). It is natural that DM1 has a fuzzy goal $\tilde{R}$ for the ratio $\tilde{\Delta}$ of satisfactory degrees in order to take into account the overall satisfactory balance between both levels as well as the satisfactory degrees of himself.

The fuzzy goal $\tilde{R}$ is expressed in words such as "the ratio $\tilde{\Delta}$ should be in the vicinity of a certain value m."

We introduce a termination condition that the maximin value between the membership value of the ratio $\tilde{\Delta}$ of satisfactory degrees and that of its fuzzy goal $\tilde{R}$ is larger than or equal to the permissible level $\hat{\delta}_{\tilde{\Delta}}$, i.e.,

$$\alpha \triangleq \max_{p} \min\{\mu_{\tilde{\Delta}}(p), \mu_{\tilde{R}}(p)\} \geqq \hat{\delta}_{\tilde{\Delta}}, \tag{4.129}$$

where $\mu_{\tilde{R}}(p)$ denote a membership function of the fuzzy goal $\tilde{R}$ for the ratio $\tilde{\Delta}$ of satisfactory degrees. The condition (4.129) is a natural extension of the second condition (4.17) in interactive fuzzy programming for two-level linear programming problems shown in subsection 4.2.

When the termination condition is not satisfied, or DM1 judges that it is desirable for himself to increase his satisfactory degree at the sacrifice of that of DM2 or the reverse is true, DM1 should update some or all of the minimal satisfactory levels. Trade-off ratio between satisfactory degrees of DM1 and DM2 would provide DM1 with important information that suggests which minimal satisfactory level should be updated. To be precise, DM1 can see the trade-off ratio $-\partial\mu_2^j(z_2^j(\boldsymbol{x}))/\partial\mu_1^i(z_1^i(\boldsymbol{x}))$ between the ith satisfactory degree of DM1 and the jth satisfactory degree of DM2 and the weighted trade-off ratios $-\partial(\sum_{j=1}^{k_2}\lambda_{2\min}^j\mu_2^j)/\partial\mu_1^i$ and $-\partial(\sum_{j=1}^{k_2}\lambda_{2\max}^j\mu_2^j)/\partial\mu_1^i$ by $\lambda_{2\min}$ and $\lambda_{2\max}$. Here, let π_i^1 and π_j^2 be the simplex multiplier corresponding to the constraints with respect to the fuzzy goals of DM1 and DM2 in problem (4.124), respectively, then the trade-off ratio can be represented by (Haimes and Chankong, 1979; Sakawa and Yano, 1988)

$$-\frac{\partial\, \mu_2^j(z_2^j(\boldsymbol{x}^*))}{\partial\, \mu_1^i(z_1^j(\boldsymbol{x}^*))} = \frac{\pi_i^2}{\pi_j^1}. \tag{4.130}$$

We are now ready to present an interactive algorithm for obtaining an overall satisfactory solution to problem (4.115), which is summarized in the following and is illustrated with a flowchart in Figure 4.13:

[Algorithm of interactive fuzzy programming for multiobjective two-level linear programming problems]

Step 0 Solve the individual problems (4.116) for all the objectives of the two decision makers. Ask the two decision makers about partial information of their preference.

Step 1 Ask DM1 to identify the membership functions $\mu_1^j(z_1^j)$, $j = 1,\dots,k_1$, and also ask DM2 to identify $\mu_2^j(z_2^j)$, $j = 1,\dots,k_2$ of the fuzzy goals for the objective functions.

Step 2 Set $l := 1$. Solve problem (4.123). If DM1 is satisfied with an optimal solution to problem (4.123), the solution becomes a satisfactory solution. Otherwise, ask DM1 to specify the minimal satisfactory levels $\hat{\delta}_j^1$, $j = 1,\dots,k_1$ by considering the current satisfaction degrees, and also ask DM2 to specify the membership values $\bar{\mu}_j^2$, $j = 1,\ldots,k_2$ representing the aspiration levels. Moreover, ask DM1 to identify the membership function $\mu_{\tilde{R}}(p)$ of the fuzzy goal for the ratio $\tilde{\Delta}$ of satisfactory degrees, and ask DM1 to specify the permissible level $\hat{\delta}_{\tilde{\Delta}}$.

Step 3 Set $l := l+1$. Solve problem (4.124) with the minimal satisfactory levels and the aspiration levels. If DM2 is satisfied with an obtained solution, go to Step 5.

Step 4 After DM2 updates the membership values representing the aspiration levels on the preference of herself, return to Step 3.

Step 5 The satisfactory degrees $\tilde{S}^i$, $i = 1,2$ and the ratio of satisfactory degrees $\tilde{\Delta}$ corresponding to an optimal solution to problem (4.124) are shown to DM1. If the solution satisfies the termination conditions and DM1 accepts the solution as a satisfactory solution, the algorithm stops.

Step 6 Ask DM1 to update some of the minimal satisfactory levels taking account of the optimal solution to problem (4.124) and the related information including trade-off ratios. Return to Step 3.

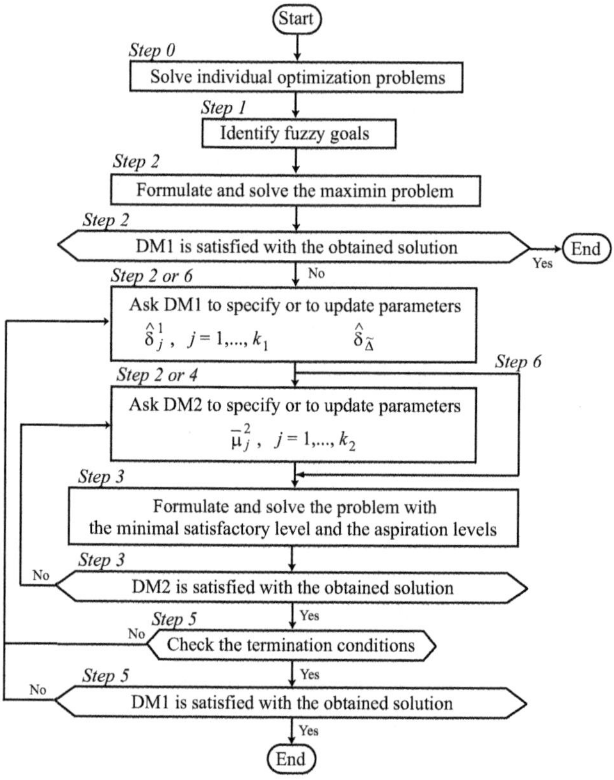

Fig. 4.13 Flowchart of the interactive fuzzy programming.

4.8.2 Numerical example

To illustrate the interactive fuzzy programming method described in the previous subsection, consider the following multiobjective two-level linear programming problem:

$$\underset{\text{for DM1}}{\text{minimize}}\ \boldsymbol{c}_{11}^1\boldsymbol{x}_1+\boldsymbol{c}_{12}^1\boldsymbol{x}_2 \tag{4.131a}$$
$$\underset{\text{for DM1}}{\text{minimize}}\ \boldsymbol{c}_{11}^2\boldsymbol{x}_1+\boldsymbol{c}_{12}^2\boldsymbol{x}_2 \tag{4.131b}$$
$$\underset{\text{for DM1}}{\text{minimize}}\ \boldsymbol{c}_{11}^3\boldsymbol{x}_1+\boldsymbol{c}_{12}^3\boldsymbol{x}_2 \tag{4.131c}$$
$$\underset{\text{for DM2}}{\text{minimize}}\ \boldsymbol{c}_{21}^1\boldsymbol{x}_1+\boldsymbol{c}_{22}^1\boldsymbol{x}_2 \tag{4.131d}$$
$$\underset{\text{for DM2}}{\text{minimize}}\ \boldsymbol{c}_{21}^2\boldsymbol{x}_1+\boldsymbol{c}_{22}^2\boldsymbol{x}_2 \tag{4.131e}$$
$$\text{subject to}\ A_1\boldsymbol{x}_1+A_2\boldsymbol{x}_2 \leqq \boldsymbol{b} \tag{4.131f}$$
$$\boldsymbol{x}_1 \geqq \boldsymbol{0},\ \boldsymbol{x}_2 \geqq \boldsymbol{0}. \tag{4.131g}$$

In problem (4.131), DM1 and DM2 have three and two objective functions, respectively, and $\boldsymbol{x}_1 = (x_{11},\dots,x_{15})^T$ and $\boldsymbol{x}_2 = (x_{21},\dots,x_{25})^T$ are the decision variable vectors of DM1 and DM2, respectively. Each entry of the 16×10 coefficient matrix A is randomly chosen in the interval $[-50,50]$; an entry of the right-hand side constant column vector $\boldsymbol{b}$ is a sum of entries of the corresponding row vector of A multiplied by 0.7; coefficients $\boldsymbol{c}_1^i$, $j=1,2,3$ and $\boldsymbol{c}_2^j$, $j=1,2$ of the objective functions are shown in Table 4.36.

Suppose that DM1 and DM2 specify the partial information of their preferences as follows:

$$\Lambda^1 \triangleq \{\boldsymbol{\lambda}^1 \in \mathbb{R}^3 \mid \lambda_1^1 \leqq \lambda_2^1 \leqq \lambda_3^1,\ \lambda_1^1 \geqq 0.2,\ \lambda_1^1+\lambda_2^1+\lambda_3^1=1,$$
$$\lambda_1^1 \geqq 0,\ \lambda_2^1 \geqq 0,\ \lambda_3^1 \geqq 0\},$$
$$\Lambda^2 \triangleq \{\boldsymbol{\lambda}^2 \in \mathbb{R}^2 \mid \lambda_1^2 \geqq \lambda_2^2,\ \lambda_2^2 \geqq 0.2,\ \lambda_1^2+\lambda_2^2=1,\ \lambda_1^2 \geqq 0,\ \lambda_2^2 \geqq 0\}.$$

Suppose that the decision makers employ the linear membership function (4.7) whose parameters are determined by the Zimmermann method (1978). The individual minima and the corresponding optimal solutions are shown in Table 4.37.

After the membership functions of the decision makers are identified, the algorithm starts in formulating the following linear programming problem which maximizes the minimum among the satisfactory degrees of the decision makers:

Table 4.36 Coefficients in problem (4.131).

$\boldsymbol{c}_1^1$	15	−46	1	34	−30	42	−18	39	46	25		
$\boldsymbol{c}_1^2$	26	27	6	1	49	−16	−45	18	41	−40		
$\boldsymbol{c}_1^3$	14	36	17	−5	−26	17	37	6	25	−4		
$\boldsymbol{c}_2^1$	23	14	12	−10	27	−11	−14	−6	−27	−7		
$\boldsymbol{c}_2^2$	12	−19	−39	−6	18	24	−9	37	16	−2		
A	−43	−4	38	−45	−18	−6	25	46	48	−20	$\boldsymbol{b}$	13
	25	−39	−38	9	47	−32	26	−45	−1	17		−20
	−39	−3	33	−27	34	−26	20	43	−29	−9		−1
	38	19	−22	−35	39	−21	30	41	34	−38		55
	13	−19	45	17	−47	10	33	−40	−5	−30		−14
	−41	49	−10	−19	−22	−23	−36	−49	−11	4		−102
	−41	38	−9	−11	12	−9	48	14	13	−39		10
	−32	−48	48	30	−16	29	−3	−35	−38	−43		−70
	−38	−11	−48	−5	41	19	−36	−28	11	−34		−83
	−26	−30	38	−36	41	−41	−33	−18	7	22		−49
	−28	40	28	−9	46	23	10	7	−44	6		51
	32	−10	18	−37	−25	36	−9	−26	34	16		18
	−47	−38	38	−7	−40	−35	−27	10	−15	36		−81
	−6	−34	−3	2	7	48	−34	16	18	26		26
	−44	−11	39	−23	−43	0	−42	−28	−29	−9		−123
	11	3	−36	25	12	3	42	25	6	47		89

$$\text{maximize } \eta \quad (4.132a)$$

$$\text{subject to } (z_1^1(\boldsymbol{x}) - 131.994)/(43.016 - 131.994) \geqq \eta \quad (4.132b)$$

$$(z_1^2(\boldsymbol{x}) - 96.824)/(23.387 - 96.824) \geqq \eta \quad (4.132c)$$

$$(z_1^3(\boldsymbol{x}) - 56.319)/(28.387 - 56.319) \geqq \eta \quad (4.132d)$$

$$(z_2^1(\boldsymbol{x}) - 60.046)/(-33.594 - 60.046) \geqq \eta \quad (4.132e)$$

$$(z_2^2(\boldsymbol{x}) - 64.013)/(-65.981 - 64.013) \geqq \eta \quad (4.132f)$$

$$\boldsymbol{x} \in S, \quad (4.132g)$$

where S denotes the feasible region of problem (4.131). An optimal solution to problem (4.132) and the related information are shown in Table 4.38.

Suppose that DM1 is not satisfied with the optimal solution to problem (4.132), and then he specifies the minimal satisfactory levels for the membership functions $\mu_1^i(z_1^i(\boldsymbol{x}))$, $i = 1,2,3$ at

$$\hat{\delta}_1^1 = 0.400, \;\; \hat{\delta}_2^1 = 0.600, \;\; \text{and} \;\; \hat{\delta}_3^1 = 0.700$$

by taking account of the minimal satisfactory degree of both levels 0.486 and the partial information Λ^1; suppose that DM2 also sets the aspiration levels to the mem-

Table 4.37 Optimal solutions to the individual problems.

$z_1^{1\min}$	43.016		z_1^{1m}	131.994	
x_1^{11o}	1.171	0.000	0.000	0.000	0.575
x_2^{11o}	0.000	0.343	0.918	0.000	0.000
$z_1^{2\min}$	23.387		z_1^{2m}	96.824	
x_1^{12o}	1.232	0.000	0.000	0.000	0.167
x_2^{12o}	0.789	0.387	0.808	0.000	0.000
$z_1^{3\min}$	28.387		z_1^{3m}	56.319	
x_1^{13o}	0.986	0.000	0.000	0.000	0.688
x_2^{13o}	0.000	0.000	1.146	0.000	0.000
$z_2^{1\min}$	−33.594		z_2^{1m}	60.046	
x_1^{21o}	0.705	0.000	0.000	0.000	0.000
x_2^{21o}	0.256	0.000	1.188	1.399	0.000
$z_2^{2\min}$	−65.981		z_2^{2m}	64.013	
x_1^{22o}	2.149	0.000	0.000	0.000	0.560
x_2^{22o}	0.000	0.000	0.750	0.000	0.000

Table 4.38 Iteration 1.

x_1	1.284	0.621	0.057	1.446	0.048
x_2	0.097	0.045	0.960	0.073	0.249
(z_1^1, z_1^2, z_1^3)	(88.751, 61.134, 42.744)				
$(\mu_1^1(z_1^1), \mu_1^2(z_1^2), \mu_1^3(z_1^3))$	(0.486, 0.486, 0.486)				
(z_2^1, z_2^2)	(14.559, 0.836)				
$(\mu_2^1(z_2^1), \mu_2^2(z_2^2))$	(0.486, 0.486)				

bership functions $\mu_2^j(z_2^j(\boldsymbol{x}))$, $j = 1,2$ at

$$\bar{\mu}_1^2 = 0.800, \text{ and } \bar{\mu}_2^2 = 0.600.$$

Moreover, suppose that DM1 thinks the ratio $\tilde{\Delta}$ should be in the vicinity of about 0.9, and identifies the membership function of the fuzzy goal $\tilde{R}$ for the ratio $\tilde{\Delta}$ of satisfactory degrees as

$$\mu_{\tilde{R}}(p) = \begin{cases} \max\{0, 10p - 8\}, & \text{if } p < 0.9 \\ \max\{0, -10p + 10\}, & \text{if } p \geqq 0.9. \end{cases} \tag{4.133}$$

The fuzzy goal $\tilde{R}$ corresponds to the fuzzy number $(0.9, 0.1)_{LL}$, $L(p) = \max\{0, 1 - |p|\}$. Suppose that DM1 determines the permissible level $\hat{\delta}_{\tilde{\Delta}}$ at 0.8, and then the termination condition is represented as $\alpha \triangleq \max_p \min\{\mu_{\tilde{\Delta}}(p), \mu_{\tilde{R}}(p)\} \geqq \hat{\delta}_{\tilde{\Delta}} = 0.8$. Then, we solve the following problem corresponding to problem (4.124):

$$
\begin{aligned}
&\text{minimize } \eta && (4.134a)\\
&\text{subject to } (z_1^1(\boldsymbol{x}) - 131.994)/(43.016 - 131.994) \geqq 0.400 && (4.134b)\\
&\qquad (z_1^2(\boldsymbol{x}) - 96.824)/(23.387 - 96.824) \geqq 0.600 && (4.134c)\\
&\qquad (z_1^3(\boldsymbol{x}) - 56.319)/(28.387 - 56.319) \geqq 0.700 && (4.134d)\\
&\qquad 0.800 - (z_2^1(\boldsymbol{x}) - 60.046)/(-33.594 - 60.046) \leqq \eta && (4.134e)\\
&\qquad 0.600 - (z_2^2(\boldsymbol{x}) - 64.013)/(-65.981 - 64.013) \leqq \eta && (4.134f)\\
&\qquad \boldsymbol{x} \in S, && (4.134g)
\end{aligned}
$$

and the result is shown in Table 4.39.

Table 4.39 Iteration 2.

	DM1	DM2
Individual satisfactory degree	$\mu_1^1 = 0.400$ $\mu_2^1 = 0.600$ $\mu_3^1 = 0.700$	$\mu_1^2 = 0.499$ $\mu_2^2 = 0.299$ —
$(\lambda^{1\min}, \lambda^{2\min})$	$\lambda_1^{1\min} = 0.332$ $\lambda_2^{1\min} = 0.333$ $\lambda_3^{1\min} = 0.334$	$\lambda_1^{2\min} = 0.500$ $\lambda_2^{2\min} = 0.500$ —
Minimum of the weighted membership value	$S_{\min}^1 = 0.567$	$S_{\min}^2 = 0.399$
$(\lambda^{1\max}, \lambda^{2\max})$	$\lambda_1^{1\max} = 0.200$ $\lambda_2^{1\max} = 0.201$ $\lambda_3^{1\max} = 0.599$	$\lambda_1^{2\max} = 0.800$ $\lambda_2^{2\max} = 0.200$ —
Maximum of the weighted membership value	$S_{\max}^1 = 0.620$	$S_{\max}^2 = 0.459$
Aggregated satisfactory degree	$\tilde{S}^1 = (0.594, 0.027)_{LL}$	$\tilde{S}^2 = (0.429, 0.030)_{LL}$
Ratio of satisfactory degrees	$\tilde{\Delta} = (0.722, 0.160)_{LL}$	
Maximin value α	$\alpha = 0.030$	

Suppose that DM2 is satisfied with the solution obtained in Iteration 2. However, from the result of Iteration 2 shown in Table 4.39, it is found that the maximin value α of the ratio of satisfactory degrees and the fuzzy goal for the ratio is 0.030 and it is smaller than the permissible level $\hat{\delta}_{\tilde{\Delta}} = 0.8$. Therefore, DM1 should update some of the minimal satisfactory levels, and the following trade-off ratios provide him with important information that suggests which minimal satisfactory level should be updated:

$$
T_1 = \begin{bmatrix} 0.138 & 0.764 \\ 0.148 & 0.818 \\ 0.319 & 1.763 \end{bmatrix}, \text{ and } T_2 = \begin{bmatrix} 0.451 & 0.264 \\ 0.483 & 0.282 \\ 1.041 & 0.608 \end{bmatrix}, \tag{4.135}
$$

where the ij element of T_1 is $-\partial\mu_2^j(z_2^j(\boldsymbol{x}))\ /\partial\mu_1^i(z_1^i(\boldsymbol{x}))$, and the $i1$ and $i2$ elements of T_2 are $-\partial(\sum_{j=1}^{k_2}\lambda_j^{2\min}\mu_2^j)/\partial\mu_1^i$ and $-\partial(\sum_{j=1}^{k_2}\lambda_j^{2\max}\mu_2^j)/\partial\mu_1^i$, respectively.

From the trade-off ratios (4.135), we can see that reducing the third minimal satisfactory level $\hat{\delta}_3^1$ is most effective to raise the satisfactory degree of DM2. Therefore, suppose that DM1 reduces the minimal satisfactory level for the third membership function by 0.050. The revised problem (4.124), in which the third minimal satisfactory level is updated from 0.700 to 0.650, is solved, and the result is shown in Table 4.40.

Table 4.40 Iteration 3.

	DM1	DM2
Individual satisfactory degree	$\mu_1^1 = 0.400$	$\mu_2^1 = 0.526$
	$\mu_1^2 = 0.600$	$\mu_2^2 = 0.326$
	$\mu_1^3 = 0.650$	—
$(\lambda^{1\min}, \lambda^{2\min})$	$\lambda_1^{1\min} = 0.332$	$\lambda_1^{2\min} = 0.500$
	$\lambda_2^{1\min} = 0.333$	$\lambda_2^{2\min} = 0.500$
	$\lambda_3^{1\min} = 0.334$	—
Minimum of the weighted membership value	$S_{\min}^1 = 0.550$	$S_{\min}^2 = 0.426$
$(\lambda^{1\max}, \lambda^{2\max})$	$\lambda_1^{1\max} = 0.200$	$\lambda_1^{2\max} = 0.800$
	$\lambda_2^{1\max} = 0.201$	$\lambda_2^{2\max} = 0.200$
	$\lambda_3^{1\max} = 0.599$	—
Maximum of the weighted membership value	$S_{\max}^1 = 0.590$	$S_{\max}^2 = 0.486$
Aggregated satisfactory degree	$\tilde{S}^1 = (0.570, 0.020)_{LL}$	$\tilde{S}^2 = (0.456, 0.030)_{LL}$
Ratio of satisfactory degrees	$\tilde{\Delta} = (0.800, 0.081)_{LL}$	
Maximin value α	$\alpha = 0.441$	

Although the aggregated satisfactory degree is improved, because the termination condition is not still satisfied at Iteration 3, DM1 must update the minimal satisfactory levels. Suppose that DM1 reduces the minimal satisfactory level for the second and the third membership functions by 0.050 with the trade-off ratios in mind. The revised problem (4.124), in which the second and the third minimal satisfactory levels are updated from 0.600 to 0.550 and from 0.650 to 0.600, respectively, is solved. The result is shown in Table 4.41.

At Iteration 4, because the termination condition is satisfied, i.e., $\alpha = 0.952 \geqq 0.8$, it follows that the obtained solution becomes a satisfactory solution if DM1 accepts the solution.

Table 4.41 Iteration 4.

	DM1	DM2
Individual satisfactory degree	$\mu_1^1 = 0.400$ $\mu_2^1 = 0.550$ $\mu_3^1 = 0.600$	$\mu_1^2 = 0.545$ $\mu_2^2 = 0.345$ —
$(\lambda^{1\min}, \lambda^{2\min})$	$\lambda_1^{1\min} = 0.332$ $\lambda_2^{1\min} = 0.333$ $\lambda_3^{1\min} = 0.334$	$\lambda_1^{2\min} = 0.500$ $\lambda_2^{2\min} = 0.500$ —
Minimum of the weighted membership value	$S_{\min}^1 = 0.517$	$S_{\min}^2 = 0.445$
$(\lambda^{1\max}, \lambda^{2\max})$	$\lambda_1^{1\max} = 0.200$ $\lambda_2^{1\max} = 0.201$ $\lambda_3^{1\max} = 0.599$	$\lambda_1^{2\max} = 0.800$ $\lambda_2^{2\max} = 0.200$ —
Maximum of the weighted membership value	$S_{\max}^1 = 0.550$	$S_{\max}^2 = 0.505$
Aggregated satisfactory degree	$\tilde{S}^1 = (0.534, 0.017)_{LL}$	$\tilde{S}^2 = (0.475, 0.030)_{LL}$
Ratio of satisfactory degrees	$\tilde{\Delta} = (0.890, 0.084)_{LL}$	
Maximin value α	$\alpha = 0.952$	

4.9 Fuzzy stochastic two-level linear programming

In this section, we deal with two-level linear programming problems with random variable coefficients in objective functions and constraints. Following the expectation and the variance optimization models with the chance constraint, we transform the two-level stochastic linear programming problems into deterministic problems, and then present interactive fuzzy programming to derive a satisfactory solution for decision makers. Finally, we describe formulations from the maximum probability model and the fractile criterion model concisely.

A two-level linear programming problem with random variable coefficients in objective functions and constraints is formulated as

$$\underset{\text{for DM1}}{\text{minimize}} \; z_1(\boldsymbol{x}_1, \boldsymbol{x}_2) = \tilde{\boldsymbol{c}}_{11}\boldsymbol{x}_1 + \tilde{\boldsymbol{c}}_{12}\boldsymbol{x}_2 \tag{4.136a}$$

$$\underset{\text{for DM2}}{\text{minimize}} \; z_2(\boldsymbol{x}_1, \boldsymbol{x}_2) = \tilde{\boldsymbol{c}}_{21}\boldsymbol{x}_1 + \tilde{\boldsymbol{c}}_{22}\boldsymbol{x}_2 \tag{4.136b}$$

$$\text{subject to } A_1\boldsymbol{x}_1 + A_2\boldsymbol{x}_2 \leqq \tilde{\boldsymbol{b}} \tag{4.136c}$$

$$\boldsymbol{x}_1 \geqq \boldsymbol{0},\ \boldsymbol{x}_2 \geqq \boldsymbol{0}, \tag{4.136d}$$

where $\boldsymbol{x}_1$ is an n_1-dimensional decision variable column vector of DM1, $\boldsymbol{x}_2$ is an n_2-dimensional decision variable column vector of DM2, $A_j,\ j = 1,2$ are $m \times n_j$ coefficient matrices, $\tilde{\boldsymbol{c}}_{lj},\ l = 1,2,\ j = 1,2$ are n_j-dimensional random variable coefficient row vectors, $\tilde{\boldsymbol{b}}$ is a random variable column vector whose elements are independent of each other.

Because problem (4.136) includes random variables in the right-hand side of the constraints, the chance constraint programming approach (Charnes and Cooper, 1963) is introduced. This means that the constraint violations are permitted up to the specified probability limits. Let α_i be a probability of the extent to which the ith constraint violation is admitted. Then, the chance constraint conditions are represented as follows:

$$P\left[A_1^i \boldsymbol{x}_1 + A_2^i \boldsymbol{x}_2 \leqq \tilde{b}_i\right] \geqq \alpha_i,\ i = 1,\ldots,m, \tag{4.137}$$

where P means a probability measure, and A_1^i, A_2^i and $\tilde{b}_i$ are the coefficients of the ith constraint. Inequality (4.137) means that the ith constraint may be violated, but at most $1-\alpha_i$ proportion of the time.

Let $F_i(\tau)$ be a distribution function of the random variable $\tilde{b}_i$. Then, because of the fact that

$$P\left[A_1^i \boldsymbol{x}_1 + A_2^i \boldsymbol{x}_2 \leqq \tilde{b}_i\right] = 1 - F(A_1^i \boldsymbol{x}_1 + A_2^i \boldsymbol{x}_2),$$

inequality (4.137) is rewritten as

$$F(A_1^i \boldsymbol{x}_1 + A_2^i \boldsymbol{x}_2) \leqq 1 - \alpha_i.$$

Let $K_{1-\alpha_i}$ be the maximal value of τ satisfying $\tau = F^{-1}(1-\alpha_i)$. Then, from the monotonicity of the distribution function $F(\tau)$, inequality (4.137) is rewritten as

$$A_1^i \boldsymbol{x}_1 + A_2^i \boldsymbol{x}_2 \leqq K_{1-\alpha_i},\ i = 1,\ldots,m, \tag{4.138}$$

and it is equivalently represented by

$$A_1 \boldsymbol{x}_1 + A_2 \boldsymbol{x}_2 \leqq \boldsymbol{K}_{1-\alpha}, \tag{4.139}$$

where $\boldsymbol{K}_{1-\alpha} = (K_{1-\alpha_1},\ldots,K_{1-\alpha_m})^T$.

With the chance constraint formulation, problem (4.136) is rewritten as the following problem with deterministic constraints:

$$\underset{\text{for DM1}}{\text{minimize}}\ z_1(\boldsymbol{x}_1,\boldsymbol{x}_2) = \tilde{\boldsymbol{c}}_{11}\boldsymbol{x}_1 + \tilde{\boldsymbol{c}}_{12}\boldsymbol{x}_2 \tag{4.140a}$$

$$\underset{\text{for DM2}}{\text{minimize}}\ z_2(\boldsymbol{x}_1,\boldsymbol{x}_2) = \tilde{\boldsymbol{c}}_{21}\boldsymbol{x}_1 + \tilde{\boldsymbol{c}}_{22}\boldsymbol{x}_2 \tag{4.140b}$$

$$\text{subject to}\ A_1\boldsymbol{x}_1 + A_2\boldsymbol{x}_2 \leqq \boldsymbol{K}_{1-\alpha} \tag{4.140c}$$

$$\boldsymbol{x}_1 \geqq \boldsymbol{0},\ \boldsymbol{x}_2 \geqq \boldsymbol{0}. \tag{4.140d}$$

4.9.1 Stochastic two-level linear programming models

To minimize the objective functions with random variable coefficients in the two-level linear programming problem, the concepts of the minimum expected value model (E-model) and the minimum variance model (V-model) are applied (Charnes and Cooper, 1963).

E-model for stochastic two-level linear programming problems

In the E-model, the means of the objective functions of DM1 and DM2 are minimized, and the deterministic problem corresponding to problem (4.140) is formulated as

$$\underset{\text{for DM1}}{\text{minimize}} \quad E[\tilde{\boldsymbol{c}}_{11}\boldsymbol{x}_1 + \tilde{\boldsymbol{c}}_{12}\boldsymbol{x}_2] = \boldsymbol{m}_{11}\boldsymbol{x}_1 + \boldsymbol{m}_{12}\boldsymbol{x}_2 \tag{4.141a}$$

$$\underset{\text{for DM2}}{\text{minimize}} \quad E[\tilde{\boldsymbol{c}}_{21}\boldsymbol{x}_1 + \tilde{\boldsymbol{c}}_{22}\boldsymbol{x}_2] = \boldsymbol{m}_{21}\boldsymbol{x}_1 + \boldsymbol{m}_{22}\boldsymbol{x}_2 \tag{4.141b}$$

$$\text{subject to} \quad A_1\boldsymbol{x}_1 + A_2\boldsymbol{x}_2 \leqq \boldsymbol{K}_{1-\alpha} \tag{4.141c}$$

$$\boldsymbol{x}_1 \geqq \boldsymbol{0},\ \boldsymbol{x}_2 \geqq \boldsymbol{0}, \tag{4.141d}$$

where $E[\tilde{f}]$ denotes the mean of $\tilde{f}$, and $(\boldsymbol{m}_{11}, \boldsymbol{m}_{12})$ and $(\boldsymbol{m}_{21}, \boldsymbol{m}_{22})$ are vectors of the means of $(\tilde{\boldsymbol{c}}_{11}, \tilde{\boldsymbol{c}}_{12})$ and $(\tilde{\boldsymbol{c}}_{21}, \tilde{\boldsymbol{c}}_{22})$, respectively.

Because problem (4.141) is an ordinary two-level linear programming problem, the interactive fuzzy programming technique for two-level linear programming problems (Sakawa, Nishizaki and Uemura, 1998, 2000a) given in subsection 4.2 is directly applicable in order to obtain a satisfactory solution for the decision makers. If the expectation minimization model is employed, we obtain a best solution in the sense of minimizing the means of the objective functions, but the decision makers are not always satisfied with the solution due to the volatility of the objective function values.

V-model for stochastic two-level linear programming problems

The solution obtained in the E-model might be undesirable for the decision maker if the dispersion of the objective function value is large. In such a situation, it is reasonable for the decision maker to employ the V-model which minimizes the variance of the objective function, and the two-level linear programming problem with random variable coefficients in the V-model is represented by

$$\text{minimize} \quad Var[\tilde{\boldsymbol{c}}_{11}\boldsymbol{x}_1 + \tilde{\boldsymbol{c}}_{12}\boldsymbol{x}_2] = \begin{bmatrix} \boldsymbol{x}_1 \\ \boldsymbol{x}_2 \end{bmatrix}^T V_1 \begin{bmatrix} \boldsymbol{x}_1 \\ \boldsymbol{x}_2 \end{bmatrix} \tag{4.142a}$$

$$\text{minimize} \quad Var[\tilde{\boldsymbol{c}}_{21}\boldsymbol{x}_1 + \tilde{\boldsymbol{c}}_{22}\boldsymbol{x}_2] = \begin{bmatrix} \boldsymbol{x}_1 \\ \boldsymbol{x}_2 \end{bmatrix}^T V_2 \begin{bmatrix} \boldsymbol{x}_1 \\ \boldsymbol{x}_2 \end{bmatrix} \tag{4.142b}$$

$$\text{subject to} \quad A_1\boldsymbol{x}_1 + A_2\boldsymbol{x}_2 \leqq \boldsymbol{K}_{1-\alpha} \tag{4.142c}$$

$$\boldsymbol{x}_1 \geqq \boldsymbol{0},\ \boldsymbol{x}_2 \geqq \boldsymbol{0}, \tag{4.142d}$$

where $Var[\tilde{f}]$ denotes the variance of $\tilde{f}$, and V_1 and V_2 are the variance-covariance matrices of $(\tilde{\boldsymbol{c}}_{11}, \tilde{\boldsymbol{c}}_{12})$ and $(\tilde{\boldsymbol{c}}_{21}, \tilde{\boldsymbol{c}}_{22})$, respectively. We assume that V_1 and V_2 are positive-definite without loss of generality.

Under the variance minimization model, the obtained solution might be undesirable in the sense of expectations of objective functions, while it accomplishes the minimization of the variances of them. In order to take account of reduction of the mean of the objective function value, in the V-model, a problem characterized by the elements governed by stochastic events is often formulated such that the variance of the objective function value is minimized under the condition that the mean of the objective function value is smaller than a certain level specified by the decision maker. Let β_1 and β_2 denote such specified levels for the means of the objective function values for DM1 and DM2, respectively. Then, the two-level linear programming problem with random variable coefficients is formulated as

$$\text{minimize } Var[\tilde{c}_{11}\boldsymbol{x}_1 + \tilde{c}_{12}\boldsymbol{x}_2] = \begin{bmatrix}\boldsymbol{x}_1\\ \boldsymbol{x}_2\end{bmatrix}^T V_1 \begin{bmatrix}\boldsymbol{x}_1\\ \boldsymbol{x}_2\end{bmatrix} \tag{4.143a}$$

$$\text{minimize } Var[\tilde{c}_{21}\boldsymbol{x}_1 + \tilde{c}_{22}\boldsymbol{x}_2] = \begin{bmatrix}\boldsymbol{x}_1\\ \boldsymbol{x}_2\end{bmatrix}^T V_2 \begin{bmatrix}\boldsymbol{x}_1\\ \boldsymbol{x}_2\end{bmatrix} \tag{4.143b}$$

$$\text{subject to } A_1\boldsymbol{x}_1 + A_2\boldsymbol{x}_2 \leqq \boldsymbol{K}_{1-\alpha} \tag{4.143c}$$

$$\boldsymbol{m}_{11}\boldsymbol{x}_1 + \boldsymbol{m}_{12}\boldsymbol{x}_2 \leqq \beta_1 \tag{4.143d}$$

$$\boldsymbol{m}_{21}\boldsymbol{x}_1 + \boldsymbol{m}_{22}\boldsymbol{x}_2 \leqq \beta_2 \tag{4.143e}$$

$$\boldsymbol{x}_1 \geqq \boldsymbol{0},\ \boldsymbol{x}_2 \geqq \boldsymbol{0}. \tag{4.143f}$$

4.9.2 Interactive fuzzy programming

After determining the probability limits α_i, $i = 1, \ldots, m$ for constraint violations, to specify the permissible level of the mean of the objective function value, it is helpful to solve the following two individual minimum and maximum of the means of the objective functions for $i = 1, 2$:

$$\text{minimize or maximize } E[\tilde{c}_{i1}\boldsymbol{x}_1 + \tilde{c}_{i2}\boldsymbol{x}_2] = \boldsymbol{m}_{i1}\boldsymbol{x}_1 + \boldsymbol{m}_{i2}\boldsymbol{x}_2 \tag{4.144a}$$

$$\text{subject to } A_1\boldsymbol{x}_1 + A_2\boldsymbol{x}_2 \leqq \boldsymbol{K}_{1-\alpha} \tag{4.144b}$$

$$\boldsymbol{x}_1 \geqq \boldsymbol{0},\ \boldsymbol{x}_2 \geqq \boldsymbol{0}. \tag{4.144c}$$

The decision makers specify the permissible levels β_1 and β_2 with the minimal and the maximal values of the objective functions in problems (4.144) in mind.

As in the two-level linear programming problems considered in the previous section, it is natural for each of the decision makers to have a fuzzy goal for the objective function when they take fuzziness of human judgments into consideration. To identify membership functions of the fuzzy goals for the objective functions, we solve the following two individual minimization variance problems for $i = 1, 2$:

$$\text{minimize} \quad Var[\tilde{\boldsymbol{c}}_{i1}\boldsymbol{x}_1 + \tilde{\boldsymbol{c}}_{i2}\boldsymbol{x}_2] = \begin{bmatrix}\boldsymbol{x}_1\\ \boldsymbol{x}_2\end{bmatrix}^T V_i \begin{bmatrix}\boldsymbol{x}_1\\ \boldsymbol{x}_2\end{bmatrix} \tag{4.145a}$$

$$\text{subject to} \quad A_1\boldsymbol{x}_1 + A_2\boldsymbol{x}_2 \leqq \boldsymbol{K}_{1-\alpha} \tag{4.145b}$$

$$\boldsymbol{x}_1 \geqq \boldsymbol{0},\ \boldsymbol{x}_2 \geqq \boldsymbol{0}. \tag{4.145c}$$

It should be noted that problems (4.145) can be solved by a convex programming method like the sequential quadratic programming method because they are ordinary single-objective convex programming problems due to the nonnegative definiteness of V_i. We use the concise notation $\boldsymbol{x} = (\boldsymbol{x}_1^T, \boldsymbol{x}_2^T)^T \in \mathbb{R}^{n_1+n_2}$ in appropriate cases hereafter.

Let $v_1(\boldsymbol{x})$ and $v_2(\boldsymbol{x})$ denote the objective functions $Var[\tilde{\boldsymbol{c}}_{11}\boldsymbol{x}_1 + \tilde{\boldsymbol{c}}_{12}\boldsymbol{x}_2]$ (4.143a) and $Var[\tilde{\boldsymbol{c}}_{21}\boldsymbol{x}_1 + \tilde{\boldsymbol{c}}_{22}\boldsymbol{x}_2]$ (4.143b), respectively. After identifying the membership functions $\mu_1(v_1(\boldsymbol{x}))$ and $\mu_2(v_2(\boldsymbol{x}))$ for the objective functions $v_1(\boldsymbol{x})$ and $v_2(\boldsymbol{x})$ of the two decision makers, problem (4.143) formulated in the variance minimization model with the permissible levels of the means of the objective function values can be interpreted as the membership function maximization problem defined by

$$\underset{\text{for DM1}}{\text{minimize}} \quad \mu_1(v_1(\boldsymbol{x})) \tag{4.146a}$$

$$\underset{\text{for DM2}}{\text{minimize}} \quad \mu_2(v_2(\boldsymbol{x})) \tag{4.146b}$$

$$\text{subject to} \quad A_1\boldsymbol{x}_1 + A_2\boldsymbol{x}_2 \leqq \boldsymbol{K}_{1-\alpha} \tag{4.146c}$$

$$\boldsymbol{m}_{11}\boldsymbol{x}_1 + \boldsymbol{m}_{12}\boldsymbol{x}_2 \leqq \beta_1 \tag{4.146d}$$

$$\boldsymbol{m}_{21}\boldsymbol{x}_1 + \boldsymbol{m}_{22}\boldsymbol{x}_2 \leqq \beta_2 \tag{4.146e}$$

$$\boldsymbol{x}_1 \geqq \boldsymbol{0},\ \boldsymbol{x}_2 \geqq \boldsymbol{0}. \tag{4.146f}$$

To derive an overall satisfactory solution to the formulated problem (4.146), we first solve the following maximin problem for obtaining a solution which maximizes the smaller degree between the satisfactory degrees of the two decision makers:

$$\text{minimize} \quad \min\{\mu_1(v_1(\boldsymbol{x})), \mu_2(v_2(\boldsymbol{x}))\} \tag{4.147a}$$

$$\text{subject to} \quad \boldsymbol{x} \in S, \tag{4.147b}$$

where S denotes the feasible region of problem (4.146). By introducing an auxiliary variable λ, this problem can be transformed into the following equivalent maximization problem:

$$\text{minimize} \quad -\lambda \tag{4.148a}$$

$$\text{subject to} \quad \mu_1(v_1(\boldsymbol{x})) \geqq \lambda \tag{4.148b}$$

$$\mu_2(v_2(\boldsymbol{x})) \geqq \lambda \tag{4.148c}$$

$$\boldsymbol{x} \in S. \tag{4.148d}$$

Assume that the membership functions $\mu_i(v_i(\boldsymbol{x}))$, $i = 1, 2$ are nonincreasing and concave. Let $g_i(\boldsymbol{x}, \lambda) = -\mu_i(v_i(\boldsymbol{x})) + \lambda$. Because $v_i(\boldsymbol{x})$, $i = 1, 2$ are convex, $g_i(\boldsymbol{x}, \lambda)$,

$i = 1,2$ are also convex. To be more specific, the following inequality holds, for any $\boldsymbol{x}^1, \boldsymbol{x}^2 \in S$, any $\gamma^1, \gamma^2 \in \mathbb{R}$, and any $\alpha \in [0,1]$,

$$\begin{aligned}
& g_i(\alpha \boldsymbol{x}^1 + (1-\alpha)\boldsymbol{x}^2, \alpha\gamma^1 + (1-\alpha)\gamma^2) \\
&= -\mu_i(v_i(\alpha \boldsymbol{x}^1 + (1-\alpha)\boldsymbol{x}^2)) + (\alpha\gamma^1 + (1-\alpha)\gamma^2) \\
&\leqq -\mu_i(\alpha v_i(\boldsymbol{x}^1) + (1-\alpha)v_i(\boldsymbol{x}^2)) + (\alpha\gamma^1 + (1-\alpha)\gamma^2) \\
&\leqq -\alpha\mu_i(v_i(\boldsymbol{x}^1)) - (1-\alpha)\mu_i(v_i(\boldsymbol{x}^2)) + (\alpha\gamma^1 + (1-\alpha)\gamma^2) \\
&= \alpha(-\mu_i(v_i(\boldsymbol{x}^1)) + \gamma^1) + (1-\alpha)(-\mu_i(v_i(\boldsymbol{x}^2)) + \gamma^2) \\
&= \alpha g_i(\boldsymbol{x}^1, \gamma^1) + (1-\alpha) g_i(\boldsymbol{x}^2, \gamma^2).
\end{aligned}$$

Therefore, problem (4.148) can be solved by a conventional convex programming method such as the sequential quadratic programming method.

We formulate the following problem which maximizes DM2's membership function under the condition that DM1's membership function $\mu_1(v_1(\boldsymbol{x}))$ is larger than or equal to the minimal satisfactory level $\hat{\delta}$ specified by DM1:

$$\text{maximize} \quad \mu_2(v_2(\boldsymbol{x})) \tag{4.149a}$$

$$\text{subject to} \quad \mu_1(v_1(\boldsymbol{x})) \geqq \hat{\delta} \tag{4.149b}$$

$$\boldsymbol{x} \in S, \tag{4.149c}$$

which can be also solved by a conventional convex programming method.

For problem (4.146) formulated in the variance minimization model with the permissible levels for the means, we can provide the termination conditions of the interactive process and the procedure for updating the minimal satisfactory level $\hat{\delta}$ which are the same as that of interactive fuzzy programming for the two-level linear programming problem (4.4), and we also give a similar algorithm with problems (4.148) and (4.149) for deriving satisfactory solutions.

4.9.3 Numerical example

In this subsection, we provide an illustrative numerical example for a stochastic two-level linear programming problem to demonstrate the feasibility of the interactive fuzzy programming method described above. Consider the following two-level linear programming problem involving random variable coefficients:

$$\underset{\text{for DM1}}{\text{minimize}}\ z_1(\boldsymbol{x}_1,\boldsymbol{x}_2)=\tilde{\boldsymbol{c}}_{11}\boldsymbol{x}_1+\tilde{\boldsymbol{c}}_{12}\boldsymbol{x}_2 \tag{4.150a}$$

$$\underset{\text{for DM2}}{\text{minimize}}\ z_2(\boldsymbol{x}_1,\boldsymbol{x}_2)=\tilde{\boldsymbol{c}}_{21}\boldsymbol{x}_1+\tilde{\boldsymbol{c}}_{22}\boldsymbol{x}_2 \tag{4.150b}$$

$$\text{subject to}\ \boldsymbol{a}_{11}\boldsymbol{x}_1+\boldsymbol{a}_{12}\boldsymbol{x}_2 \leqq \tilde{b}_1 \tag{4.150c}$$

$$\boldsymbol{a}_{21}\boldsymbol{x}_1+\boldsymbol{a}_{22}\boldsymbol{x}_2 \leqq \tilde{b}_2 \tag{4.150d}$$

$$\boldsymbol{a}_{31}\boldsymbol{x}_1+\boldsymbol{a}_{32}\boldsymbol{x}_2 \leqq \tilde{b}_3 \tag{4.150e}$$

$$\boldsymbol{x}_1=(x_{11},\ldots,x_{15})^T \geqq \boldsymbol{0},\ \boldsymbol{x}_2=(x_{21},\ldots,x_{23})^T \geqq \boldsymbol{0}, \tag{4.150f}$$

where $\tilde{b}_1$, $\tilde{b}_2$, and $\tilde{b}_3$ are Gaussian random variables expressed by $N(220,4^2)$, $N(145,3^2)$, and $N(-18,5^2)$, respectively, where $N(m,s^2)$ stands for a Gaussian random variable with the mean m and the variance s^2. Each element of $\tilde{\boldsymbol{c}}_{i1}$ and $\tilde{\boldsymbol{c}}_{i2}$, $i=1,2$ is also a Gaussian random variable whose mean is shown in Table 4.42, and each element of the coefficient vectors $\boldsymbol{a}_{i1}$ and $\boldsymbol{a}_{i2}$, $i=1,2,3$ is shown in Table 4.43.

Table 4.42 Mean of each element of $\boldsymbol{c}_{ij}$ in problem (4.150).

$\boldsymbol{m}_{11}$	5	2	1	2	1	$\boldsymbol{m}_{12}$	6	1	3
$\boldsymbol{m}_{21}$	10	−7	1	−2	−5	$\boldsymbol{m}_{22}$	3	−4	6

Table 4.43 Each element of $\boldsymbol{a}_{ij}$ in problem (4.150).

$\boldsymbol{a}_{11}$	7	2	6	9	11	$\boldsymbol{a}_{12}$	4	3	8
$\boldsymbol{a}_{21}$	5	6	−4	3	3	$\boldsymbol{a}_{22}$	−7	−1	−3
$\boldsymbol{a}_{31}$	−4	−7	−2	−6	−8	$\boldsymbol{a}_{32}$	−3	−5	−6

Furthermore, the covariance matrices V_l, $l=1,2$ are given as:

$$V_1=\begin{bmatrix} 16.0 & -1.6 & 1.8 & -3.5 & 1.3 & -2.0 & 4.0 & -1.4 \\ -1.6 & 25.0 & -2.2 & 1.6 & -0.7 & 0.5 & -1.3 & 2.0 \\ 1.8 & -2.2 & 25.0 & -2.0 & 5.0 & -2.4 & 1.2 & -2.1 \\ -3.5 & 1.6 & -2.0 & 16.0 & -2.0 & 3.0 & 2.2 & 2.8 \\ 1.3 & -0.7 & 5.0 & -2.0 & 4.0 & -1.0 & 0.8 & -2.0 \\ -2.0 & 0.5 & -2.4 & 3.0 & -1.0 & 1.0 & -1.5 & 0.6 \\ 4.0 & -1.3 & 1.2 & 2.2 & 0.8 & -1.5 & 4.0 & -2.3 \\ -1.4 & 2.0 & -2.1 & 2.8 & -2.0 & 0.6 & -2.3 & 4.0 \end{bmatrix}, \tag{4.151a}$$

$$V_2 = \begin{bmatrix} 4.0 & -1.4 & 0.8 & 0.2 & 1.6 & 1.0 & 1.2 & 2.0 \\ -1.4 & 4.0 & 0.2 & -1.0 & -2.2 & 0.8 & 0.9 & 1.8 \\ 0.8 & 0.2 & 9.0 & 0.2 & -1.5 & 1.5 & 1.0 & 0.6 \\ 0.2 & -1.0 & 0.2 & 36.0 & 0.8 & 0.4 & -1.5 & 0.7 \\ 1.6 & -2.2 & -1.5 & 0.8 & 25.0 & 1.2 & -0.2 & 2.0 \\ 1.0 & 0.8 & 1.5 & 0.4 & 1.2 & 25.0 & 0.5 & 1.4 \\ 1.2 & 0.9 & 1.0 & -1.5 & -0.2 & 0.5 & 9.0 & 0.8 \\ 2.0 & 1.8 & 0.6 & 0.7 & 2.0 & 1.4 & 0.8 & 16.0 \end{bmatrix}. \tag{4.151b}$$

Suppose that the decision makers specify the probabilities α_i, $i = 1,2,3$ of the extent to which the constraint violations are admitted at $\alpha_1 = 0.85$, $\alpha_2 = 0.95$, and $\alpha_3 = 0.90$, and the corresponding $K_{1-\alpha_i}$, $i = 1,2,3$ are calculated as $K_{1-\alpha_1} = 215.85$, $K_{1-\alpha_2} = 140.07$, and $K_{1-\alpha_3} = 24.41$. After solving problems (4.144) corresponding to the numerical example (4.150), suppose that they specify the permissible levels for the means of the objective function values at $\beta_1 = 60.0$ and $\beta_2 = 35.0$ by taking account of the minimal and the maximal values of the objective functions in problems (4.144): $E[\tilde{\boldsymbol{c}}_{11}\boldsymbol{x}_1 + \tilde{\boldsymbol{c}}_{12}\boldsymbol{x}_2]^{\min} = 3.053$, $E[\tilde{\boldsymbol{c}}_{21}\boldsymbol{x}_1 + \tilde{\boldsymbol{c}}_{22}\boldsymbol{x}_2]^{\min} = -425.621$, $E[\tilde{\boldsymbol{c}}_{11}\boldsymbol{x}_1 + \tilde{\boldsymbol{c}}_{12}\boldsymbol{x}_2]^{\max} = 323.784$, and $E[\tilde{\boldsymbol{c}}_{21}\boldsymbol{x}_1 + \tilde{\boldsymbol{c}}_{22}\boldsymbol{x}_2]^{\max} = 304.479$. Then, the following problem is formulated as:

$$\underset{\text{for DM1}}{\text{minimize}} \quad v_1(\boldsymbol{x}) = Var[\tilde{\boldsymbol{c}}_{11}\boldsymbol{x}_1 + \tilde{\boldsymbol{c}}_{12}\boldsymbol{x}_2] = \begin{bmatrix} \boldsymbol{x}_1 \\ \boldsymbol{x}_2 \end{bmatrix}^T V_1 \begin{bmatrix} \boldsymbol{x}_1 \\ \boldsymbol{x}_2 \end{bmatrix} \tag{4.152a}$$

$$\underset{\text{for DM2}}{\text{minimize}} \quad v_2(\boldsymbol{x}) = Var[\tilde{\boldsymbol{c}}_{21}\boldsymbol{x}_1 + \tilde{\boldsymbol{c}}_{22}\boldsymbol{x}_2] = \begin{bmatrix} \boldsymbol{x}_1 \\ \boldsymbol{x}_2 \end{bmatrix}^T V_2 \begin{bmatrix} \boldsymbol{x}_1 \\ \boldsymbol{x}_2 \end{bmatrix} \tag{4.152b}$$

$$\text{subject to} \quad \boldsymbol{a}_{11}\boldsymbol{x}_1 + \boldsymbol{a}_{12}\boldsymbol{x}_2 \leqq 215.85 \tag{4.152c}$$

$$\boldsymbol{a}_{21}\boldsymbol{x}_1 + \boldsymbol{a}_{22}\boldsymbol{x}_2 \leqq 140.07 \tag{4.152d}$$

$$\boldsymbol{a}_{31}\boldsymbol{x}_1 + \boldsymbol{a}_{32}\boldsymbol{x}_2 \leqq 24.41 \tag{4.152e}$$

$$\boldsymbol{m}_{11}\boldsymbol{x}_1 + \boldsymbol{m}_{12}\boldsymbol{x}_2 \leqq 60.0 \tag{4.152f}$$

$$\boldsymbol{m}_{21}\boldsymbol{x}_1 + \boldsymbol{m}_{22}\boldsymbol{x}_2 \leqq 35.0 \tag{4.152g}$$

$$\boldsymbol{x}_1 = (x_{11}, \ldots, x_{15})^T \geqq \boldsymbol{0}, \; \boldsymbol{x}_2 = (x_{21}, \ldots, x_{23})^T \geqq \boldsymbol{0}. \tag{4.152h}$$

Moreover, suppose that the decision makers employ the following linear membership function whose parameters are determined by the Zimmermann method (1978):

$$\mu_i(v_i(\boldsymbol{x})) = \begin{cases} 0 & \text{if } v_i(\boldsymbol{x}) > v_i^0 \\ \dfrac{v_i(\boldsymbol{x}) - v_i^0}{v_i^1 - v_i^0} & \text{if } v_i^1 < v_i(\boldsymbol{x}) \leqq v_i^0 \\ 1 & \text{if } v_i(\boldsymbol{x}) \leqq v_i^1, \end{cases}$$

where $v_1^1 = 0.818$, $v_1^0 = v_1^{\mathrm{m}} = 129.623$, $v_2^1 = 19.524$, and $v_2^0 = v_2^{\mathrm{m}} = 206.620$. Then, the maximin problem (4.54) for this numerical example can be formulated as

$$\begin{aligned}
&\text{maximize } \lambda && (4.153a)\\
&\text{subject to } (v_1(\boldsymbol{x}) - 129.623)/(0.818 - 129.623) \geqq \lambda && (4.153b)\\
&\qquad (v_2(\boldsymbol{x}) - 206.620)/(19.524 - 206.620) \geqq \lambda && (4.153c)\\
&\qquad \boldsymbol{x} \in S, && (4.153d)
\end{aligned}$$

where S denotes the feasible region of problem (4.152). The result of the first iteration including an optimal solution to problem (4.153) is shown in Table 4.44, where m_1^1 and m_2^1 denote the means $E[\tilde{\boldsymbol{c}}_{11}\boldsymbol{x}_1 + \tilde{\boldsymbol{c}}_{12}\boldsymbol{x}_2]$ and $E[\tilde{\boldsymbol{c}}_{21}\boldsymbol{x}_1 + \tilde{\boldsymbol{c}}_{22}\boldsymbol{x}_2]$ of objective functions, respectively.

Table 4.44 Iteration 1.

$\boldsymbol{x}_1^1$	0.410	0.632	0.067	0.270	0.801
$\boldsymbol{x}_2^1$	0.276	0.973	0.749		
mean: (m_1^1, m_2^1)	$(9.598, -3.372)$				
variance: (v_1^1, v_2^1)	$(20.654, 48.355)$				
membership value: $(\mu_1(v_1^1), \mu_2(v_2^1))$	$(0.846, 0.846)$				
λ^1	0.846				
Δ^1	1.000				

Suppose that DM1 is not satisfied with the solution obtained in Iteration 1, and then he specifies the minimal satisfactory level at $\hat{\delta} = 0.9$ and the bounds of the ratio at the interval $[\Delta_{\min}, \Delta_{\max}] = [0.65, 0.75]$, taking account of the result of Iteration 1. Then, the problem with the minimal satisfactory level (4.149) is formulated as

$$\begin{aligned}
&\text{maximize } \mu_2(v_2(\boldsymbol{x})) && (4.154a)\\
&\text{subject to } (v_1(\boldsymbol{x}) - 129.623)/(0.818 - 129.623) \geqq 0.9 && (4.154b)\\
&\qquad \boldsymbol{x} \in S. && (4.154c)
\end{aligned}$$

The result of the second iteration including an optimal solution to problem (4.154) is shown in Table 4.45.

Table 4.45 Iteration 2.

$\boldsymbol{x}_1^2$	0.264	0.459	0.034	0.200	0.876
$\boldsymbol{x}_2^2$	0.366	1.064	0.907		
mean: (m_1^2, m_2^2)	$(9.529, -3.035)$				
variance: (v_1^2, v_2^2)	$(13.699, 57.881)$				
membership value: $(\mu_1(v_1^2), \mu_2(v_2^2))$	$(0.900, 0.795)$				
λ^2	0.795				
Δ^2	0.884				

At Iteration 2, the satisfactory degree $\mu_1^2 = 0.900$ of DM1 becomes equal to the minimal satisfactory level $\hat{\delta} = 0.900$, but the ratio $\Delta^2 = 0.884$ of satisfactory degrees is not in the valid interval $[0.65, 0.75]$ of the ratio. Therefore, this solution does not satisfy the second condition of termination of the interactive procedure. Suppose that DM1 updates the minimal satisfactory level at $\hat{\delta} = 0.95$. Then, the problem with the revised minimal satisfactory level (4.149) is solved, and the result of the third iteration is shown in Table 4.46.

Table 4.46 Iteration 3.

$\boldsymbol{x}_1^3$	0.092	0.285	0.000	0.053	0.946
$\boldsymbol{x}_2^3$	0.538	1.193	1.097		
mean: (m_1^3, m_2^3)	$(9.794, -2.487)$				
variance: (v_1^3, v_2^3)	$(7.258, 73.170)$				
membership value: $(\mu_1(v_1^3), \mu_2(v_2^3))$	$(0.950, 0.714)$				
λ^3	0.714				
Δ^3	0.751				

At Iteration 3, the ratio $\Delta^3 = 0.751$ of satisfactory degrees is slightly larger than the upper bound of the valid interval $[0.65, 0.75]$ of the ratio. Therefore, this solution does not satisfy the second condition of termination of the interactive procedure. Suppose that DM1 raises the minimal satisfactory level a little bit to $\hat{\delta} = 0.96$. Then, the problem with the revised minimal satisfactory level (4.149) is solved, and the result of the fourth iteration is given in Table 4.47.

Table 4.47 Iteration 4.

$\boldsymbol{x}_1^4$	0.053	0.251	0.000	0.007	0.950
$\boldsymbol{x}_2^4$	0.599	1.232	1.140		
mean: (m_1^4, m_2^4)	$(9.977, -2.282)$				
variance: (v_1^4, v_2^4)	$(5.970, 77.725)$				
membership value: $(\mu_1(v_1^4), \mu_2(v_2^4))$	$(0.960, 0.689)$				
λ^4	0.689				
Δ^4	0.718				

At Iteration 4, the satisfactory degree $\mu_1^4 = 0.960$ of DM1 becomes equal to the minimal satisfactory level $\hat{\delta} = 0.96$, and the ratio $\Delta^4 = 0.718$ of satisfactory degrees is in the valid interval $[0.65, 0.75]$ of the ratio. Therefore, this solution satisfies the termination conditions of the interactive procedure and becomes a satisfactory solution for both decision makers if DM1 accepts the solution.

4.9.4 Alternative stochastic models

In this subsection, we give a summary of the maximum probability model (Charnes and Cooper, 1963) and the fractile criterion model (Kataoka, 1963; Geoffrion, 1967) for two-level linear programming problems with random variables.

Maximum probability model for stochastic two-level linear programming problems

In the maximum probability model referred to as the P-model, we restrict representation of the random variable coefficients $\tilde{\boldsymbol{c}}_{ij} = (\tilde{c}_{ij1}, \ldots, \tilde{c}_{ijn_j})$ in the objective functions to a specific form in order to reduce difficulties of computation. To be more specific, assume that the random variables are represented by

$$\tilde{c}_{ijk_j} = c_{ijk_j} + \bar{c}_{ijk_j}\tilde{t}_i,\ i=1,2,\ j=1,2,\ k_j=1,\ldots,n_j, \tag{4.155}$$

where c_{ijk} and $\bar{c}_{ijk}$ are constants, and $\tilde{t}_i$ is a random variable, and the two random variables $\tilde{t}_1$ and $\tilde{t}_2$ are mutually independent; thus, it is supposed that each objective function value is governed by a single and independent stochastic event. Let T_i denote a distribution function of the random variable $\tilde{t}_i$, and it is assumed that T_i is a continuous and strictly monotonically increasing function.

Then, a stochastic two-level linear programming problem is represented as

$$\underset{\text{for DM1}}{\text{minimize}}\ z_1(\boldsymbol{x}_1,\boldsymbol{x}_2;\tilde{t}_1) = (\boldsymbol{c}_{11}+\bar{\boldsymbol{c}}_{11}\tilde{t}_1)\boldsymbol{x}_1 + (\boldsymbol{c}_{12}+\bar{\boldsymbol{c}}_{12}\tilde{t}_1)\boldsymbol{x}_2 \tag{4.156a}$$

$$\underset{\text{for DM2}}{\text{minimize}}\ z_2(\boldsymbol{x}_1,\boldsymbol{x}_2;\tilde{t}_2) = (\boldsymbol{c}_{21}+\bar{\boldsymbol{c}}_{21}\tilde{t}_2)\boldsymbol{x}_1 + (\boldsymbol{c}_{22}+\bar{\boldsymbol{c}}_{22}\tilde{t}_2)\boldsymbol{x}_2 \tag{4.156b}$$

$$\text{subject to}\ \boldsymbol{x} \in S, \tag{4.156c}$$

where S is a feasible region with the chance constraints.

In the P-model, the probability that the objective function value is smaller than a ceratin target value is maximized, and then the two-level programming problem in the P-model corresponding to problem (4.156) is represented as

$$\underset{\text{for DM1}}{\text{minimize}}\ P[(\boldsymbol{c}_{11}+\bar{\boldsymbol{c}}_{11}\tilde{t}_1)\boldsymbol{x}_1 + (\boldsymbol{c}_{12}+\bar{\boldsymbol{c}}_{12}\tilde{t}_1)\boldsymbol{x}_2 \leqq f_1] \tag{4.157a}$$

$$\underset{\text{for DM2}}{\text{minimize}}\ P[(\boldsymbol{c}_{21}+\bar{\boldsymbol{c}}_{21}\tilde{t}_2)\boldsymbol{x}_1 + (\boldsymbol{c}_{22}+\bar{\boldsymbol{c}}_{22}\tilde{t}_1)\boldsymbol{x}_2 \leqq f_2] \tag{4.157b}$$

$$\text{subject to}\ \boldsymbol{x} \in S, \tag{4.157c}$$

where f_1 and f_2 are the target values for the objective functions of DM1 and DM2, respectively.

Additionally, we assume that $\bar{\boldsymbol{c}}_{11}\boldsymbol{x}_1 + \bar{\boldsymbol{c}}_{12}\boldsymbol{x}_2 > 0$ for all $\boldsymbol{x} \in S$. From the assumption of the distribution function T_i of the random variable $\tilde{t}_i$, one finds the followings:

$$
\begin{aligned}
& P\left[(\boldsymbol{c}_{i1}+\bar{\boldsymbol{c}}_{i1}\tilde{t}_i)\boldsymbol{x}_1+(\boldsymbol{c}_{i2}+\bar{\boldsymbol{c}}_{i2}\tilde{t}_i)\boldsymbol{x}_2 \leqq f_i\right] \\
&= P\left[\tilde{t}_i \leqq \frac{f_i-(\boldsymbol{c}_{i1}\boldsymbol{x}_1+\boldsymbol{c}_{i2}\boldsymbol{x}_2)}{\bar{\boldsymbol{c}}_{i1}\boldsymbol{x}_1+\bar{\boldsymbol{c}}_{i2}\boldsymbol{x}_2}\right] \\
&= T_i\left(\frac{f_i-(\boldsymbol{c}_{i1}\boldsymbol{x}_1+\boldsymbol{c}_{i2}\boldsymbol{x}_2)}{\bar{\boldsymbol{c}}_{i1}\boldsymbol{x}_1+\bar{\boldsymbol{c}}_{i2}\boldsymbol{x}_2}\right).
\end{aligned}
$$

Because maximizing $T_i\left(\frac{f_i-(\boldsymbol{c}_{i1}+\boldsymbol{c}_{i2})}{\bar{\boldsymbol{c}}_{i1}+\bar{\boldsymbol{c}}_{i2}}\right)$ is equivalent to maximizing $\frac{f_i-(\boldsymbol{c}_{i1}+\boldsymbol{c}_{i2})}{\bar{\boldsymbol{c}}_{i1}+\bar{\boldsymbol{c}}_{i2}}$, problem (4.157) can be transformed into

$$
\begin{aligned}
& \underset{\text{for DM1}}{\text{maximize}} \quad \frac{f_1-(\boldsymbol{c}_{11}\boldsymbol{x}_1+\boldsymbol{c}_{12}\boldsymbol{x}_2)}{\bar{\boldsymbol{c}}_{11}\boldsymbol{x}_1+\bar{\boldsymbol{c}}_{12}\boldsymbol{x}_2} && (4.158a) \\
& \underset{\text{for DM2}}{\text{maximize}} \quad \frac{f_2-(\boldsymbol{c}_{21}\boldsymbol{x}_1+\boldsymbol{c}_{22}\boldsymbol{x}_2)}{\bar{\boldsymbol{c}}_{21}\boldsymbol{x}_1+\bar{\boldsymbol{c}}_{22}\boldsymbol{x}_2} && (4.158b) \\
& \text{subject to} \quad \boldsymbol{x} \in S. && (4.158c)
\end{aligned}
$$

Then, problem (4.158) is reduced to a two-level linear fractional programming problem (4.50) dealt with in section 4.4, and we can directly apply the interactive programming method for two-level linear fractional programming problems to the two-level programming problem (4.158) in the P-model.

Fractile criterion model for stochastic two-level linear programming problems

The fractile criterion model is considered as complementary to the P-model; a target variable to the objective function is minimized under the condition that the probability that the objective function value is smaller than the target variable is larger than a given assurance level. Then, a stochastic two-level linear programming problem is represented as

$$
\begin{aligned}
& \underset{\text{for DM1}}{\text{minimize}} \quad f_1 && (4.159a) \\
& \underset{\text{for DM2}}{\text{minimize}} \quad f_2 && (4.159b) \\
& \text{subject to} \quad P\left[\tilde{\boldsymbol{c}}_{11}\boldsymbol{x}_1+\tilde{\boldsymbol{c}}_{12}\boldsymbol{x}_2 \leqq f_1\right] \geqq \alpha_1 && (4.159c) \\
& \qquad\qquad\quad P\left[\tilde{\boldsymbol{c}}_{21}\boldsymbol{x}_1+\tilde{\boldsymbol{c}}_{22}\boldsymbol{x}_2 \leqq f_2\right] \geqq \alpha_2 && (4.159d) \\
& \qquad\qquad\quad \boldsymbol{x} \in S, && (4.159e)
\end{aligned}
$$

where f_1 and f_2 are the target variables to the objective functions of DM1 and DM2, respectively; α_1 and α_2 are the assurance levels specified by DM1 and DM2, respectively, for the probabilities that the objective function values are smaller than the target variables. Furthermore, we assume that $\tilde{\boldsymbol{c}}_{ij}$, $i=1,2$, $j=1,2$ are n_j-dimensional Gaussian random variable coefficient row vectors; and let $\bar{c}_{ijk}$ denote the mean of the Gaussian random variable $\tilde{c}_{ijk_j}$, $i=1,2$, $j=1,2$, $k_j=1,\dots,n_j$.

From the assumption of the random variable $\tilde{c}_{ij}$, the constraint (4.159c) or the constraint (4.159d)

$$P\left[\tilde{\boldsymbol{c}}_{i1}\boldsymbol{x}_1+\tilde{\boldsymbol{c}}_{i2}\boldsymbol{x}_2 \leqq f_i\right] \geqq \alpha_i \tag{4.160}$$

is transformed into

$$P\left[\frac{(\tilde{\boldsymbol{c}}_{i1}\boldsymbol{x}_1+\tilde{\boldsymbol{c}}_{i2}\boldsymbol{x}_2)-(\bar{\boldsymbol{c}}_{i1}\boldsymbol{x}_1+\bar{\boldsymbol{c}}_{i2}\boldsymbol{x}_2)}{\sqrt{\begin{bmatrix}\boldsymbol{x}_1\\ \boldsymbol{x}_2\end{bmatrix}^T V_i \begin{bmatrix}\boldsymbol{x}_1\\ \boldsymbol{x}_2\end{bmatrix}}} \leqq \frac{f_i-(\bar{\boldsymbol{c}}_{i1}\boldsymbol{x}_1+\bar{\boldsymbol{c}}_{i2}\boldsymbol{x}_2)}{\sqrt{\begin{bmatrix}\boldsymbol{x}_1\\ \boldsymbol{x}_2\end{bmatrix}^T V_i \begin{bmatrix}\boldsymbol{x}_1\\ \boldsymbol{x}_2\end{bmatrix}}}\right] \geqq \alpha_i, \tag{4.161}$$

where V_i is the variance-covariance matrix of $(\tilde{\boldsymbol{c}}_{i1},\tilde{\boldsymbol{c}}_{i2})$. Because the left-hand side of the inequality in the square bracket of the probability $P[\cdot]$ in (4.161) is the standard Gaussian random variable where the mean is 0 and the variance is 1, the inequality (4.161) is also transformed into

$$\Phi\left(\frac{f_i-(\bar{\boldsymbol{c}}_{i1}\boldsymbol{x}_1+\bar{\boldsymbol{c}}_{i2}\boldsymbol{x}_2)}{\sqrt{\begin{bmatrix}\boldsymbol{x}_1\\ \boldsymbol{x}_2\end{bmatrix}^T V_i \begin{bmatrix}\boldsymbol{x}_1\\ \boldsymbol{x}_2\end{bmatrix}}}\right) \geqq \alpha_i \tag{4.162a}$$

$$\Longleftrightarrow \frac{f_i-(\bar{\boldsymbol{c}}_{i1}\boldsymbol{x}_1+\bar{\boldsymbol{c}}_{i2}\boldsymbol{x}_2)}{\sqrt{\begin{bmatrix}\boldsymbol{x}_1\\ \boldsymbol{x}_2\end{bmatrix}^T V_i \begin{bmatrix}\boldsymbol{x}_1\\ \boldsymbol{x}_2\end{bmatrix}}} \geqq K_{\alpha_i} \tag{4.162b}$$

$$\Longleftrightarrow f_i \geqq \bar{\boldsymbol{c}}_{i1}\boldsymbol{x}_1+\bar{\boldsymbol{c}}_{i2}\boldsymbol{x}_2+K_{\alpha_i}\sqrt{\begin{bmatrix}\boldsymbol{x}_1\\ \boldsymbol{x}_2\end{bmatrix}^T V_i \begin{bmatrix}\boldsymbol{x}_1\\ \boldsymbol{x}_2\end{bmatrix}}, \tag{4.162c}$$

where $\Phi(\cdot)$ is the distribution function of the standard Gaussian random variable, and $K_{\alpha_i}=\Phi^{-1}(\alpha_i)$.

Furthermore, because the minimum of f_i is attained at equality in (4.162c), i.e.,

$$f_i = \bar{\boldsymbol{c}}_{i1}\boldsymbol{x}_1+\bar{\boldsymbol{c}}_{i2}\boldsymbol{x}_2+K_{\alpha_i}\sqrt{\begin{bmatrix}\boldsymbol{x}_1\\ \boldsymbol{x}_2\end{bmatrix}^T V_i \begin{bmatrix}\boldsymbol{x}_1\\ \boldsymbol{x}_2\end{bmatrix}}, \tag{4.163}$$

eventually, problem (4.159) is transformed into

$$\underset{\text{for DM1}}{\text{minimize}}\ \zeta_1(\boldsymbol{x}_1,\boldsymbol{x}_2)=\bar{\boldsymbol{c}}_{11}\boldsymbol{x}_1+\bar{\boldsymbol{c}}_{12}\boldsymbol{x}_2+K_{\alpha_1}\sqrt{\begin{bmatrix}\boldsymbol{x}_1\\ \boldsymbol{x}_2\end{bmatrix}^T V_1 \begin{bmatrix}\boldsymbol{x}_1\\ \boldsymbol{x}_2\end{bmatrix}} \tag{4.164a}$$

$$\underset{\text{for DM2}}{\text{minimize}}\ \zeta_2(\boldsymbol{x}_1,\boldsymbol{x}_2)=\bar{\boldsymbol{c}}_{21}\boldsymbol{x}_1+\bar{\boldsymbol{c}}_{22}\boldsymbol{x}_2+K_{\alpha_2}\sqrt{\begin{bmatrix}\boldsymbol{x}_1\\ \boldsymbol{x}_2\end{bmatrix}^T V_2 \begin{bmatrix}\boldsymbol{x}_1\\ \boldsymbol{x}_2\end{bmatrix}} \tag{4.164b}$$

$$\text{subject to } \boldsymbol{x} \in S. \tag{4.164c}$$

Employing an algorithm similar to that of two-level linear programming problems given in section 4.2, we solve the maximin problem

$$\begin{aligned}
&\text{minimize} \quad -\lambda && (4.165a)\\
&\text{subject to} \quad \lambda - \mu_1(\zeta_1(\boldsymbol{x}_1,\boldsymbol{x}_2)) \leqq 0 && (4.165b)\\
&\qquad \lambda - \mu_2(\zeta_2(\boldsymbol{x}_1,\boldsymbol{x}_2)) \leqq 0 && (4.165c)\\
&\qquad \boldsymbol{x} \in S, && (4.165d)
\end{aligned}$$

and the problem with the minimal satisfactory level

$$\begin{aligned}
&\text{minimize} \quad -\mu_2(\zeta_2(\boldsymbol{x}_1,\boldsymbol{x}_2)) && (4.166a)\\
&\text{subject to} \quad \hat{\delta} - \mu_2(\zeta_2(\boldsymbol{x}_1,\boldsymbol{x}_2)) \leqq 0 && (4.166b)\\
&\qquad \boldsymbol{x} \in S. && (4.166c)
\end{aligned}$$

Because problems (4.165) and (4.166) are convex problems when μ_i is a monotonically increasing concave function, they can be solved by a conventional convex programming method such as the sequential quadratic programming method.

Chapter 5
Some applications

In this chapter, we apply cooperative and noncooperative decision making methods to real-world decision making problems in decentralized organizations. First, we deal with decision making problems on production and work force assignment in a housing material manufacturer and a subcontract company. We formulate two kinds of two-level programming problems: one is a profit maximization problem of both the housing material manufacturer and the subcontract company, and the other is a profitability maximization problem of them. Applying the interactive fuzzy programming methods for two-level linear and linear fractional programming problems, we derive satisfactory solutions to the problems. Second, we treat a transportation problem in the housing material manufacturer and derive a satisfactory solution to the problem by taking into account not only the degree of satisfaction with respect to objectives of the housing material manufacturer but also those of two forwarding agents to which the housing material manufacturer entrusts transportation of products. Third, as a noncooperative decision making problem, we deal with a purchase problem for food retailing, and formulate a two-level linear programming problem with a food retailer and a distributer. The food retailer deals with vegetables and fruits which are purchased from the distributer; the distributer buys vegetables and fruits ordered from the food retailer at the central wholesale markets in several cities, and transports them by truck from each of the central wholesaler markets to the food retailer's storehouse. We compute the Stackelberg solution to the two-level linear programming problem in which the profits of the food retailer and the distributer are maximized.

5.1 Two-level production and work force assignment problem

A housing material manufacturer usually places orders with other companies for processing or making some parts of its products, and then it finishes the end products in its own factories. Such a production procedure is often employed in many industries. In general, a company tries to maximize or minimize its profit, profitabil-

M. Sakawa and I. Nishizaki (eds.), *Cooperative and Noncooperative Multi-Level Programming*, Operations Research/Computer Science Interfaces Series 48, DOI: 10.1007/978-1-4419-0676-2_5,

ity, or production cost under some conditions such as resource conditions. However, when a parent company orders to a subsidiary or there exists a business relationship extending over a long period of time between the two companies, it is natural that the two companies not only optimize their own objectives but also make decisions cooperatively by balancing satisfaction of one company with that of the other.

In this section, we consider production and work force assignment problems in the housing material manufacturer and the company contracting to process some parts of the products of the housing material manufacturer (Sakawa, Nishizaki and Uemura, 2001). Because the housing material manufacturer has had dealings with the subcontracting company in the long term, decisions of the two companies are made cooperatively. For simplicity, we call the housing material manufacturer the upper level company and the company contracting to process the parts the lower level company hereafter.

The upper level company determines the production numbers of goods and orders processing some parts to the lower level company in compliance with the production planning. The lower level company employs some workers who are classified into several skill stages according to work experience, and assigns them tasks ordered by the upper level company. We formulate this decision making problem as two-level programming problems.

The concept of Stackelberg solutions has been employed as a solution concept when decision making problems are modeled as two-level programming problems (Bard and Falk, 1982; Bard, 1983a; Bialas and Karwan, 1984; White and Anandalingam, 1993; Shimizu, Ishizuka and Bard, 1997), whether there is a cooperative relationship between the decision makers or not. It should be noted that Stackelberg solutions do not always satisfy Pareto optimality because of their noncooperative nature.

Due to the cooperative relations between the two companies, we do not formulate the production and work force assignment problem of the two companies as a Stackelberg problem, in which the upper level company first decides the production numbers of the goods, and then the lower level company determines the assignment of the workers with full knowledge of the production numbers specified by the upper level company in a noncooperative manner, but employ the interactive fuzzy programming shown in chapter 4 to derive a satisfactory solution in which the satisfactory degrees of the two companies are taken into account simultaneously (Lai, 1996; Shih, Lai and Lee, 1996; Sakawa, Nishizaki and Uemura, 1998, 2000a; Sakawa and Nishizaki, 2001a; Sakawa, Nishizaki and Uemura, 2000b).

First, we formulate a two-level linear programming problem in which both of the upper level and the lower level companies maximize their profits, and derive a satisfactory solution satisfying the minimal satisfactory level of the upper level company and maximizing the satisfactory degree of the lower level company by applying the interactive fuzzy programming for two-level linear programming problems described in section 4.2 of chapter 4 to the production and work force assignment problems. Second, formulating a two-level linear fractional programming problem in which the two companies maximize the profitability represented by the ratio of profit to sales, we apply the interactive fuzzy programming for two-level linear fractional programming problems described in section 4.4 of chapter 4 to the profitability maximization problem. Finally, after comparing the two problems, we discuss the results of the applications, and examine actual planning of the production and work forth assignment of the two companies to be implemented.

5.1.1 Problem formulation

The upper level company determines the production numbers of m kinds of goods and orders processing some parts of the goods to the lower level company. The lower level company assigns workers, who are divided into the n skill stages, the jobs ordered by the upper level company. In the upper level company, a standard output X_i, $i = 1,\ldots,m$ of each product is determined by the company's policy, and the production numbers of the goods must be within the limits of possible outputs corresponding the standard outputs. For given α and β, $0 \leqq \alpha, \beta \leqq 1$, the output x_i of product i, $i = 1,\ldots,m$ is an integer, and it is constrained as follows:

$$(1-\beta)X_i \leqq x_i \leqq (1+\alpha)X_i,\ i = 1,\ldots,m. \tag{5.1}$$

Let a_i and b_i denote a finishing cost of product i at the upper level company and its processing cost at the lower level company, respectively. The upper limits of the costs, which mean budgets of the two companies, are $\sum_{i=1}^{m} a_i X_i$ and $\sum_{i=1}^{m} b_i X_i$, and the constraints on the budgets are represented by

$$\sum_{i=1}^{m} a_i x_i \leqq \sum_{i=1}^{m} a_i X_i, \tag{5.2a}$$

$$\sum_{i=1}^{m} b_i x_i \leqq \sum_{i=1}^{m} b_i X_i. \tag{5.2b}$$

The lower level company receives orders from the upper level company, assigns the workers who treat the parts, and then earns an income $\sum_{i=1}^{m} b_i x_i$. It is assumed that there are the n skill stages of workers according to the level of work experience. The skill of a worker becomes greater as the number of stage increases, and thus, skill stage n is the most skillful stage. The difference of skill between two stages can be found clearly in processing high-grade products, but any worker has almost equal ability in processing low-grade products. Let d_j denote a labor cost of a worker in skill stage j, and the greater the labor cost grows, the higher the skill level becomes. Let y_j denote the number of assigned workers in skill stage j, and e_{ij} the ability of processing one unit of product i by a worker in skill stage j. To meet the orders from the upper level company, the lower level company must assign workers so as to satisfy the condition:

$$\sum_{j=1}^{n} e_{ij} y_j \geqq x_i,\ i = 1,\cdots,m. \tag{5.3}$$

The number of assigned workers is not necessary to be an integral number because the labor force corresponding to the fractional portion can be supplemented with part time jobs and overtime works. From the restriction on employment of the lower level company, there exist the upper and the lower limits Y_j^L and Y_j^U of the number of workers in each skill stage and the upper limit Y of assignable total workers. Therefore, the numbers of assigned workers are constrained as follows:

$$Y_j^L \leqq y_j \leqq Y_j^U,\ j = 1,\ldots,n, \tag{5.4a}$$

$$\sum_{j=1}^{n} y_j \leqq Y. \tag{5.4b}$$

We first consider the two-level linear programming problem in which both of the upper level and the lower level companies maximize their profits. Let c_i denote a price of one unit of product i. Its profit of one unit is $c_i - a_i - b_i$, and the gross profits of the upper level and the lower level companies become $\sum_{i=1}^{m}(c_i - a_i - b_i)x_i$ and $\sum_{i=1}^{m} b_i x_i - \sum_{j=1}^{n} d_j y_j$, respectively. Then, the two-level linear programming problem can be formulated as:

$$\underset{\text{for upper level company}}{\text{maximize}} \quad z_1(\boldsymbol{x},\boldsymbol{y}) = \sum_{i=1}^{m}(c_i - a_i - b_i)x_i \tag{5.5a}$$

$$\underset{\text{for lower level company}}{\text{maximize}} \quad z_2(\boldsymbol{x},\boldsymbol{y}) = \sum_{i=1}^{m} b_i x_i - \sum_{j=1}^{n} d_j y_j \tag{5.5b}$$

$$\text{subject to} \quad \sum_{i=1}^{m} a_i x_i \leqq \sum_{i=1}^{m} a_i X_i \tag{5.5c}$$

$$\sum_{i=1}^{m} b_i x_i \leqq \sum_{i=1}^{m} b_i X_i \tag{5.5d}$$

$$\sum_{j=1}^{n} e_{ij} y_j \geqq x_i,\ i = 1,\ldots,m \tag{5.5e}$$

$$(1-\beta)X_i \leqq x_i \leqq (1+\alpha)X_i,\ i = 1,\ldots,m \tag{5.5f}$$

$$Y_j^L \leqq y_j \leqq Y_j^U,\ j = 1,\ldots,n \tag{5.5g}$$

$$\sum_{j=1}^{n} y_j \leqq Y. \tag{5.5h}$$

For the sake of simplicity, the feasible region of problem (5.5) is denoted by S hereafter.

Profitability is also an important index of business efficiency as well as profit. The profitability of the two companies is represented by the ratios of profit to sales:

$$f_1(\boldsymbol{x},\boldsymbol{y}) = \Big(\sum_{i=1}^{m}(c_i - b_i - a_i)x_i\Big) \Big/ \Big(\sum_{i=1}^{m} c_i x_i\Big), \tag{5.6a}$$

$$f_2(\boldsymbol{x},\boldsymbol{y}) = \Big(\sum_{i=1}^{m} b_i x_i - \sum_{j=1}^{n} d_j y_j\Big) \Big/ \Big(\sum_{i=1}^{m} b_i x_i\Big). \tag{5.6b}$$

Then, the following two-level linear fractional programming problem is formulated when both of the upper level and the lower level companies maximize their profitability:

$$\underset{\text{for upper level company}}{\text{maximize}} \quad f_1(\boldsymbol{x},\boldsymbol{y}) = \frac{\sum_{i=1}^{m}(c_i - b_i - a_i)x_i}{\sum_{i=1}^{m} c_i x_i} \tag{5.7a}$$

$$\underset{\text{for lower level company}}{\text{maximize}} \quad f_2(\boldsymbol{x},\boldsymbol{y}) = \frac{\sum_{i=1}^{m} b_i x_i - \sum_{j=1}^{n} d_j y_j}{\sum_{i=1}^{m} b_i x_i} \tag{5.7b}$$

$$\text{subject to} \quad (\boldsymbol{x},\boldsymbol{y}) \in S. \tag{5.7c}$$

5.1.2 Maximization of profit

In this subsection, we examine the production and work force assignment problem of the two companies from the viewpoint of profit maximization. The examination starts with maximization of individual objective functions under the constraints described in the previous subsection, and then the fuzzy interactive programming is applied to the two-level linear production and work force assignment problem. The data of the problem are shown in Table 5.1. All the products are divided into five grades: high, semi-high, medium, semi-low, and low. The grades of products are also shown in Table 5.1, where high, semi-high, medium, semi-low, and low grades are denoted by the numbers: 5, 4, 3, 2, and 1, respectively. The parameters α and β of the limits of the output x_i are specified at $\alpha = \beta = 0.3$.

The differences of skill among five stages are defined by the number of processing products in a unit of time, and they are represented by:

$$e_{ij} = 0.1 k_{ij} X_i, \tag{5.8}$$

where k_{ij} denotes a ratio of skill to a worker in skill stage 3, who has the standard skill level. We assume that a worker with the standard skill level can process 10% of the standard output X_i in a unit of time.

The ability and the labor cost of a worker in each skill stage are shown in Table 5.2. As seen in Table 5.2, the difference between skill stages can be recognized obviously in high-grade products, but it is hardly seen in low-grade products. The grades of products are defined by the processing costs, and there are four products in each grade.

The upper and the lower limits Y_j^L and Y_j^U of the numbers of workers in each skill stage and the upper limit Y of assignable total workers are specified as follows:

Table 5.1 Data of the production and work force assignment problem.

product number i	standard output X_i	price c_i	finishing cost a_i	processing cost b_i	profit $c_i - a_i - b_i$	grade
1	800	700	67	33	600	4
2	60	700	459	41	200	5
3	1500	600	124	26	450	2
4	1600	600	317	33	250	3
5	120	500	159	41	300	5
6	2000	500	374	26	100	2
7	1300	400	67	33	300	3
8	150	400	209	41	150	5
9	6500	300	124	26	150	1
10	2300	300	167	33	100	3
11	250	200	9	41	150	4
12	20	200	79	51	70	5
13	9000	100	17	23	60	1
14	2200	100	29	31	40	3
15	15000	650	277	23	350	1
16	8000	550	69	31	450	2
17	700	450	312	38	100	4
18	6000	350	77	23	250	1
19	1200	250	49	31	170	2
20	160	150	62	38	50	4

Table 5.2 Ability and labor cost of a worker.

skill stage j	1	2	3	4	5
high-grade k_{1j}	0.6	0.8	1.0	1.3	1.7
semi-high-grade k_{2j}	0.7	0.85	1.0	1.25	1.6
medium-grade k_{3j}	0.8	0.9	1.0	1.2	1.5
semi-low-grade k_{4j}	0.9	0.95	1.0	1.15	1.4
low-grade k_{5j}	1.0	1.0	1.0	1.1	1.3
labor cost d_j	101978	114715	127473	152967	191209

$$0 \leqq y_j \leqq 3,\ j = 1, \ldots, 5,$$
$$\sum_{j=1}^{5} y_j \leqq 10.$$

Before applying the interactive fuzzy programming to the two-level production and work force assignment problem, we solve the following maximization problems of the profits of the upper level and the lower level companies in order to understand the characteristics of the two-level linear programming problem:

$$\text{maximize}\ z_1(\boldsymbol{x},\boldsymbol{y}) = \sum_{i=1}^{m}(c_i - a_i - b_i)x_i \tag{5.9a}$$
$$\text{subject to}\ (\boldsymbol{x},\boldsymbol{y}) \in S, \tag{5.9b}$$

and

$$\text{maximize } z_2(\boldsymbol{x},\boldsymbol{y}) = \sum_{i=1}^{m} b_i x_i - \sum_{j=1}^{n} d_j y_j \quad (5.10a)$$

$$\text{subject to } (\boldsymbol{x},\boldsymbol{y}) \in S. \quad (5.10b)$$

An optimal solution to problem (5.9) of the upper level company is shown in Table 5.3. The relation between profitability and an output of each product is also shown in Figure 5.1, where the profitability of each production means a ratio of profit to price, $(c_i - a_i - b_i)/c_i$, and the whole profitability is $f_1(\boldsymbol{x},\boldsymbol{y})$ defined by (5.6a).

Table 5.3 Optimal solution to the profit maximization problem of the upper level company.

	the numbers of products $x_1,\ldots,x_{20}$			
992	42	1740	1120	153
1400	1560	105	5721	1610
310	14	7609	1540	16800
9280	490	6720	1392	112
	the numbers of assigned workers $y_1,\ldots,y_5$			
0	1	3	3	3
	profit z_i	profitability f_i	sales $\boldsymbol{cx}$ or processing cost $\boldsymbol{bx}$	
upper level company	15948260	0.613	$\boldsymbol{cx} = 26022600$	
lower level company	18169	0.012	$\boldsymbol{bx} = 1547831$	

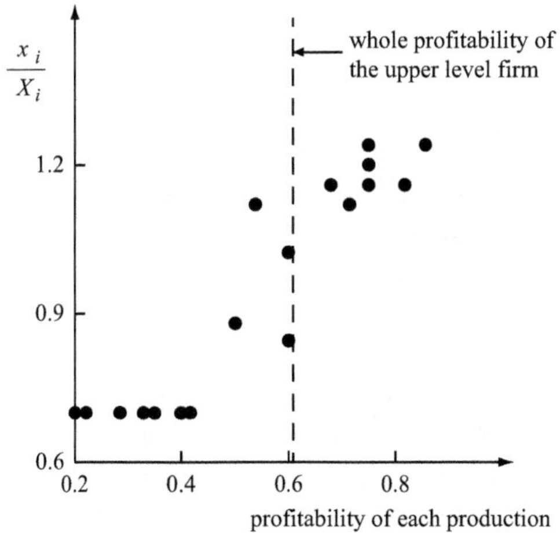

Fig. 5.1 Profitability and output of each product in the profit maximization problem of the upper level company.

As seen in Table 5.3 and Figure 5.1, because the profit of the upper level company is maximized under the budget constraints (5.2a) and (5.2b), in the optimal solution, the numbers of products with higher profitability reach the upper limits as long as the budget constraints are satisfied. The lower and the upper limits of the number x_i of product i are $0.7X_i$ and $\min\{1.3X_i, \sum_{j=1}^{5} e_{ij}y_j\}$, respectively, because the value of x_i is constrained by the conditions (5.1) and (5.3). The maximized profit z_1 of the upper level company is 15,948,260, and the corresponding profitability f_1 becomes 0.613.

Similarly, the result of the profit maximization of the lower level company is shown in Table 5.4. When the objective function of the lower level company is maximized, all the constraints (5.3) for 20 products become active. The reason why they are all active is as follows. Let x_i^* denote the output of product i. Then, by determining y_j^* so as to satisfy $x_i^* = \sum_{j=1}^{5} e_{ij}y_j^*$, the lower level company can finish the order from the upper level company at the minimal cost.

Moreover, as seen in Table 5.2, because the cost can be most reduced when workers with lower skill process lower grade products, the workers with lower skill are assigned fully in the optimal solution to the profit maximization problem of the lower level company, and the optimal solution is shown in the upper panel of Table 5.4. We also see that the lower grade products are made more than the higher grade ones in Figure 5.2, which shows the relation between a grade and an output of each product. It is noted that because ratios of the four products in the same grade are the same values, only the five points can be found in Figure 5.2.

Table 5.4 Optimal solution to the profit maximization problem of the lower level company.

		the numbers of products $x_1,\ldots,x_{20}$		
712	51	1455	1488	102
1940	1209	127	6565	2139
222	17	9090	2046	15150
7760	623	6060	1164	142
		the numbers of assigned workers $y_1,\ldots,y_5$		
3	3	3	1	0

	profit z_i	profitability f_i	sales $\boldsymbol{cx}$ or processing cost $\boldsymbol{bx}$
upper level company	14507660	0.593	$\boldsymbol{cx} = 24457050$
lower level company	344068	0.225	$\boldsymbol{bx} = 1529533$

To apply the interactive fuzzy programming to the profit maximization problem, we should first identify fuzzy goals for the profits of the upper level and the lower level companies. Suppose that membership functions of the fuzzy goals are linear and the parameters of the membership functions are specified by the Zimmermann method (1978). Then, the parameters of the linear membership function (4.7) are determined as $z_1^1 = 15948260$, $z_1^0 = 14507660$, $z_2^1 = 344068$, and $z_2^0 = 18169$ from the optimal solutions to the profit maximization problems of the upper level and the lower level companies which are shown in Tables 5.3 and 5.4, and the membership functions are represented as

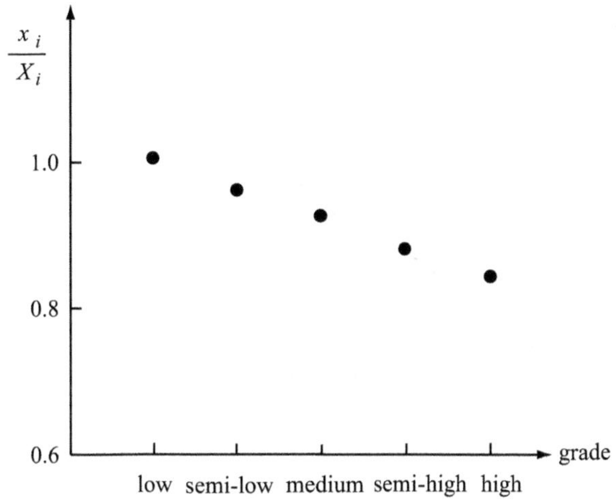

Fig. 5.2 Grade and output of each product in the profit maximization problem of the lower level company.

$$\mu_1^z(z_1(\boldsymbol{x},\boldsymbol{y})) = (z_1(\boldsymbol{x},\boldsymbol{y}) - 14507660)/(15948260 - 14507660), \tag{5.11a}$$
$$\mu_2^z(z_2(\boldsymbol{x},\boldsymbol{y})) = (z_2(\boldsymbol{x},\boldsymbol{y}) - 18169)/(344068 - 18169). \tag{5.11b}$$

The interactive procedure starts to solve the following problem which is to find a solution maximizing the minimum between the satisfactory degrees of the two companies:

$$\text{maximize } \lambda \tag{5.12a}$$
$$\text{subject to } \mu_1^z(z_1(\boldsymbol{x},\boldsymbol{y})) \geqq \lambda \tag{5.12b}$$
$$\mu_2^z(z_2(\boldsymbol{x},\boldsymbol{y})) \geqq \lambda \tag{5.12c}$$
$$(\boldsymbol{x},\boldsymbol{y}) \in S. \tag{5.12d}$$

In problem (5.12), the degrees of satisfaction of the two companies are evaluated equally, and an optimal solution to the problem is shown in Table 5.5.

As seen in Table 5.5, the degree of satisfaction of the upper level company and that of the lower level company are equal, and it is 0.609. The profit $z_1 = 15,384,810$ of the upper level company is smaller than its maximal profit $z_1 = 15,948,260$ while it is larger than the profit $z_1 = 14,507,660$ in the case where the profit of the lower level company is maximized. Similarly, the profit $z_2 = 216,653$ of the lower level company is smaller than its maximal profit $z_2 = 344,068$ while it is larger than the profit $z_2 = 18,169$ in the case where the profit of the upper level company is maximized. The numbers of assigned workers $(y_1,y_2,y_3,y_4,y_5) = (3,2.43,2,0,2.57)$ in the lower level company are not integer, but the labor forces corresponding to the fractional portion can be supplemented with part time jobs or overtime works.

Table 5.5 Maximin solution in the profit maximization problem.

the numbers of products $x_1,\dots,x_{20}$				
822	42	1590	1541	121
1400	1357	105	7000	1610
256	14	8549	1540	16155
8484	490	6462	1272	112
the numbers of assigned workers $y_1,\dots,y_5$				
3	2.43	2	0	2.57
	profit z_i	$\mu_i^z(z_i)$	sales $\boldsymbol{cx}$ or processing cost $\boldsymbol{bx}$	
upper level company	15384810	0.609	$\boldsymbol{cx} = 25458550$	
upper level company	216653	0.609	$\boldsymbol{bx} = 1547698$	
ratio of satisfaction: $\Delta = \dfrac{\mu_2^z(z_2)}{\mu_1^z(z_1)} = 1.00$				

To compare the optimal solution of the maximin problem with that of the maximization problem of the upper level company, we show a relation between profitability and an output of each product in Figure 5.3. The products with lower profitability are made at the lower limit of output $0.7X_i$ both in the maximin problem and in the maximization problem of the upper level company. For the products with higher profitability in the maximization problem of the upper level company, the higher the profitability of a product is, the larger the output of the product becomes. In contrast, if the profitability is larger than 0.4, the outputs of such products range from $0.9X_i$ to $1.1X_i$ in the maximin problem, regardless of the profitability of each product.

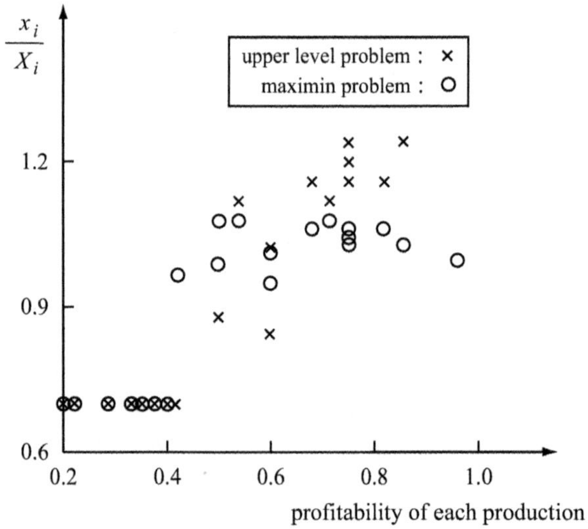

Fig. 5.3 Profitability and output of each product in the maximin problem and the profit maximization problem of the upper level company.

We compare the optimal solution of the maximin problem with that of the maximization problem of the lower level company. In Figure 5.4, a relation between a grade and an output of each product is shown. For the products with higher profitability both in the maximin problem and in the maximization problem of the lower level company, the higher the grade of a product is, the smaller the output of the product becomes. For the products with lower profitability, the outputs of them are at the lower limit $0.7X_i$, regardless of their grades.

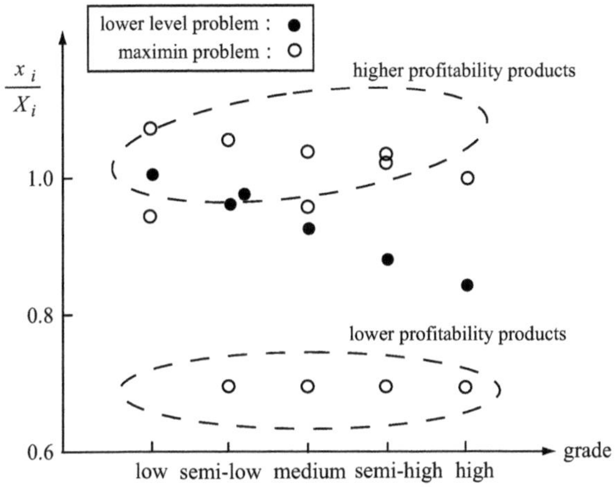

Fig. 5.4 Profitability and output of each product in the maximin problem and the profit maximization problem of the lower level company.

From the above analysis, it is found that the optimal solution to the maximin problem, where the satisfactory degrees with respect to the profits of the two companies are treated impartially, has a part of the characteristics of the optimal solution to the profit maximization problem of the upper level company and that of the lower level company in addition to the fact that the satisfactory degrees of the two company are equal.

After obtaining the optimal solution to the maximin problem, we will derive a satisfactory solution to the two-level linear production and work force assignment problem through the interactive fuzzy programming method for two-level programming problems in which the upper level company is guaranteed a certain satisfactory degree specified by itself and the overall satisfactory balance between both decision makers is taken into account; the procedure of the method is shown diagrammatically in Figure 4.3 of chapter 4.

Suppose that the upper level company judges that it is desirable for itself to increase its satisfactory degree at the sacrifice of that of the lower level company. With the optimal solution to the maximin problem in mind, suppose that the upper level company specifies the minimal satisfactory level at $\hat{\delta} = 0.8$ and the bounds of the ratio of the satisfactory degrees of the two companies as $[0.6, 1.0]$. In the interactive fuzzy programming that we apply to the production and work force assignment problem, the following two conditions are required as the termination conditions already given in section 4.2 of chapter 4:

Condition 1 The satisfactory degree $\mu_1^z(z_1)$ of the upper level company is larger than or equal to the minimal satisfactory level $\hat{\delta}$: $\mu_1(z_1) \geqq \hat{\delta}$.

Condition 2 The ratio $\Delta = \mu_2^z(z_2)/\mu_1^z(z_1)$ between the satisfactory degrees of the two companies lies in the closed interval between the lower and the upper bounds specified by the upper level company: $\Delta \in [\Delta_{\min}, \Delta_{\max}]$.

The satisfactory degree of the lower level company is maximized under the condition that the satisfactory degree of the upper level company is larger than or equal to 0.8 and the constraints of the problem given in the previous section. Then, the following problem is solved:

$$\begin{aligned} &\text{maximize } \mu_2^z(z_2(\boldsymbol{x},\boldsymbol{y})) &\quad (5.13\text{a})\\ &\text{subject to } \mu_1^z(z_1(\boldsymbol{x},\boldsymbol{y})) \geqq 0.8 &\quad (5.13\text{b})\\ &\qquad (\boldsymbol{x},\boldsymbol{y}) \in S, &\quad (5.13\text{c}) \end{aligned}$$

and the result is shown in Table 5.6.

Table 5.6 Iteration 1 (profit maximization).

		the numbers of products $x_1, \ldots, x_{20}$		
888	42	1657	1120	133
1400	1440	105	6831	1610
277	14	7909	1540	16538
8841	490	6615	1326	112
		the numbers of assigned workers $y_1, \ldots, y_5$		
3	0.74	2	1.26	3

	profit z_i	$\mu_i^z(z_i)$	sales $\boldsymbol{cx}$ or processing cost $\boldsymbol{bx}$
upper level company	15659340	0.80	$\boldsymbol{cx} = 25733400$
upper level company	135638	0.36	$\boldsymbol{bx} = 1547772$
Δ	$-\dfrac{\partial z_2(\boldsymbol{x},\boldsymbol{y})}{\partial z_1(\boldsymbol{x},\boldsymbol{y})}$	$-\dfrac{\partial \mu_2^z(z_2(\boldsymbol{x},\boldsymbol{y}))}{\partial \mu_1^z(z_1(\boldsymbol{x},\boldsymbol{y}))}$	
0.45	0.35	1.55	

At Iteration 1, because while the satisfactory degree of the upper level company is 0.8, the ratio of the satisfactory degrees is $\Delta = 0.45$, *Condition 2* of the termination conditions is not satisfied. Thus, the upper level company should update the minimal satisfactory level $\hat{\delta}$. For this end, the information of trade-off between the two objective functions, z_1 and z_2, or between the two membership functions, $\mu_1(z_1)$ and $\mu_2(z_2)$, is useful; it is provided in Table 5.6 as trade-off values $-\partial z_2(\boldsymbol{x},\boldsymbol{y})/\partial z_1(\boldsymbol{x},\boldsymbol{y})$ and $-\partial \mu_2^z(z_2(\boldsymbol{x},\boldsymbol{y}))/\partial \mu_1^z(z_1(\boldsymbol{x},\boldsymbol{y}))$ in the relaxed linear programming problem.

For a tentative solution obtained in Iteration 1, because of $-\partial z_2(\boldsymbol{x},\boldsymbol{y})/\partial z_1(\boldsymbol{x},\boldsymbol{y}) = 0.35$ and $-\partial\mu_2^z(z_2(\boldsymbol{x},\boldsymbol{y}))/\partial\mu_1^z(z_1(\boldsymbol{x},\boldsymbol{y})) = 1.55$, a small change of the profit of the upper level company does not affect the profit of the lower level company seriously, but a small change of the minimal satisfactory level has a strong influence on that of the lower level company. For example, if the upper level company reduces 0.1 of the minimal satisfactory level, an increase of about 0.15 of satisfactory degree of the lower level company is expected.

Suppose that the upper level company judges that its own satisfactory degree is relatively higher, and the minimal satisfactory level should be reduced by 0.1. Then, the upper level company specifies it at $\hat{\delta} = 0.7$, taking account of the trade-off information. Problem (5.13) with the minimal satisfactory level updated from 0.8 to 0.7 is solved, and the result is shown in Table 5.7.

Table 5.7 Iteration 2 (profit maximization).

		the numbers of products $x_1,\ldots,x_{20}$		
848	42	1620	1269	126
1400	1391	105	7086	1610
265	14	8172	1540	16353
8643	490	6541	1296	112
		the numbers of assigned workers $y_1,\ldots,y_5$		
3	1.97	2	0.02	3
	profit z_i	$\mu_i^z(z_i)$	sales $\boldsymbol{cx}$ or processing cost $\boldsymbol{bx}$	
upper level company	15513920	0.70	$\boldsymbol{cx} = 25587350$	
upper level company	184110	0.51	$\boldsymbol{bx} = 1547665$	
Δ	$-\dfrac{\partial z_2(\boldsymbol{x},\boldsymbol{y})}{\partial z_1(\boldsymbol{x},\boldsymbol{y})}$	$-\dfrac{\partial\mu_2^z(z_2(\boldsymbol{x},\boldsymbol{y}))}{\partial\mu_1^z(z_1(\boldsymbol{x},\boldsymbol{y}))}$		
0.72	0.31	1.37		

At Iteration 2, because the satisfactory degree of the upper level company is 0.7 and the ratio of the satisfactory degrees is $\Delta = 0.72$, the two termination conditions are satisfied. If the upper level company accepts the solution, it follows that the final satisfactory solution is derived.

5.1.3 Maximization of profitability

In this subsection, we formulate the two-level linear fractional programming problem (5.7) in which both of the upper level company and the lower level company maximize profitability represented by the ratio of profit to sales. After examining the two individual maximization problems on profitability, the fuzzy interactive programming is applied to the two-level linear fractional production and work force assignment problem.

To understand the characteristics of the two-level linear fractional programming problem, we solve the following two individual problems of the upper level and the lower level companies:

$$\text{maximize} \;\; f_1(\boldsymbol{x},\boldsymbol{y}) = \frac{\sum_{i=1}^{m}(c_i - b_i - a_i)x_i}{\sum_{i=1}^{m} c_i x_i} \qquad (5.14a)$$

$$\text{subject to} \;\; (\boldsymbol{x},\boldsymbol{y}) \in S, \qquad (5.14b)$$

and

$$\text{maximize} \;\; f_2(\boldsymbol{x},\boldsymbol{y}) = \frac{\sum_{i=1}^{m} b_i x_i - \sum_{j=1}^{n} d_j y_j}{\sum_{i=1}^{m} b_i x_i} \qquad (5.15a)$$

$$\text{subject to} \;\; (\boldsymbol{x},\boldsymbol{y}) \in S. \qquad (5.15b)$$

An optimal solution to problem (5.14) and the related information are shown in Table 5.8, and the relation between profitability and an output of each product is also shown in Figure 5.5.

Table 5.8 Optimal solution to the profitability maximization problem of the upper level company.

		the numbers of products $x_1,\ldots,x_{20}$		
992	42	1740	1120	84
1400	1560	105	4550	1610
310	14	6300	1540	10500
9280	490	6720	1392	112
		the numbers of assigned workers $y_1,\ldots,y_5$		
0	1	3	3	3
	profitability f_i	profit z_i	sales $\boldsymbol{cx}$ or processing cost $\boldsymbol{bx}$	
upper level company	0.629	13468370	$\boldsymbol{cx} = 21410900$	
lower level company	−0.140	−190117	$\boldsymbol{bx} = 1339549$	

In the optimal solution to problem (5.14), products with higher profitability are made at the upper limit, $\min\{1.3X_i, \sum_{j=1}^{5} e_{ij}y_j\}$, and products with lower profitability are made at the lower limit, $0.7X_i$, as in the profit maximization problem (5.9) of the upper level company. However, although in the profit maximization problem (5.9), the products with profitability smaller than the whole profitability 0.613 in the profit maximization problem are made to some extent, in the profitability maximization problem (5.14), the products with profitability smaller than the whole profitability 0.629 in the profitability maximization problem are made no more than the lower limit, $0.7X_i$.

An optimal solution to problem (5.15) of the lower level company and the related information are shown in Table 5.9, and the relation between a grade and an output of each product is also shown in Figure 5.6.

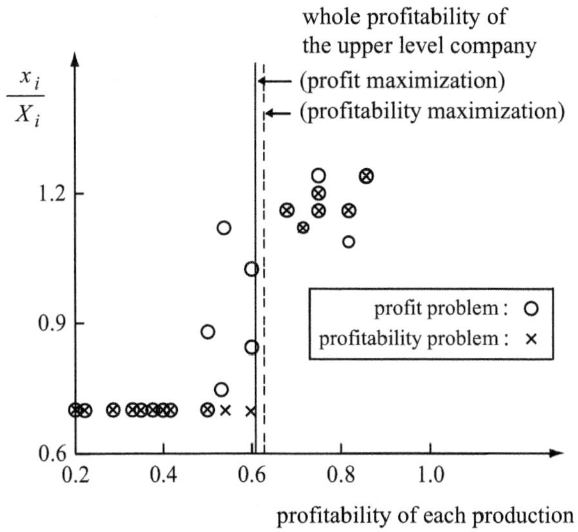

Fig. 5.5 Profitability and output of each product in the profitability maximization problem of the upper level company.

Table 5.9 Optimal solution to the profitability maximization problem of the lower level company.

		the numbers of products $x_1,\dots,x_{20}$		
596	42	1252	1264	84
1670	1027	105	5720	1817
186	14	7920	1738	13200
6680	521	5280	1002	119
		the numbers of assigned workers $y_1,\dots,y_5$		
3	3	2.8	0	0

	profitability f_i	profit z_i	sales $\boldsymbol{cx}$ or processing cost $\boldsymbol{bx}$
upper level company	0.593	12546740	$\boldsymbol{cx} = 21154900$
lower level company	0.238	314140	$\boldsymbol{bx} = 1321143$

It can be seen in Figure 5.6 that the lower grade products are made more than the higher grade products as seen also in the profit maximization problem of the lower level company, but the sum of the outputs of all the products in the profitability maximization problem is smaller than that of the profit maximization problem because products with lower profitability are not always produced as many as possible.

To apply the interactive fuzzy programming method for the two-level linear fractional problem maximizing the profitability of the two companies, we should first identify fuzzy goals for the profitability of the two companies. Suppose that membership functions of the fuzzy goals are linear and the parameters of the membership functions are specified by the Zimmermann method (1978). Then, the parameters

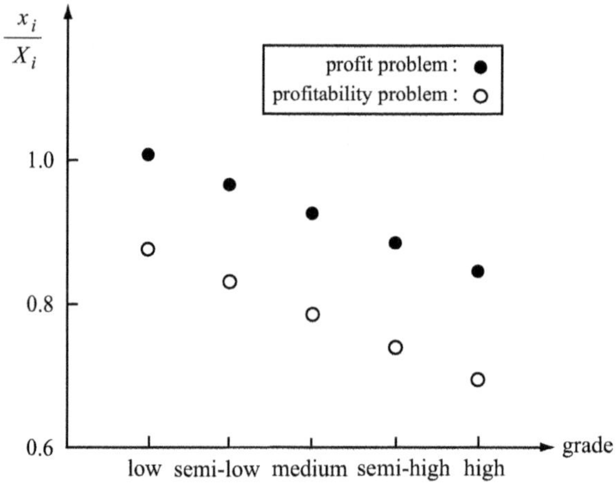

Fig. 5.6 Grade and output of each product in the profitability maximization problem of the lower level company.

of the linear membership function (4.7) are determined as $z_1^1 = 0.629$, $z_1^0 = 0.593$, $z_2^1 = 0.238$, and $z_2^0 = -0.140$ from the solutions shown in Tables 5.8 and 5.9, the membership functions are represented as

$$\mu_1^f(f_1(\boldsymbol{x},\boldsymbol{y})) = (f_1(\boldsymbol{x},\boldsymbol{y}) - 0.593)/(0.629 - 0.593), \tag{5.16a}$$

$$\mu_2^f(f_2(\boldsymbol{x},\boldsymbol{y})) = (f_2(\boldsymbol{x},\boldsymbol{y}) + 0.140)/(0.238 + 0.140). \tag{5.16b}$$

At the beginning of the interactive fuzzy programming method, the following problem which is to find a solution maximizing the minimum between the satisfactory degrees of the two companies is solved:

$$\text{maximize } \lambda \tag{5.17a}$$

$$\text{subject to } \mu_1^f(f_1(\boldsymbol{x},\boldsymbol{y})) \geqq \lambda \tag{5.17b}$$

$$\mu_2^f(f_2(\boldsymbol{x},\boldsymbol{y})) \geqq \lambda \tag{5.17c}$$

$$(\boldsymbol{x},\boldsymbol{y}) \in S, \tag{5.17d}$$

and the result is shown in Table 5.10.

As seen in Table 5.10, both the degree of satisfaction of the upper level company and that of the lower level company are equal, and it is 0.63. The profitability $f_1 = 0.616$ of the upper level company is smaller than its maximal profitability $f_1 = 0.629$ while it is larger than the profitability $f_1 = 0.593$ in the case where the profitability of the lower level company is maximized. Similarly, the profitability $f_2 = 0.097$ of the lower level company is smaller than its maximal profitability $f_2 = 0.238$ while it is larger than the profitability $f_2 = -0.140$ in the case where the profitability of the upper level company is maximized.

Table 5.10 Maximin solution in the profitability maximization problem.

the numbers of products $x_1,\ldots,x_{20}$				
788	42	1552	1120	115
1400	1313	105	6621	1610
246	19	9540	2222	10500
8280	490	6360	1242	157
the numbers of assigned workers $y_1,\ldots,y_5$				
3	3	2	0	2

	profitability f_i	$\mu_i^f(f_i)$	profit z_i	sales $\boldsymbol{cx}$ or processing cost $\boldsymbol{bx}$
upper level company	0.616	0.63	13156400	$\boldsymbol{cx} = 21366950$
upper level company	0.097	0.63	138525	$\boldsymbol{bx} = 1425968$
ratio of satisfaction: $\Delta = \dfrac{\mu_2^f(f_2)}{\mu_1^f(f_1)} = 1.00$				

To compare the optimal solution of the maximin problem (5.12) on profit with that of the maximin problem (5.17) on profitability, we show the relation between profitability and an output of each product in Figure 5.7 and the relation between a grade and an output of each product in Figure 5.8.

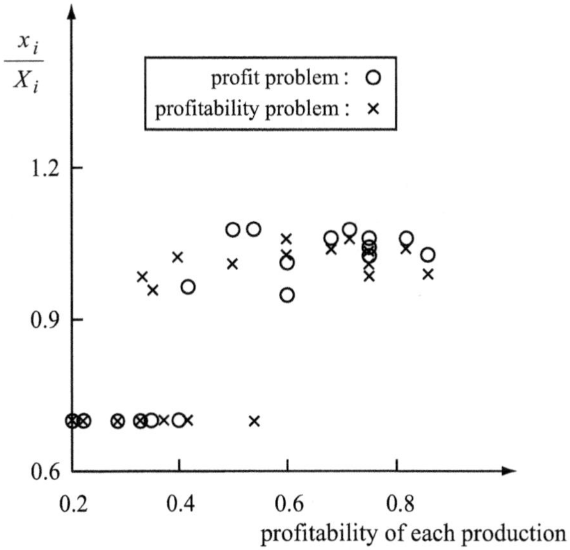

Fig. 5.7 Profitability and output of each product in the maximin problems on profit and profitability.

As seen in Figures 5.7 and 5.8, products with lower profitability are made at the lower limit of output, $0.7X_i$, in the maximin problems both on profit and on

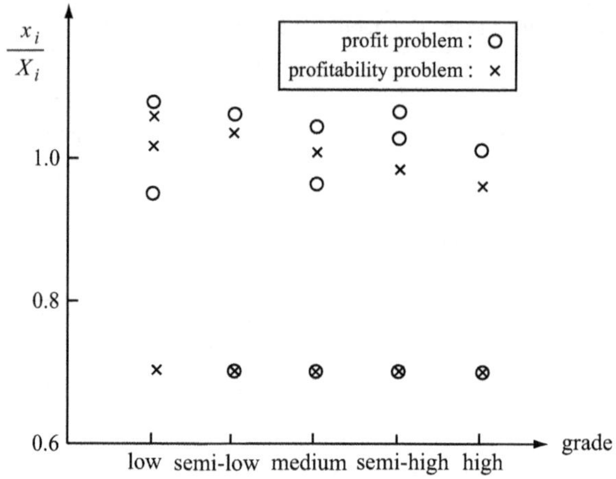

Fig. 5.8 Grade and output of each product in the maximin problems on profit and profitability.

profitability while the outputs of products with higher profitability in the maximin problem on profitability are slightly smaller than those on profit. In the profitability maximization problems (5.14) and (5.15) of the upper level and the lower level companies, we can also see similar characteristics.

Suppose that the upper level company judges that it is desirable for itself to increase its satisfactory degree at the sacrifice of that of the lower level company. With the optimal solution to the maximin problem in mind, suppose that the upper level company specifies the minimal satisfactory level at $\hat{\delta} = 0.8$ and the bounds of the ratio of the satisfactory degrees of the two companies at $[0.6, 1.0]$.

The following problem maximizing the satisfactory degree of the lower level company is solved under the original constraints and the condition that the satisfactory degree of the upper level company is larger than or equal to 0.8:

$$\text{maximize } \mu_2^f(f_2(\boldsymbol{x},\boldsymbol{y})) \tag{5.18a}$$

$$\text{subject to } \mu_1^f(f_1(\boldsymbol{x},\boldsymbol{y})) \geqq 0.8 \tag{5.18b}$$

$$(\boldsymbol{x},\boldsymbol{y}) \in S. \tag{5.18c}$$

An optimal solution to problem (5.18) and the related information are shown in Table 5.11.

At Iteration 1, because while the satisfactory degree of the upper level company is 0.8, the ratio of the satisfactory degrees is $\Delta = 0.578$, *Condition 2* of the termination conditions is not satisfied. Thus, the upper level company must update the minimal satisfactory level $\hat{\delta}$. For this end, the information of trade-off between the two objective functions, f_1 and f_2, or between the two membership functions, $\mu_1^f(f_1)$ and $\mu_2^f(f_2)$, is utilized; it is given in Table 5.11 as trade-off values

Table 5.11 Iteration 1 (profitability maximization).

the numbers of products $x_1, \ldots, x_{20}$				
842	42	1613	1120	125
1400	1383	105	4550	1610
263	21	9785	2341	10500
8606	490	6523	1290	168
the numbers of assigned workers $y_1, \ldots, y_5$				
3	2.09	2	0	2.91

	profitability f_i	$\mu_i^f(f_i)$	sales $\boldsymbol{cx}$ or processing cost $\boldsymbol{bx}$
upper level company	0.622	0.80	$\boldsymbol{cx} = 21143250$
upper level company	0.033	0.463	$\boldsymbol{bx} = 1404094$
Δ	$-\dfrac{\partial f_2(\boldsymbol{x},\boldsymbol{y})}{\partial f_1(\boldsymbol{x},\boldsymbol{y})}$	$-\dfrac{\partial \mu_2^f(f_2(\boldsymbol{x},\boldsymbol{y}))}{\partial \mu_1^f(f_1(\boldsymbol{x},\boldsymbol{y}))}$	
0.578	10.49	1.45	

$-\partial f_2(\boldsymbol{x},\boldsymbol{y})/\partial f_1(\boldsymbol{x},\boldsymbol{y})$ and $-\partial \mu_2^f(f_2(\boldsymbol{x},\boldsymbol{y}))/\partial \mu_1^f(f_1(\boldsymbol{x},\boldsymbol{y}))$ in the relaxed linear fractional programming problem.

Suppose that the upper level company judges that its own satisfactory degree is relatively higher, the minimal satisfactory level should be reduced by 0.1. Then, the upper level company specifies it at $\hat{\delta} = 0.7$, with the trade-off information in mind. Problem (5.18) with the minimal satisfactory level updated from 0.8 to 0.7 is solved, and the result is shown in Table 5.12.

Table 5.12 Iteration 2 (profitability maximization).

the numbers of products $x_1, \ldots, x_{20}$				
788	42	1552	1120	115
1400	1313	105	5097	1610
246	19	9540	2222	10500
8280	490	6360	1242	157
the numbers of assigned workers $y_1, \ldots, y_5$				
3	3	2	0	2

	profitability f_i	$\mu_i^f(f_i)$	sales $\boldsymbol{cx}$ or processing cost $\boldsymbol{bx}$
upper level company	0.618	0.70	$\boldsymbol{cx} = 20909750$
upper level company	0.0713	0.562	$\boldsymbol{bx} = 1386344$
Δ	$-\dfrac{\partial f_2(\boldsymbol{x},\boldsymbol{y})}{\partial f_1(\boldsymbol{x},\boldsymbol{y})}$	$-\dfrac{\partial \mu_2^f(f_2(\boldsymbol{x},\boldsymbol{y}))}{\partial \mu_1^f(f_1(\boldsymbol{x},\boldsymbol{y}))}$	
0.802	10.26	1.42	

At Iteration 2, because the satisfactory degree of the upper level company is 0.7 and the ratio of the satisfactory degrees becomes $\Delta = 0.802$, the two termination conditions are satisfied. If the upper level company accepts the solution, it follows that the final satisfactory solution is obtained.

5.1.4 Discussions and implementation

In this subsection, we summarize the characteristics of the profit maximization problem and the profitability maximization problem which we have analyzed above, and examine an actual planning of the production and work forth assignment of the upper and the lower level companies to be implemented.

To compare the profit maximization problem and the profitability maximization problem, we show the related information such as sales, profit, profitability, etc. in Tables 5.13 and 5.14, and describe the characteristics of the problems as follows:

Profit maximization of the upper level company: The production volume of products with higher profitability reaches the upper limits as long as the budget constraint is satisfied, but conversely the other products with relatively lower profitability are produced at the lower limit.

Profitability maximization of the upper level company: Products with higher profitability are made at the upper limits, and products with lower profitability are made at the lower limits in a way similar to the profit maximization of the upper level company. However, it is not always true that the products are made as many as possible up to the budget limits.

Profit maximization of the lower level company: The minimal labor forces are assigned from the most efficient workers on cost per unit time as long as the amount of orders received can be produced.

Profitability maximization of the lower level company: Although the minimal labor forces are assigned from the most efficient workers on cost per unit time as long as the amount of orders received can be produced, it is not always true that the products are made as many as possible up to the budget limits.

The above characteristics are found out clearly in the individual problems of the upper level and the lower level companies, and some characteristics of the upper level and the lower level companies are revealed simultaneously in the maximin problems. As for the satisfactory (final) solutions, because the upper level company specifies the minimal satisfactory level of itself, the characteristics of the upper level company can be seen rather than those of the lower level company.

We can suggest the following criteria for deciding which formulation to be employed. If it is desirable to implement the budget fully, we should formulate the profit maximization problem. In contrast, in the case where the higher profitability is expected even if the budget is left over, the profitability should be chosen as an objective function.

The above analysis provides us a good guide to implementation of planning for the production and the work forth assignment. From the fact that there exists a possibility that maximization of the profitability would reduce the scale of the business,

Table 5.13 Profit maximization problems.

		upper level	lower level	maximin	final solution
upper level company	sales ($\times 10^3$)	26023	24457	25459	25587
	profit ($\times 10^3$)	15948	14508	15385	15514
	profitability	0.613	0.593	0.604	0.606
	cost ($\times 10^3$)	8527	8420	8530	8526
	satisfactory degree	1	0	0.61	0.7
lower level company	sales ($\times 10^3$)	1548	1530	1548	1548
	profit ($\times 10^3$)	18	344	217	183
	profitability	0.012	0.224	0.141	0.118
	satisfactory degree	0	1	0.61	0.51

Table 5.14 Profitability maximization problems.

		upper level	lower level	maximin	final solution
upper level company	sales ($\times 10^3$)	21411	21155	21367	20910
	profit ($\times 10^3$)	13468	12547	13156	12928
	profitability	0.629	0.593	0.616	0.618
	cost ($\times 10^3$)	6603	7687	6785	6596
	satisfactory degree	1	0	0.630	0.7
lower level company	sales ($\times 10^3$)	1339	1321	1426	1386
	profit ($\times 10^3$)	−190	314	139	99
	profitability	−0.142	0.238	0.097	0.0713
	satisfactory degree	0	1	0.630	0.561

because the housing material manufacturer expects to enlarge the scale of business, we recommend to formulate the problems on the production and work force assignment as the profit maximization problem. Through interactive process of the fuzzy programming for two-level programming problems, the housing material manufacturer and the subcontracting company can obtain a satisfactory solution for planning of the production and the work forth assignment such as the solution seen in Table 5.7.

5.2 Decentralized two-level transportation problem

In this section, we deal with a transportation problem in the housing material manufacturer (Sakawa, Nishizaki and Uemura, 2002). The housing material manufacturer does not transport its products from its factory or warehouse to customers on its own account, but entrusts the transportation to forwarding agents. There are two kinds of forwarding agents: one handles the regular transportation, and the other handles the small lot transportation. Minimizing the transportation cost and the opportunity loss with respect to transportation time, the housing material manufacturer gives the two forwarding agents orders for the transportation of the products to the customers. The two forwarding agents assign work force so as to maximize their profits, taking the ability of drivers into account. Such a transportation planning and work force assignment problem can be formulated as a decentralized two-level integer programming problem.

In this decision making problem, the housing material manufacturer has been connected in business with each of the forwarding agents over a long period of time, and there exists cooperative relationship between them. For this cooperative relationship, it is natural that the housing material manufacturer does not optimize only its own objectives, but makes decisions cooperatively by balancing its own satisfaction with those of the two forwarding agents. To do so, we employ interactive fuzzy programming approach to the transportation planning and work force assignment problem.

Especially, the interactive fuzzy programming method for a decentralized two-level linear programming problem, which was developed by Sakawa and Nishizaki (2002a) and described in chapter 4, is applicable to the transportation problem in the housing material manufacturer by revising it partly because the transportation problem in the housing material manufacturer is formulated as a decentralized two-level integer programming problem with the two objectives of the upper level decision maker.

In the subsequent subsections, we formulate the transportation problem in the housing material manufacturer, and solve four individual programming problems with a single objective function which are two problems with two single different objectives of the housing material manufacturer and the other two problems of the two forwarding agents in order to understand the characteristics of the transportation problem and to specify parameters of membership functions of fuzzy goals. The fuzzy goals for the objective functions are identified, and in particular, the two fuzzy goals of the housing material manufacturer are aggregated by three methods. Then, a satisfactory solution is derived for the transportation problem by applying the interactive fuzzy programming method consisting of the two interactive phases.

5.2.1 Problem formulation

The transportation problem in the housing material manufacturer is formulated as a decentralized two-level integer programming problem. A decision maker at the upper level is the housing material manufacturer and there are two decision makers at the lower level who are the two forwarding agents handling the regular transportation and the small lot transportation. The housing material manufacturer determines allocation of the transportation tasks to five customers between the two forwarding agents, and they assign work force, taking the three levels of ability of drivers into account. The housing material manufacturer and the two forwarding agents have the following objectives.

5.2.1.1 Objective functions

To describe concisely, let Company U, Company L1 and Company L2 denote the housing material manufacturer, the forwarding agent handling the regular transportation and the forwarding agent handling the small lot transportation, respectively.

The first objective function of Company U

Coordinating orders to Companies L1 and L2, Company U minimizes the transportation cost:

$$z_{11}(\boldsymbol{y},\boldsymbol{z}) = \sum_{i=1}^{5}(a_i y_i + b_i z_i), \tag{5.19}$$

where a_i and b_i are unit costs of the regular and the small lot transportation to customer i, respectively, and they are shown in Table 5.15; y_i and z_i denote decision variables which are the numbers of the products to be sent to customer i by Company L1 and Company L2, respectively. Because the decision variables $\boldsymbol{y} = (y_1,\dots,y_5)$ and $\boldsymbol{z} = (z_1,\dots,z_5)$ of Company U are the numbers of the products to be sent to the five customers, they are integer variables.

The second objective function of Company U

Company U also minimizes the opportunity loss with respect to the transportation time which is a delay from the earliest delivery time:

$$z_{12}(\boldsymbol{u},\boldsymbol{v}) = \sum_{i=1}^{5}\sum_{j=1}^{3}(e_{ij}u_{ij} + f_{ij}v_{ij}), \tag{5.20}$$

where e_{ij} and f_{ij} are the values of the opportunity loss when drivers with ability level j in Companies L1 and L2 transport the products to customer i, respectively, and they are defined by $e_{ij} = o_{ij} - o_{i3}$ and $f_{ij} = p_{ij} - p_{i3}$; o_{ij} and p_{ij} are transportation times when the drivers with ability level j of Companies L1 and L2 transport the products to customer i. We use a larger index j for drivers with higher ability, and therefore, o_{i3} and p_{i3} are the smallest values. The values of o_{ij} and p_{ij} are shown in Tables 5.16 and 5.17, respectively. The symbols u_{ij} and v_{ij} denote decision variables which are the numbers of the assigned drivers with ability level j in Companies L1 and L2 which transform the products to customer i. Because the decision variables $\boldsymbol{u} = (u_{11},\dots,u_{53})$ and $\boldsymbol{v} = (v_{11},\dots,v_{53})$ of Companies L1 and L2 are the numbers of drivers, they are also integral.

The objective function of Company L1

By assigning drivers efficiently, Company L1 maximizes the following profit defined by taking the necessary expenses and the costs on drivers from the money paid by Company U.

$$z_{21}(\boldsymbol{y},\boldsymbol{u}) = \sum_{i=1}^{5} \left\{ a_i y_i - \left(0.3 a_i y_i + \sum_{j=1}^{3} s_{ij} u_{ij} \right) \right\} \tag{5.21}$$

The necessary expense is defined as 30% of the income from Company U. In the objective function (5.21), s_{ij} denotes the cost on a driver with ability level j who transports the products to customer i, and values of s_{ij} are shown in Table 5.16.

The objective function of Company L2

Similarly to Company L1, Company L2 also maximizes the following profit defined by taking the necessary expenses and the costs on drivers from the money paid by Company U.

$$z_{22}(\boldsymbol{z},\boldsymbol{v}) = \sum_{i=1}^{5} \left\{ b_i z_i - \left(0.3 b_i z_i + \sum_{j=1}^{3} t_{ij} v_{ij} \right) \right\}, \tag{5.22}$$

where t_{ij} denotes the cost on a driver with ability level j who transports the products to customer i, and values of t_{ij} are shown in Table 5.17.

5.2.1.2 Constraints

Constraints on the amount of transportation

Because the amount of transportation to customer i is fixed at X_i, the sum $y_i + z_i$ of the numbers of the products to be sent to customer i by Companies L1 and L2 equals X_i, and then the constraints on the amount of transportation are represented as

$$X_i = y_i + z_i, \ i = 1, \ldots, 5. \tag{5.23}$$

The values of X_i are shown in Table 5.15.

Constraints on the minimal orders

Company U must request Companies L1 and L2 to transport more than 30 and 20 products to each customer, respectively. The constraints on the minimal orders are represented as

$$y_i \geqq 30,\ i = 1,\dots,5, \tag{5.24a}$$
$$z_i \geqq 20,\ i = 1,\dots,5. \tag{5.24b}$$

Constraints on assignment of work force

Companies L1 and L2 must transport the products to the customers before the specified date of delivery, and therefore they cannot assign transportation of the products to drivers who take longer time than the due date. The constraints on assignment of work force are represented as

$$y_i \leqq \sum_{j \in K_i} q_j u_{ij},\ i = 1,\dots,5, \tag{5.25a}$$
$$z_i \leqq \sum_{j \in L_i} r_j v_{ij},\ i = 1,\dots,5, \tag{5.25b}$$

where q_j and r_j are capacities of transportation of a driver with ability level j in Companies L1 and L2, respectively; Let $K_i = \{j \mid o_{ij} \leqq c_i\}$ and $L_i = \{j \mid p_{ij} \leqq d_i\}$, where c_i and d_i denote the due date to customer i for Companies L1 and L2, respectively, and they are shown in Table 5.15.

Constraints on the numbers of drivers

For each ability level, Companies L1 and L2 must assign at least one driver but no more than three drivers, and the constraints on the numbers of drivers are represented as

$$1 \leqq \sum_{i=1}^{5} u_{ij} \leqq 3,\ j = 1,2,3, \tag{5.26a}$$
$$1 \leqq \sum_{i=1}^{5} v_{ij} \leqq 3,\ j = 1,2,3. \tag{5.26b}$$

Constraints on redundant assignment of work force

To prevent redundant assignment of work force, the following constraints are imposed, and they mean that the total amount of actual transportation to each customer must be smaller than the sum of the order volume and the capacity of transportation of a driver with the lowest ability level.

$$\sum_{j \in K_i} q_j u_{ij} \leqq y_i + q_1 - 1,\ i = 1,\dots,5, \tag{5.27a}$$
$$\sum_{j \in L_i} r_j v_{ij} \leqq z_i + r_1 - 1,\ i = 1,\dots,5. \tag{5.27b}$$

Table 5.15 Data on transportation of Company U.

customer	X_i	a_i (¥/product)	b_i (¥/product)	c_i (days)	d_i (days)
1	300	1,000	2,000	10	5
2	150	2,000	4,000	2	1
3	250	2,200	3,000	2	1
4	100	800	1,400	7	3
5	100	2,000	3,400	2	1

Table 5.16 Drivers data of Company L1.

	Level 1	Level 2	Level 3
q_j	100	120	150
o_{1j}, e_{1j} (days)	10, 2	9, 1	8, 0
o_{2j}, e_{2j} (days)	3*, –	2, 1	1, 0
o_{3j}, e_{3j} (days)	3*, –	2, 0	2, 0
o_{4j}, e_{4j} (days)	7, 2	6, 1	5, 0
o_{5j}, e_{5j} (days)	3*, –	2, 1	1, 0
s_{1j} (¥)	40,000	57,000	72,000
s_{2j} (¥)	—	36,000	44,000
s_{3j} (¥)	—	36,000	44,000
s_{4j} (¥)	34,000	48,000	60,000
s_{5j} (¥)	—	36,000	44,000

An asterisk $*$ denotes drivers who cannot be assigned.

Table 5.17 Drivers data of Company L2.

	Level 1	Level 2	Level 3
r_j	60	72	90
p_{1j}, e_{1j} (days)	5, 2	4, 1	3, 0
p_{2j}, e_{2j} (days)	1, 0	1, 0	1, 0
p_{3j}, e_{3j} (days)	3*, –	2, 0	2, 0
p_{4j}, e_{4j} (days)	2*, –	1, 0	1, 0
p_{5j}, e_{5j} (days)	1, 0	1, 0	1, 0
t_{1j} (¥)	40,000	52,000	62,000
t_{2j} (¥)	32,000	43,000	54,000
t_{3j} (¥)	—	43,000	54,000
t_{4j} (¥)	—	49,000	58,000
t_{5j} (¥)	32,000	43,000	54,000

An asterisk $*$ denotes drivers who cannot be assigned.

5.2.2 Interactive fuzzy programming

We apply the interactive fuzzy programming method to the transportation planning and work force assignment problem, which is formulated as the decentralized two-level integer programming problem, because the housing material manufacturer denoted by Company U makes decisions by taking into account not only its own satisfaction but also the satisfaction of the two forwarding agents denoted by Companies L1 and L2. The problem can be formally formulated as:

$$\underset{\text{for Company U}}{\text{minimize}} \quad z_{11}(\boldsymbol{y},\boldsymbol{z}) \tag{5.28a}$$

$$\underset{\text{for Company U}}{\text{minimize}} \quad z_{12}(\boldsymbol{u},\boldsymbol{v}) \tag{5.28b}$$

$$\underset{\text{for Company L1}}{\text{minimize}} \quad z_{21}(\boldsymbol{y},\boldsymbol{u}) \tag{5.28c}$$

$$\underset{\text{for Company L2}}{\text{minimize}} \quad z_{22}(\boldsymbol{z},\boldsymbol{v}) \tag{5.28d}$$

$$\text{subject to} \quad (\boldsymbol{y},\boldsymbol{z},\boldsymbol{u},\boldsymbol{v}) \in S, \tag{5.28e}$$

where S denotes the feasible region satisfying the constraints (5.23)–(5.27).

To support cooperative decision making in decentralized organizations effectively, the interactive fuzzy programming methods are developed. In the methods, it is assumed that multiple decision makers can arrive at an agreement to coordinate their actions, and the interactive process is performed by a delegate of them or an analyst who makes contact with the decision makers. Consequently, although the delegate or the analyst controls all of the decision variables, the overall satisfactory balance between both decision makers is taken into account.

Before examining the application, we show an outline of the interactive fuzzy programming method which is described in section 4.5 of chapter 4. The interactive fuzzy programming method by Sakawa and Nishizaki (2002a) was developed for obtaining a satisfactory solution to a decentralized two-level linear programming problem in which there are a single decision maker at the upper level and two or more decision makers at the lower level, and each decision maker has a single objective function. The algorithm is composed of two phases. In the first phase, the decision makers at both levels identify membership functions of their fuzzy goals for the objective functions. Let μ_0 denote the membership function of the fuzzy goal for the objective function of the decision maker at the upper level, and μ_i, $i = 1, \ldots, k$ the membership functions of those for the k decision makers at the lower level. To take into consideration overall satisfactory balance between the two levels, the following ratio of satisfactory degrees between the two levels is utilized:

$$\Delta = \frac{\min\{\mu_1(z_1), \ldots, \mu_k(z_k)\}}{\mu_0(z_0)}. \tag{5.29}$$

In the interactive process, the decision maker at the upper level specifies the minimal satisfactory level $\hat{\delta}$ and updates it if necessary, and then a tentative solution is obtained. In this phase, the decision makers at the lower level are treated impartially, and therefore they can be regarded as a group.

In the second phase, with the ratios of satisfaction between the decision maker at the upper level and each of the decision makers at the lower level in mind, the decision maker at the upper level specifies minimal or maximal satisfactory levels to some of the decision makers at the lower level and updates them if necessary. By coordinating the satisfactory degrees of the decision makers individually, the final satisfactory solution can be derived. The interactive procedure is depicted by a flowchart in Figure 4.7 in section 4.5 of chapter 4.

5.2.2.1 Individual optimization problems

To incorporate fuzziness of human judgment of the decision makers, membership functions of fuzzy goals for their objective functions are identified. For this end, the four single-objective programming problems with the single objective functions shown in the previous subsection are optimized separately, and parameters of the membership functions are determined by taking account to the optimal values and solutions of the four problems.

At the beginning, we solve the problem minimizing the transportation cost z_{11} of Company U:

$$\text{minimize } z_{11}(\boldsymbol{y},\boldsymbol{z}) \tag{5.30a}$$
$$\text{subject to } (\boldsymbol{y},\boldsymbol{z},\boldsymbol{u},\boldsymbol{v}) \in S. \tag{5.30b}$$

Problem (5.30) is solved by using the "lp_solve" developed by Berkelaar, which is a software for solving mixed integer programming problems based on the branch-and-bound techniques, and subsequent problems formulated in this subsection are also solved by using the lp_solve. The obtained optimal solution to problem (5.30) is shown in Table 5.18.

Table 5.18 Optimal solution to the problem minimizing z_{11}.

customer i	1	2	3	4	5
orders to Company L1: y_i	280	130	229	70	80
orders to Company L2: z_i	20	20	21	30	20
assigned drivers (Level 1): u_{i1}	0	0	0	1	0
in Company L1 (Level 2): u_{i2}	0	0	2	0	1
(Level 3): u_{i3}	2	1	0	0	0
assigned drivers (Level 1): v_{i1}	1	1	0	0	1
in Company L2 (Level 2): v_{i2}	0	0	1	0	0
(Level 3): v_{i3}	0	0	0	1	0

	z_{11}	z_{12}	z_{21}	z_{22}
objective function value	(¥) 1,552,800	7 (days)	(¥) 551,860	(¥) 100

In the optimal solution shown in Table 5.18, the transportation planning is one-sided to the regular transportation of Company L1, and Company L2 receives only the minimum orders. Therefore, the profit z_{22} of Company L2 is extremely small.

Similarly, the results of the remaining three problems, which minimize the opportunity loss z_{12} of Company U and maximize the profits of Companies L1 and L2, are summarized in Tables 5.19, 5.20 and 5.21.

Table 5.19 Optimal solution to the problem minimizing z_{12}.

customer i	1	2	3	4	5
orders to Company L1: y_i	150	130	150	70	80
orders to Company L2: z_i	150	20	100	30	20
assigned drivers (Level 1): u_{i1}	0	0	0	1	0
in Company L1 (Level 2): u_{i2}	0	0	0	0	1
(Level 3): u_{i3}	1	1	1	0	0
assigned drivers (Level 1): v_{i1}	0	1	0	0	1
in Company L2 (Level 2): v_{i2}	0	0	2	0	0
(Level 3): v_{i3}	2	0	0	1	0

	z_{11}	z_{12}	z_{21}	z_{22}
objective function value	(¥) 1,746,000	3 (days)	(¥) 439,260	(¥) 221,000

Table 5.20 Optimal solution to the problem minimizing z_{21}.

customer i	1	2	3	4	5
orders to Company L1: y_i	250	130	230	80	80
orders to Company L2: z_i	50	20	20	20	20
assigned drivers (Level 1): u_{i1}	1	0	0	1	0
in Company L1 (Level 2): u_{i2}	0	0	2	0	1
(Level 3): u_{i3}	1	1	0	0	0
assigned drivers (Level 1): v_{i1}	0	1	0	0	1
in Company L2 (Level 2): v_{i2}	0	0	1	1	0
(Level 3): v_{i3}	1	0	0	0	0

	z_{11}	z_{12}	z_{21}	z_{22}
objective function value	(¥) 1,576,000	8 (days)	(¥) 570,000	(¥) 17,200

For the result of minimization of the opportunity loss z_{12} of Company U shown in Table 5.19, the opportunity loss of the small lot transportation in Company L2 is zero while that of the regular transportation in Company L1 becomes three days because of the constraints (5.27a). Moreover, because the objective function (5.20) does not include the decision variables $\boldsymbol{y}$ and $\boldsymbol{z}$, it seems that the transportation planning is not one-sided to either the regular transportation or the small lot transportation in the planning shown in Table 5.19. Therefore, this planning produces impartial profits to Companies L1 and L2.

In the result of maximization of the profit z_{21} of Company L1 shown in Table 5.20, because the transportation is planned such that the income $\sum_{i=1}^{5} a_i y_i$ from

Table 5.21 Optimal solution to the problem minimizing z_{22}.

customer i	1	2	3	4	5
orders to Company L1: y_i	120	30	30	30	40
orders to Company L2: z_i	180	120	220	70	60
assigned drivers (Level 1): u_{i1}	0	0	0	1	0
in Company L1 (Level 2): u_{i2}	0	1	1	0	1
(Level 3): u_{i3}	1	0	0	0	0
assigned drivers (Level 1): v_{i1}	0	2	0	0	1
in Company L2 (Level 2): v_{i2}	0	0	2	1	0
(Level 3): v_{i3}	2	0	1	0	0

	z_{11}	z_{12}	z_{21}	z_{22}
objective function value	(¥) 2,152,000	6 (days)	(¥) 31,000	(¥) 852,400

Company U becomes larger, the solution is one-sided to the regular transportation. On the other hand, to keep expenses to the drivers as low as possible, Company L1 assigns the transportation to drivers with relatively lower ability, and consequently the transportation time and the opportunity loss of Company U become large.

Similarly, for maximization of the profit z_{22} of Company L2 shown in Table 5.21, the solution is one-sided to the small lot transportation. Because the capacity of transportation of Company L2 is not so large, the transportation to customer 1 by Company L1 is more than its minimal order, i.e., $y_1 = 120 > 30$.

The membership functions of the fuzzy goals for the objective functions are identified by taking account of the obtained solutions and the related information of the four individual programming problems with single objective functions. Suppose that the membership functions are linear (4.7), and the Zimmermann method (1978) is employed for determining parameters of the membership functions; the parameter z^1 is specified as the optimal value of the corresponding individual programming problem and the parameter z^0 is specified as the worst objective function value among the optimal solutions to the other individual programming problems. The obtained parameters are shown in Table 5.22.

Table 5.22 Parameters z^1 and z^0 of the linear membership functions.

	z_{11}	z_{12}	z_{21}	z_{22}
z^1	1,552,800	3	570,000	852,400
z^0	2,152,000	8	31,000	100

Next, by using the interactive fuzzy programming method (Sakawa and Nishizaki, 2002a), we try to derive a satisfactory solution, in which the satisfaction of Company U and those of Companies L1 and L2 are well-balanced, by updating the satisfactory degree of Company U.

5.2.2.2 Aggregation of the fuzzy goals of Company U

In our problem, Company U at the upper level has the two objective function, while each of Companies L1 and L2 at the lower level has a single objective function. Therefore, to apply the interactive fuzzy programming method directly, we aggregate the fuzzy goals for the two objective functions of Company U.

In this section, we examine three aggregation methods: the aggregation by a minimal component, the aggregation by weighting coefficients, and the aggregation by a distance from aspiration levels. The three aggregation methods are known as methods for scalarizing multiobjective mathematical programming problems (Chankong and Haimes, 1983; Yu, 1985; Steuer, 1986; Sakawa, 1993). The characteristics of the three methods are summarized as follows.

In the aggregation by a minimal component, the contours of the aggregated function are quadrate in a two-dimensional problem, and all of the objective functions are impartially maximized because the values of the aggregated function correspond to the Tchebycheff metric. Furthermore, the aggregation by a minimal component is also regarded as the fuzzy decision rule (Bellman and Zadeh, 1970) in decision making under fuzzy environments.

Concerning the aggregation by weighting coefficients, each of the weights for the objectives is interpreted as a relative degree of importance between the objective functions. Because the contours of the aggregated function by weighting coefficients are straight lines in a two-dimensional problem, the values of the objective functions may be biased for some of the objectives.

In the aggregation by a distance from aspiration levels, it follows that a solution closest to the aspiration levels specified by a decision maker is found. When we employ the Tchebycheff metric as a measure of distance, this aggregation has characteristics similar to the aggregation by a minimal component in which solutions farthest from the origin are searched.

Aggregation by a minimal component

The membership function aggregated by a minimal component is represented as

$$\mu_1^m(z_{11}, z_{12}) = \min\{\mu_{11}(z_{11}(\boldsymbol{y}, \boldsymbol{z})), \mu_{12}(z_{12}(\boldsymbol{u}, \boldsymbol{v}))\}. \tag{5.31}$$

Employing this aggregation implies that Company U maximizes the smaller degree between the satisfactory degrees of the fuzzy goals for the objective functions.

Aggregation by weighting coefficients

When Company U tries to maximize the weighted sum of the satisfactory degrees after specifying weighting coefficients for the two fuzzy goals, the aggregated membership function is represented by

$$\mu_1^w(z_{11}, z_{12}) = \alpha\mu_{11}(z_{11}(\boldsymbol{y}, \boldsymbol{z})) + (1-\alpha)\mu_{12}(z_{12}(\boldsymbol{u}, \boldsymbol{v})), \tag{5.32}$$

where α is a weighting coefficient of the first fuzzy goal μ_{11} and then $(1-\alpha)$ is that of the second one.

Aggregation by a distance from aspiration levels

When Company U has aspiration levels to the satisfactory degrees of the fuzzy goals for the two objective functions, the two membership functions are aggregated by using a distance from the aspiration levels as follows:

$$z_1(\mu_{11}, \mu_{12}) = \max\{\bar{\mu}_{11} - \mu_{11}(z_{11}(\boldsymbol{y}, \boldsymbol{z})), \bar{\mu}_{12} - \mu_{12}(z_{12}(\boldsymbol{u}, \boldsymbol{v}))\}, \tag{5.33}$$

where $\bar{\mu}_{11}$ and $\bar{\mu}_{12}$ are membership values representing the aspiration levels to the fuzzy goals prescribed by the membership functions μ_{11} and μ_{12}, respectively. Moreover, we interpret $z_1(\mu_{11}, \mu_{12})$ as the newly defined objective function and identify the linear membership function $\mu_1^d(z_{11}, z_{12})$ in a similar way.

5.2.2.3 First phase of the interactive fuzzy programming

Examining the three models of aggregation presented above, we apply the interactive fuzzy programming method to the transportation planning and work force assignment problem in the housing material manufacturer. In the first phase, Company U specifies a minimal satisfactory level subjectively. Taking into consideration overall satisfactory balance between the two levels, Company U updates the minimal satisfactory level if necessary, and a tentative solution is derived.

Model with the aggregation by a minimal component

For the problem with the fuzzy goal of Company U aggregated by a minimal component, the first phase of the interactive fuzzy programming method starts to solve the following problem for obtaining a solution which maximizes the smaller degree between the satisfactory degrees of the two levels:

$$\text{maximize} \quad \min\{\mu_1^m(z_{11}, z_{12}), \min\{\mu_{21}(z_{21}(\boldsymbol{y}, \boldsymbol{u})), \mu_{22}(z_{22}(\boldsymbol{z}, \boldsymbol{v}))\}\} \tag{5.34a}$$

$$\text{subject to} \quad (\boldsymbol{y}, \boldsymbol{z}, \boldsymbol{u}, \boldsymbol{v}) \in S. \tag{5.34b}$$

Accordingly, problem (5.34) is a problem maximizing the minimum among the four satisfactory degrees, i.e., the two membership functions of Company U and the two membership functions of Companies L1 and L2. An optimal solution to problem (5.34) is calculated, and it is shown in Table 5.23.

If Company U is not satisfied with the obtained solution, with the related information of the solution in mind, Company U specifies the minimal satisfactory

Table 5.23 Optimal solution to problem (5.34).

customer i	1	2	3	4	5
orders to Company L1: y_i	210	90	139	31	50
orders to Company L2: z_i	90	60	111	69	50
assigned drivers (Level 1): u_{i1}	1	0	0	1	0
in Company L1 (Level 2): u_{i2}	1	0	0	0	0
(Level 3): u_{i3}	0	1	1	0	1
assigned drivers (Level 1): v_{i1}	0	1	0	0	1
in Company L2 (Level 2): v_{i2}	0	0	2	0	0
(Level 3): v_{i3}	1	0	0	1	0

	z_{11}	z_{12}	z_{21}	z_{22}
objective function value	1,840,200	5	311,420	443,720
satisfactory degree $\mu(z)$	0.52	0.60	0.52	0.52
aggregated satisfactory degree	0.52		—	—

level $\hat{\delta}$ and formulates the following problem, in which the smaller degree between the satisfactory degrees of Companies L1 and L2 is maximized under the condition that the aggregated satisfactory degree of Company U is larger than or equal to the minimal satisfactory level:

$$\text{maximize} \quad \min\{\mu_{21}(z_{21}(\boldsymbol{y},\boldsymbol{u})),\mu_{22}(z_{22}(\boldsymbol{z},\boldsymbol{v}))\} \tag{5.35a}$$

$$\text{subject to} \quad \mu_1^m(z_{11},z_{12}) \geqq \hat{\delta} \tag{5.35b}$$

$$(\boldsymbol{y},\boldsymbol{z},\boldsymbol{u},\boldsymbol{v}) \in S. \tag{5.35c}$$

Table 5.24 Optimal solution to problem (5.35).

customer i	1	2	3	4	5
orders to Company L1: y_i	262	130	88	32	40
orders to Company L2: z_i	38	20	162	68	60
assigned drivers (Level 1): u_{i1}	0	0	0	1	0
in Company L1 (Level 2): u_{i2}	0	0	1	0	1
(Level 3): u_{i3}	2	1	0	0	0
assigned drivers (Level 1): v_{i1}	0	1	0	0	1
in Company L2 (Level 2): v_{i2}	0	0	1	0	0
(Level 3): v_{i3}	1	0	1	1	0

	z_{11}	z_{12}	z_{21}	z_{22}
objective function value	1,762,400	4	280,840	377,840
satisfactory degree $\mu(z)$	0.65	0.8	0.46	0.44
aggregated satisfactory degree	0.65		—	—

An optimal solution to problem (5.35) with $\hat{\delta} = 0.65$ is shown in Table 5.24. Comparing this solution with that of the maximin problem shown in Table 5.23, to satisfy the condition of the minimal satisfactory level, the satisfactory degrees of the two fuzzy goals of Company U are increased, and conversely those of Companies

L1 and L2 are decreased. Moreover, more drivers with higher ability are assigned in this solution, compared to the maximin solution. The ratio between the satisfactory degrees of the two levels becomes $\Delta = 0.68$, and it follows that a tentative solution is derived if the ratio is in the interval $[\Delta_{\min}, \Delta_{\max}]$ specified by Company U, and Company U accepts the solution.

To examine the characteristics of the model with the aggregation by a minimal component, we repeatedly solve problem (5.35), varying the parameter $\hat{\delta}$. Suppose that Company U specifies the bounds of the ratio of satisfaction at $[\Delta_{\min}, \Delta_{\max}] = [0.6, 0.8]$. The result is shown in Figure 5.9 depicting the relation between the satisfactory degrees and the minimal satisfactory level $\hat{\delta}$.

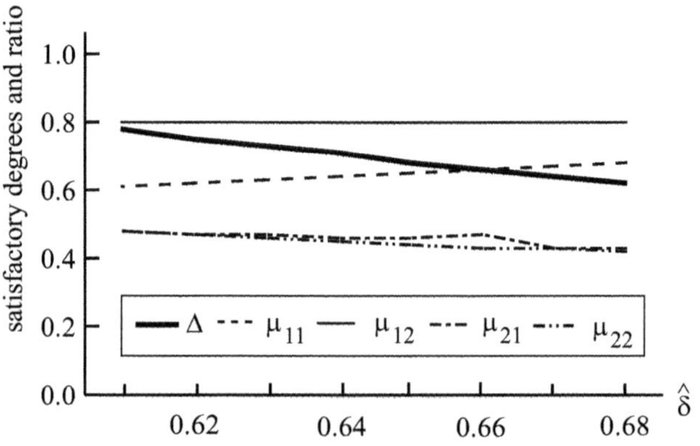

Fig. 5.9 The relation between the satisfactory degrees and $\hat{\delta}$ in the model with the aggregation by a minimal component.

The range $[0.61, 0.68]$ of the horizontal axis shown in Figure 5.9 is the interval in which the ratio Δ is in the bounds $[0.6, 0.8]$, i.e., if Company U specifies the minimal satisfactory level $\hat{\delta}$ in the interval $[0.61, 0.68]$, the ratio Δ is in the bounds $[\Delta_{\min}, \Delta_{\max}] = [0.6, 0.8]$. As seen in Figure 5.9, the satisfactory degree μ_{12} of the fuzzy goal for the opportunity loss with respect to the transportation time takes a fixed value of 0.8 for any $\hat{\delta} \in [0.61, 0.68]$. Because the aggregated satisfactory degree is represented by (5.31) and $\mu_{11} < \mu_{12} = 0.8$ holds for any $\hat{\delta} \in [0.61, 0.68]$, the aggregated satisfactory degree is always equal to the satisfactory degree μ_{11}, and therefore, it follows that the satisfactory degree μ_{12} for the second objective function cannot affect the aggregated satisfactory degree $\min\{\mu_{11}, \mu_{12}\}$ at all.

Model with the aggregation by weighting coefficients

We apply the first phase of the interactive fuzzy programming method to the problem with the fuzzy goal of Company U aggregated by weighting coefficients. In a way similar to the model with the aggregation by a minimal component, the first phase of the interactive fuzzy programming method starts to solve the following problem for obtaining a solution which maximizes the smaller degree between the satisfactory degrees of the two levels:

$$\begin{aligned} &\text{maximize} \;\; \min\{\mu_1^w(z_{11},z_{12}), \min\{\mu_{21}(z_{21}(\boldsymbol{y},\boldsymbol{u})), \mu_{22}(z_{22}(\boldsymbol{z},\boldsymbol{v}))\}\} && (5.36a)\\ &\text{subject to} \;\; (\boldsymbol{y},\boldsymbol{z},\boldsymbol{u},\boldsymbol{v}) \in S. && (5.36b)\end{aligned}$$

An optimal solution to Problem (5.36) with the weighting coefficients $(\alpha, 1-\alpha) = (0.5, 0.5)$ of the fuzzy goals μ_{11} and μ_{12} in (5.32) is shown in Table 5.25.

Table 5.25 Optimal solution to problem (5.36).

customer i	1	2	3	4	5
orders to Company L1: y_i	120	30	230	32	62
orders to Company L2: z_i	180	120	20	68	38
assigned drivers (Level 1): u_{i1}	0	0	0	1	0
in Company L1 (Level 2): u_{i2}	1	0	2	0	0
(Level 3): u_{i3}	0	1	0	0	1
assigned drivers (Level 1): v_{i1}	0	2	0	0	1
in Company L2 (Level 2): v_{i2}	0	0	1	0	0
(Level 3): v_{i3}	2	0	0	1	0

	z_{11}	z_{12}	z_{21}	z_{22}
objective function value	1,960,000	4	325,920	466,080
satisfactory degree $\mu(z)$	0.32	0.8	0.55	0.55
aggregated satisfactory degree	0.56		—	—

If Company U is not satisfied with the obtained solution, with the related information of the solution in mind, after specifying the minimal satisfactory level $\hat{\delta}$, Company U formulates the following problem, in which the smaller degree between the satisfactory degrees of Companies L1 and L2 is maximized under the condition that the aggregated satisfactory degree of Company U is larger than or equal to the minimal satisfactory level $\hat{\delta}$:

$$\begin{aligned} &\text{maximize} \;\; \min\{\mu_{21}(z_{21}(\boldsymbol{y},\boldsymbol{u})), \mu_{22}(z_{22}(\boldsymbol{z},\boldsymbol{v}))\} && (5.37a)\\ &\text{subject to} \;\; \mu_1^w(z_{11},z_{12}) \geqq \hat{\delta} && (5.37b)\\ &\qquad\qquad (\boldsymbol{y},\boldsymbol{z},\boldsymbol{u},\boldsymbol{v}) \in S. && (5.37c)\end{aligned}$$

An optimal solution to problem (5.37) with $\hat{\delta} = 0.72$ is shown in Table 5.26. Comparing the solution in Table 5.26 with the solution in Table 5.25, the small lot transportation of Company L2 becomes larger, and Company L1 decreases the

Table 5.26 Optimal solution to problem (5.37).

customer i	1	2	3	4	5
orders to Company L1: y_i	120	91	122	62	75
orders to Company L2: z_i	180	59	128	38	25
assigned drivers (Level 1): u_{i1}	0	0	0	1	0
in Company L1 (Level 2): u_{i2}	1	0	0	0	0
(Level 3): u_{i3}	0	1	1	0	1
assigned drivers (Level 1): v_{i1}	0	1	0	0	1
in Company L2 (Level 2): v_{i2}	0	0	2	0	0
(Level 3): v_{i3}	2	0	0	1	0

	z_{11}	z_{12}	z_{21}	z_{22}
objective function value	1,888,200	3	316,000	450,740
satisfactory degree $\mu(z)$	0.44	1.0	0.53	0.53
aggregated satisfactory degree	0.72		—	—

number of assigned drivers to cope with the reduction of transportation ordered by Company U.

To examine the characteristics of the model with the aggregation by weighting coefficients, we solve problem (5.37), varying the parameter $\hat{\delta}$. Assume that Company U specifies the bounds of the ratio of satisfaction at $[\Delta_{\min}, \Delta_{\max}] = [0.6, 0.8]$. The result is shown in Figure 5.10 depicting the relation between the satisfactory degrees and the minimal satisfactory level $\hat{\delta}$.

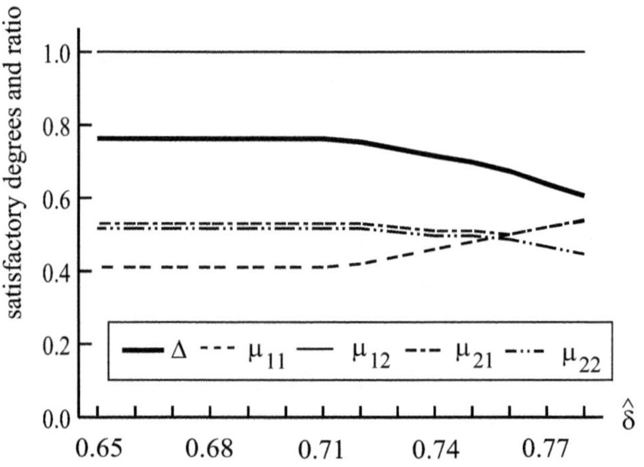

Fig. 5.10 The relation between the satisfactory degrees and $\hat{\delta}$ in the model with the aggregation by weighting coefficients.

The range $[0.65, 0.78]$ of the horizontal axis shown in Figure 5.10 is the interval in which the ratio Δ is in the bounds $[0.6, 0.8]$ as in Figure 5.9. As seen in Figure 5.10, the satisfactory degree μ_{12} of the fuzzy goal for the opportunity loss with respect to the transportation time takes a fixed value of 1.0 for any $\hat{\delta} \in [0.65, 0.78]$, and the satisfactory degree μ_{11} of the cost of transportation is in the interval $\mu_{11} \in [0.4, 0.6]$. The value of μ_{11} is smaller than that of the model with the aggregation by a minimal component, and there is a larger difference between the two satisfactory degrees μ_{11} and μ_{12} in this model. Moreover, the interval of $\hat{\delta}$ that the ratio Δ is in the specified bounds $[0.6, 0.8]$ is calculated as $[0.65, 0.78]$, and one finds that it is wider than that of the previous model.

Model with the aggregation by a distance from aspiration levels

We apply the first phase of the interactive fuzzy programming method to the problem with the fuzzy goal of Company U aggregated by a distance from aspiration levels. In a way similar to the previous two models, the first phase of the interactive fuzzy programming method starts to solve the following problem for obtaining a solution which maximizes the smaller degree between the satisfactory degrees of the two levels:

$$\text{maximize} \quad \min\{\mu_1^d(z_{11}, z_{12}), \min\{\mu_{21}(z_{21}(\boldsymbol{y}, \boldsymbol{u})), \mu_{22}(z_{22}(\boldsymbol{z}, \boldsymbol{v}))\}\} \qquad (5.38a)$$

$$\text{subject to} \quad (\boldsymbol{y}, \boldsymbol{z}, \boldsymbol{u}, \boldsymbol{v}) \in S. \qquad (5.38b)$$

Assume that Company U specifies the aspiration levels in (5.33) at $\bar{\mu}_{11} = 1.0$ and $\bar{\mu}_{12} = 0.8$, and the parameters of the linear membership function are determined as $z^1 = 0.12$ and $z^0 = 1.0$ based on the Zimmermann method (1978). Then, an optimal solution to problem (5.38) with these parameters is shown in Table 5.27. In particular, it should be noted that the difference between the two satisfactory degrees of Company U, μ_{11} and μ_{12}, is smaller than those of the previous two models.

Table 5.27 Optimal solution to problem (5.38).

customer i	1	2	3	4	5
orders to Company L1: y_i	120	122	107	73	48
orders to Company L2: z_i	180	28	143	27	52
assigned drivers (Level 1): u_{i1}	0	0	0	1	0
in Company L1 (Level 2): u_{i2}	1	0	1	0	1
(Level 3): u_{i3}	0	1	0	0	0
assigned drivers (Level 1): v_{i1}	0	1	0	0	1
in Company L2 (Level 2): v_{i2}	0	0	2	1	0
(Level 3): v_{i3}	2	0	0	0	0

	z_{11}	z_{12}	z_{21}	z_{22}
objective function value	1,869,400	6	320,660	457,920
satisfactory degree $\mu(z)$	0.47	0.4	0.54	0.54
aggregated satisfactory degree	0.54		—	—

If Company U is not satisfied with the obtained solution, taking account of the related information of the solution, after specifying the minimal satisfactory level $\hat{\delta}$,

Company U formulates the following problem, in which the smaller degree between the satisfactory degrees of Companies L1 and L2 is maximized under the condition that the aggregated satisfactory degree of Company U is larger than or equal to the minimal satisfactory level:

$$\text{maximize} \quad \min\{\mu_{21}(z_{21}(\boldsymbol{y},\boldsymbol{u})),\mu_{22}(z_{22}(\boldsymbol{z},\boldsymbol{v}))\} \tag{5.39a}$$

$$\text{subject to} \quad \mu_1^d(z_{11},z_{12}) \geqq \hat{\delta} \tag{5.39b}$$

$$(\boldsymbol{y},\boldsymbol{z},\boldsymbol{u},\boldsymbol{v}) \in S. \tag{5.39c}$$

An optimal solution to problem (5.39) with $\hat{\delta} = 0.72$ is shown in Table 5.28.

Table 5.28 Optimal solution to problem (5.39).

customer i	1	2	3	4	5
orders to Company L1: y_i	249	128	88	40	41
orders to Company L2: z_i	51	22	162	60	59
assigned drivers (Level 1): u_{i1}	1	0	0	0	0
in Company L1 (Level 2): u_{i2}	0	0	0	1	1
(Level 3): u_{i3}	1	1	1	0	0
assigned drivers (Level 1): v_{i1}	0	1	0	0	1
in Company L2 (Level 2): v_{i2}	1	0	1	0	0
(Level 3): v_{i3}	0	0	1	1	0

	z_{11}	z_{12}	z_{21}	z_{22}
objective function value	1,773,200	5	284,820	401,420
satisfactory degree $\mu(z)$	0.63	0.6	0.47	0.47
aggregated satisfactory degree	0.72		—	—

In the solution in Table 5.28, to satisfy the condition of the minimal satisfactory level, some parts of the regular transportation to customers 1 and 2 are switched to the small lot transportation, and some parts of the small lot transportation to customers 3, 4 and 5 are switched to the regular transportation, compared to the solution in Table 5.27. Moreover, drivers with higher ability level are newly assigned for the increased orders in both forwarding agents. It should be noted that the difference between μ_{11} and μ_{12} is still small.

To examine the characteristics of the model with the aggregation by a distance from aspiration levels, we solve problem (5.39), varying the parameter $\hat{\delta}$. Assume that Company U specifies the bounds of the ratio of satisfactory degrees at $[\Delta_{\min},\Delta_{\max}] = [0.6,0.8]$. The result is shown in Figure 5.11 depicting the relation between the satisfactory degrees and the minimal satisfactory level $\hat{\delta}$.

The range $[0.66,0.75]$ of the horizontal axis shown in Figure 5.11 means the interval in which the ratio Δ is in the bounds $[0.6,0.8]$ as in Figures 5.9 and 5.10. As seen in Figure 5.11, the satisfactory degree μ_{11} increases within the interval $[0.58,0.7]$, and the satisfactory degree μ_{12} for the opportunity loss is smaller than μ_{11} for the cost of transportation. Whenever $\hat{\delta}$ is the interval $[0.66,0.75]$, the ratio

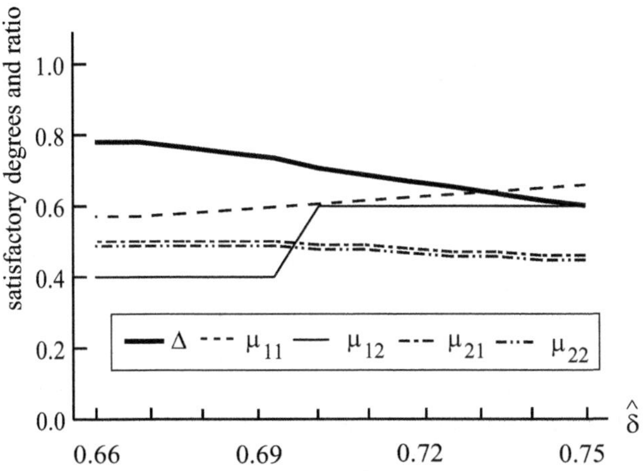

Fig. 5.11 The relation between the satisfactory degrees and $\hat{\delta}$ in the model with the aggregation by a distance from aspiration levels.

Δ is in the specified bounds $[0.6, 0.8]$, and the interval $[0.66, 0.75]$ is wider than the corresponding interval in the model with the aggregation by a minimal component.

5.2.2.4 Second phase of the interactive fuzzy programming

In the first phase of the interactive fuzzy programming, Companies L1 and L2 are treated impartially, and therefore, they can be regarded as a single group. However, if Company U wants to give either of Companies L1 and L2 special consideration, Company U might specify different intervals $[\Delta^i_{\min}, \Delta^i_{\max}]$, $i = 1, 2$ for them separately. In such a case, it is necessary to execute the second phase of the interactive fuzzy programming. In the second phase, by coordinating the aggregated satisfactory degree of Company U and those of Companies L1 and L2, the final satisfactory solution is derived.

First, we calculate the following individual ratios between the aggregated satisfactory degree of Company U and the satisfactory degree of Company L1 or that of Company L2 with respect to the solution obtained in the first phase:

$$\Delta^i = \frac{\mu_{2i}(z_{2i})}{\mu_1(z_{11}, z_{12})}, \; i = 1, 2, \tag{5.40}$$

where $\mu_1(z_{11}, z_{12})$ denotes the aggregated satisfactory degree of Company U.

For a forwarding agent whose ratio of satisfactory degrees with Company U is not in the interval $[\Delta^i_{\min}, \Delta^i_{\max}]$ assigned by Company U, after specifying a minimal satisfactory level $\bar{\delta}^i_{\min}$ or a maximal satisfactory level $\bar{\delta}^i_{\max}$ additionally, Company U formulates the following problem for $i = 1, 2$, $j \neq i$:

$$\begin{aligned}
&\text{maximize} && \mu_{2j}(z_{2j}) && \text{(5.41a)}\\
&\text{subject to} && \mu_1(z_{11}, z_{12}) \geqq \hat{\delta} && \text{(5.41b)}\\
&&& \bar{\delta}^i_{\min} \leqq \mu_{2i}(z_{2i}) \leqq \bar{\delta}^i_{\max} && \text{(5.41c)}\\
&&& (\boldsymbol{y}, \boldsymbol{z}, \boldsymbol{u}, \boldsymbol{v}) \in S. && \text{(5.41d)}
\end{aligned}$$

Assume that Company U adopts a policy that Company U gives favorable treatment to Company L1 dealing with the regular transportation, and specifies the bounds of the ratios of satisfactory degrees for Companies L1 and L2 at $[\Delta^1_{\min}, \Delta^1_{\max}] = [0.7, 0.8]$ and $[\Delta^2_{\min}, \Delta^2_{\max}] = [0.6, 0.7]$, respectively.

From the examination described above, because it is found that, in the model with the aggregation by a minimal component, the satisfactory degree for the second objective function does not affect the aggregated satisfactory degree at all, we will deal with only the model with the aggregation by weighting coefficients and the model with the aggregation by a distance from aspiration levels in the second phase of the interactive fuzzy programming method.

Model with the aggregation by weighting coefficients

Assume that, in the first phase, Company U is satisfied with the tentative solution which is obtained by solving problem (5.37) with $\hat{\delta} = 0.72$. The corresponding individual ratios of satisfactory degrees are $\Delta^1 = \Delta^2 = 0.74$, and one finds that $\Delta^1 \in [0.7, 0.8]$ and $\Delta^2 \notin [0.6, 0.7]$. Because the ratio Δ^2 is over the upper bound of 0.7, Company U should specify the permissible minimal level $\bar{\delta}^2_{\min}$ for μ_{22}. Then, assume that the following problem is formulated:

$$\begin{aligned}
&\text{maximize} && \mu_{21}(z_{21}(\boldsymbol{y}, \boldsymbol{u})) && \text{(5.42a)}\\
&\text{subject to} && \mu^w_1(z_{11}, z_{12}) \geqq \hat{\delta} = 0.72 && \text{(5.42b)}\\
&&& 0.504 = \bar{\delta}^2_{\min} \leqq \mu_{22}(z_{22}(\boldsymbol{z}, \boldsymbol{v})), && \text{(5.42c)}\\
&&& (\boldsymbol{y}, \boldsymbol{z}, \boldsymbol{u}, \boldsymbol{v}) \in S, && \text{(5.42d)}
\end{aligned}$$

where the permissible minimal level $\bar{\delta}^2_{\min}$ for μ_{22} is determined at $\bar{\delta}^2_{\min} = 0.72 \cdot 0.7 = 0.504$. An optimal solution to problem (5.42) is shown in Table 5.29.

As seen in Table 5.29, compared to the transportation planning corresponding to the tentative solution obtained in the first phase which is shown in Table 5.26, some part of the small lot transportation to customer 3 is switched to the regular transportation, and conversely some part of the regular transportation to customer 4 is switched to the small lot transportation. From these facts, it is thought that the satisfactory degree of Company L1 increases, and then the ratio Δ^2 moves into the valid interval.

To examine the characteristics of this model in the second phase of the interactive fuzzy programming, we solve problem (5.42), varying the parameter $\bar{\delta}^2_{\min}$. The result

Table 5.29 Optimal solution to problem (5.42).

customer i	1	2	3	4	5
orders to Company L1: y_i	120	90	150	32	74
orders to Company L2: z_i	180	60	100	68	26
assigned drivers (Level 1): u_{i1}	0	0	0	1	0
in Company L1 (Level 2): u_{i2}	1	0	0	0	0
(Level 3): u_{i3}	0	1	1	0	1
assigned drivers (Level 1): v_{i1}	0	1	0	0	1
in Company L2 (Level 2): v_{i2}	0	0	2	0	0
(Level 3): v_{i3}	2	0	0	1	0

	z_{11}	z_{12}	z_{21}	z_{22}
objective function value	1,887,200	3	339,520	426,520
satisfactory degree $\mu(z)$	0.44	1.0	0.57	0.50
ratio of satisfactory degrees	$\mu_1^w = 0.72$		$\Delta^1 = 0.79$	$\Delta^2 = 0.69$

is shown in Figure 5.12 depicting the relation between the ratios of satisfactory degrees and the permissible minimal level $\bar{\delta}^2_{\min}$.

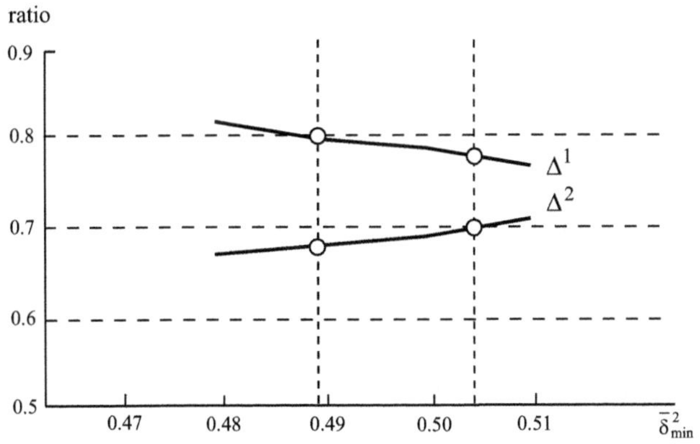

Fig. 5.12 The relation between the ratios of satisfactory degrees and $\bar{\delta}^2_{\min}$ in the model with the aggregation by weighting coefficients.

As seen in Figure 5.12, because Δ^1 decreases and Δ^2 increases as $\bar{\delta}^2_{\min}$ increases, it is found that there exists some conflict between μ_{21} and μ_{22}. Moreover, a value of $\bar{\delta}^2_{\min}$ in the interval $[0.489, 0.504]$ can lead to a possible satisfactory solution in the second phase, and therefore, it is found that the range of the parameter $\bar{\delta}^2_{\min}$ which the decision maker can control is very narrow.

Model with the aggregation by a distance from aspiration levels

Assume that, in the first phase, Company U is satisfied with the tentative solution which is obtained by solving problem (5.39) with $\hat{\delta} = 0.72$. The corresponding individual ratios of satisfactory degrees are $\Delta^1 = \Delta^2 = 0.65$, and one finds that $\Delta^1 \notin [0.7, 0.8]$ and $\Delta^2 \in [0.6, 0.7]$. Because the ratio Δ^1 is smaller than the lower bound of 0.7, Company U should specify the permissible minimal level $\bar{\delta}^1_{\min}$ for μ_{21}. Then, suppose that the following problem is formulated:

$$
\begin{aligned}
\text{maximize} \quad & \mu_{22}(z_{22}(\boldsymbol{z}, \boldsymbol{v})) && \text{(5.43a)}\\
\text{subject to} \quad & \mu_1^d(z_{11}, z_{12}) \geqq \hat{\delta} = 0.72 && \text{(5.43b)}\\
& 0.504 = \bar{\delta}^1_{\min} \leqq \mu_{21}(z_{21}(\boldsymbol{y}, \boldsymbol{u})) && \text{(5.43c)}\\
& (\boldsymbol{y}, \boldsymbol{z}, \boldsymbol{u}, \boldsymbol{v}) \in S, && \text{(5.43d)}
\end{aligned}
$$

where the permissible minimal level $\bar{\delta}^1_{\min}$ for μ_{21} is determined at $\bar{\delta}^1_{\min} = 0.72 \cdot 0.7 = 0.504$. An optimal solution to problem (5.43) is shown in Table 5.30.

Table 5.30 Optimal solution to problem (5.43).

customer i	1	2	3	4	5
orders to Company L1: y_i	249	120	106	45	40
orders to Company L2: z_i	51	30	144	55	60
assigned drivers (Level 1): u_{i1}	1	0	0	0	0
in Company L1 (Level 2): u_{i2}	0	0	0	1	1
(Level 3): u_{i3}	1	1	1	0	0
assigned drivers (Level 1): v_{i1}	0	1	0	0	1
in Company L2 (Level 2): v_{i2}	1	0	2	0	0
(Level 3): v_{i3}	0	0	0	1	0

	z_{11}	z_{12}	z_{21}	z_{22}
objective function value	1,773,200	5	302,740	394,500
satisfactory degree $\mu(z)$	0.63	0.60	0.50	0.46
ratio of satisfactory degrees	$\mu_1^d = 0.72$		$\Delta^1 = 0.70$	$\Delta^2 = 0.64$

As seen in Table 5.30, compared to the transportation planning corresponding to the tentative solution obtained in the first phase which is shown in Table 5.28, by increasing some amount of the order to the regular transportation, the satisfactory degree of Company L1 can be increased with the small reduction of the satisfactory degree of Company L2 so as to satisfy the conditions of the ratios of satisfactory degrees. It is also seen that the orders to the regular transportation are larger than those of the transportation planning corresponding to the solution of the model with the aggregation by weighting coefficients shown in Table 5.29.

To examine the characteristics of this model in the second phase of the interactive fuzzy programming, we solve problem (5.43), varying the parameter $\bar{\delta}^1_{\min}$. The result is shown in Figure 5.13 depicting the relation between the ratios of satisfactory degrees and the permissible minimal level $\bar{\delta}^1_{\min}$.

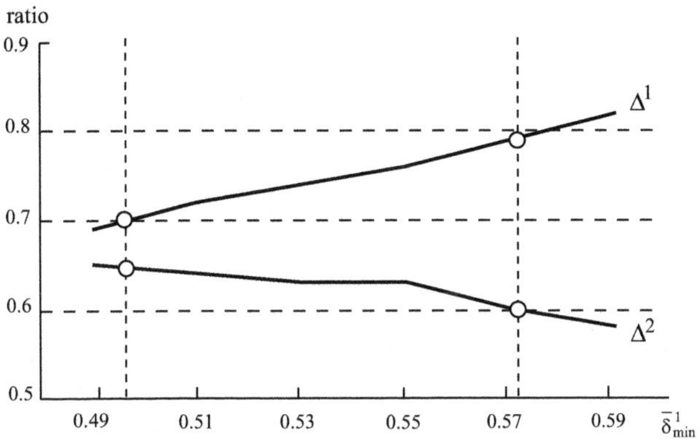

Fig. 5.13 The relation between the ratios of satisfactory degrees and $\bar{\delta}^1_{\min}$ in the model with the aggregation by a distance from aspiration levels.

As seen in Figure 5.13, because Δ^1 increases and Δ^2 decreases as $\bar{\delta}^1_{\min}$ increases, one finds that there also exists some conflict between μ_{21} and μ_{22}. Moreover, a value of $\bar{\delta}^1_{\min}$ in the interval $[0.504, 0.573]$ can lead to a possible satisfactory solution in the second phase, and therefore, it is found that the range of the parameter $\bar{\delta}^1_{\min}$ which the decision maker can control is wider than that of the model with the aggregation by weighting coefficients.

From the analysis described above, in the model with the aggregation by a distance from aspiration levels, it is easier to control the two satisfactory degrees of Company U and the range of the parameter $\bar{\delta}^i_{\min}$ that Company U can control is wider than the others. It seems reasonable to conclude, from these facts, that employing the model with the aggregation by a distance from aspiration levels is appropriate for the decentralized two-level transportation problem in the housing material manufacturer.

5.3 Two-level purchase problem for food retailing

In this section, we deal with a purchase problem for food retailing, and formulate a two-level linear programming problem with a food retailer and a distributer under a noncooperative decision making environment. Many people in Japan buy vegetables and fruits in food supermarkets, and the food supermarkets usually purchase such fresh produce from distributers who obtain them in central wholesale markets. In Japan, 80% of vegetables and 60% of fruits are distributed by way of wholesale markets (Ministry of Agriculture, Forestry and Fisheries of Japan, 2008), and this fact means that the wholesale markets have been fulfilling as an efficient intermediary role connecting consumers and farm producers.

Because Japanese consumers tend to buy small amounts of vegetables and fruits frequently, food retailers such as supermarkets must provide a wide range of fresh products every day. To cope with Japanese consumer behavior, in most situations, food retailers do not obtain vegetables and fruits by themselves but contract with distributers to purchase them. This method of purchasing decreases the transaction cost and enables distributers to supply a wider range of fresh products for customers in a timely manner (Kidachi, 2006).

To take into account the mutual interdependence of a food retailer and a distributer, we formulates a decision problem on the purchase of food for retailing as a two-level linear programming problem with self-interested decision makers where the profits of the food retailer and the distributer are maximized. In this problem, the food retailer first specifies the order quantities of vegetables and fruits, and after receiving the order from the food retailer, the distributer determines purchase volumes of them at each of the central wholesale markets in several major cities in Japan. Although the food retailer and the distributer in this application are hypothetical decision makers, data used in the mathematical modeling are realistic.

5.3.1 Problem formulation

The food retailer deals with n kinds of vegetables and fruits which are purchased from the distributer. The distributer buys vegetables and fruits ordered from the food retailer at the central wholesale markets in s cities, and transports them by truck from each of the central wholesaler markets to the food retailer's storehouse in Tokyo. The two decision makers make an agreement that the distributer has an obligation to transport the foods to the storehouse, but the cost of the transportation is paid by the food retailer.

Let x_i, $i = 1\ldots,n$ denote an order quantity of food i specified by the food retailer to the distributer, and let y_{ji}, $j = 1,\ldots,s$, $i = 1\ldots,n$ denote a purchase volume of food i at the central wholesale market in city j. For concise representation, on occasion the decision variables are expressed by $\boldsymbol{x}^T = (x_1,\ldots,x_n)$ and $\boldsymbol{y}^T = (\boldsymbol{y}_1^T,\ldots,\boldsymbol{y}_s^T)$, $\boldsymbol{y}_j^T = (y_{j1},\ldots,y_{jn})$, $j = 1,\ldots,s$.

Objective functions

The profit of the food retailer is represented by

$$z_1(\boldsymbol{x},\boldsymbol{y}) = \sum_{i=1}^{n} a_i x_i - \sum_{j=1}^{s}\sum_{i=1}^{n} b_{ji} y_{ji}, \tag{5.44}$$

where a_i is the margin per unit of food i, and b_{ji} is the transportation cost per unit of food i from city j.

The profit of the distributer is represented by

$$z_2(\boldsymbol{x},\boldsymbol{y}) = \sum_{i=1}^{n} c_i x_i - \sum_{j=1}^{s}\sum_{i=1}^{n} d_{ji} y_{ji}, \tag{5.45}$$

where c_i is the selling price of food i to the food retailer, and d_{ji} is the purchase price of food i at the central wholesale market in city j.

Constraints

Let W be the capacity of the storehouse of the food retailer, and let v_i be the cubic volume per unit of food i. The constraint for the storehouse is represented by

$$\sum_{i=1}^{n} v_i x_i \leqq W. \tag{5.46}$$

For any food i, an order quantity of food i is specified by the food retailer between the lower limit D_i^L and the upper limit D_i^U, taking into account the volume of inventories. Then, the constraints for the upper and lower limits are represented by

$$D_i^L \leqq x_i \leqq D_i^U,\ i = 1,\ldots,n. \tag{5.47}$$

The distributer buys food i at one or more central wholesale markets, and then the total volume of food i must be larger than or equal to the quantity ordered by the food retailer. Thus, the constraints for order quantities are represented by

$$\sum_{j=1}^{s} y_{ji} \geqq x_i,\ i = 1,\ldots,n. \tag{5.48}$$

Moreover, there are constraints on financial resources of the distributer for purchasing foods at the central wholesaler markets, and they are expressed by

$$\sum_{i=1}^{n} d_{ji} y_{ji} \leqq o_j,\ j = 1,\ldots,s, \tag{5.49}$$

where o_j is the budget cap in city j.

Two-level linear programming problem

A two-level linear programming problem for purchase in food retailing, in which the objective functions (5.44) and (5.45) are maximized under the constraints described above (5.46)–(5.49), is formulated as follows:

$$\text{maximize } z_1(\boldsymbol{x},\boldsymbol{y}) = \sum_{i=1}^{n} a_i x_i - \sum_{j=1}^{s}\sum_{i=1}^{n} b_{ji} y_{ji} \tag{5.50a}$$

where $\boldsymbol{y}$ solves

$$\text{maximize } z_2(\boldsymbol{x},\boldsymbol{y}) = \sum_{i=1}^{n} c_i x_i - \sum_{j=1}^{s}\sum_{i=1}^{n} d_{ji} y_{ji} \tag{5.50b}$$

$$\text{subject to } \sum_{i=1}^{n} v_i x_i \leqq W,\ i = 1,\dots,n \tag{5.50c}$$

$$D_i^L \leqq x_i \leqq D_i^U,\ i = 1,\dots,n \tag{5.50d}$$

$$\sum_{j=1}^{s} y_{ji} \geqq x_i,\ i = 1,\dots,n \tag{5.50e}$$

$$\sum_{i=1}^{n} d_{j2i} y_{ji} \leqq o_j,\ j = 1,\dots,s \tag{5.50f}$$

$$\boldsymbol{x} \geqq \boldsymbol{0},\ \boldsymbol{y} \geqq \boldsymbol{0}. \tag{5.50g}$$

5.3.2 Parameter setting and Stackelberg solution

We assume that the food retailer sells 16 vegetables and fruits, i.e., $n = 16$, and the distributer purchases them at central wholesale markets in 8 cities, i.e., $s = 8$. The retail and the purchase prices of the food retailer's 16 items are shown in Table 5.31, and the margin per unit a_i of food i is the difference between the retail price and the purchase price c_i. Foods i, $i = 1,\dots,16$ represent onions, potatoes, cabbage, Japanese radish, Chinese cabbage, carrots, cucumbers, lettuce, tomatoes, spinach, eggplant, apples, bananas, strawberries, mandarin oranges, and lemons, respectively; and cities j, $j = 1,\dots,8$ stand for Sapporo, Sendai, Niigata, Kanazawa, Tokyo, Osaka, Hiroshima, and Miyazaki, respectively. The retail prices are specified such that the cost to sales ratios range from 50% to 75%, and the average cost to sales ratios of the 16 items is about 60%. The purchase prices of the food retailer corresponding to the selling prices of the distributer are about 95% of the wholesale prices at the central wholesale market in Tokyo. The wholesale prices d_{ji} in each city are shown in Table 5.32, and these prices are the averages of prices in March, 2008 at the central wholesale markets.

The fresh foods are transported from each of the 8 cities to the storehouse of the food retailer in Tokyo by truck. The transportation cost per unit b_{ji} of food i from city j to the storehouse is given in Table 5.33, and it is calculated under the assumption that the capacity of a truck is 8 tons, express toll highways are utilized, and the cost of fuel is ¥116 per liter. The capacity of the storehouse is 150 [m^2] $\times$ 2 [m], and the cubic volumes of food i per kilogram are shown in Table 5.34.

The lower limit D_i^L of an order quantity of food i is determined by reference to the demand of 10,000 households, and the upper limit D_i^U is set from 1.1 to 1.4

Table 5.31 Retail and purchase prices of fresh foods [yen/kg].

	food 1	food 2	food 3	food 4	food 5	food 6
retail	150.417	158.785	197.6	136.167	191.727	256.500
purchase c_i	90.25	111.15	98.8	81.7	105.45	179.55
	food 7	food 8	food 9	food 10	food 11	food 12
retail	370.500	269.167	533.462	392.214	651.182	377.077
purchase c_i	259.35	161.5	346.75	274.55	358.15	245.1
	food 13	food 14	food 15	food 16		
retail	279.300	1183.066	282.077	500.909		
purchase c_i	139.65	887.3	183.35	275.5		

Table 5.32 Wholesale prices in each city [yen/kg].

	food 1	food 2	food 3	food 4	food 5	food 6
city 1 d_{1i}	55	57	100	102	104	156
city 2 d_{2i}	78	87	113	95	115	187
city 3 d_{3i}	73	90	98	85	114	169
city 4 d_{4i}	83	105	103	83	113	178
city 5 d_{5i}	95	117	104	86	111	189
city 6 d_{6i}	111	110	88	71	97	189
city 7 d_{7i}	92	81	87	72	104	179
city 8 d_{8i}	85	106	72	60	88	151
	food 7	food 8	food 9	food 10	food 11	food 12
city 1 d_{1i}	288	229	349	339	421	221
city 2 d_{2i}	270	168	394	284	336	250
city 3 d_{3i}	274	186	312	335	342	231
city 4 d_{4i}	276	188	429	296	373	226
city 5 d_{5i}	273	170	365	289	377	258
city 6 d_{6i}	260	173	317	287	368	274
city 7 d_{7i}	248	138	300	257	315	265
city 8 d_{8i}	217	93	249	242	260	249
	food 13	food 14	food 15	food 16		
city 1 d_{1i}	157	926	195	294		
city 2 d_{2i}	165	867	198	353		
city 3 d_{3i}	149	743	168	283		
city 4 d_{4i}	115	872	159	290		
city 5 d_{5i}	147	934	193	290		
city 6 d_{6i}	147	939	156	310		
city 7 d_{7i}	176	693	168	301		
city 8 d_{8i}	186	782	150	231		

times the quantities of the lower limit D_i^L; these figures are shown in Table 5.35. The budget caps o_j on purchases in 8 cities are given in Table 5.36.

We computed the Stackelberg solution to problem (5.50) with parameters shown in Tables 5.31–5.36 by using the kth best method (Bialas and Karwan, 1984) and the Hansen, Jaumard and Savard method (1992). The solution is given in Table 5.37. We used a PC with Intel Pentium IV 2.80 GHz, and the computational times of the kth best method and the Hansen, Jaumard and Savard method were 2186.296 seconds and 5.531 seconds, respectively.

Table 5.33 Transportation costs [yen/kg].

	food 1	food 2	food 3	food 4	food 5	food 6
city 1 b_{1i}	12.476020	7.984653	12.476020	2.694820	9.980816	5.988490
city 2 b_{2i}	2.834936	1.814359	2.834936	0.612346	2.267949	1.360769
city 3 b_{3i}	2.837123	1.815758	2.837123	0.612818	2.269698	1.361819
city 4 b_{4i}	3.882100	2.484544	3.882100	0.838534	3.105680	1.863408
city 5 b_{5i}	0.202730	0.129747	0.202730	0.043790	0.162184	0.097310
city 6 b_{6i}	4.553846	2.914462	4.553846	0.983631	3.643077	2.185846
city 7 b_{7i}	6.225852	3.984545	6.225852	1.344784	4.980682	2.988409
city 8 b_{8i}	10.273461	6.575015	10.273461	2.219068	8.218769	4.931261
	food 7	food 8	food 9	food 10	food 11	food 12
city 1 b_{1i}	2.495204	59.884896	4.990408	39.923264	24.952040	9.980816
city 2 b_{2i}	0.566987	13.607693	1.133974	9.071796	5.669872	2.267949
city 3 b_{3i}	0.567425	13.618188	1.134849	9.078792	5.674245	2.269698
city 4 b_{4i}	0.776420	18.634078	1.552840	12.422719	7.764199	3.105680
city 5 b_{5i}	0.040546	0.973104	0.081092	0.648736	0.405460	0.162184
city 6 b_{6i}	0.910769	21.858463	1.821539	14.572308	9.107693	3.643077
city 7 b_{7i}	1.245170	29.884090	2.490341	19.922727	12.451704	4.980682
city 8 b_{8i}	2.054692	49.312615	4.109385	32.875076	20.546923	8.218769
	food 13	food 14	food 15	food 16		
city 1 b_{1i}	2.495204	4.990408	3.742806	3.742806		
city 2 b_{2i}	0.566987	1.133974	0.850481	0.850481		
city 3 b_{3i}	0.567425	1.134849	0.851137	0.851137		
city 4 b_{4i}	0.776420	1.552840	1.164630	1.164630		
city 5 b_{5i}	0.040546	0.081092	0.060819	0.060819		
city 6 b_{6i}	0.910769	1.821539	1.366154	1.366154		
city 7 b_{7i}	1.245170	2.490341	1.867756	1.867756		
city 8 b_{8i}	2.054692	4.109385	3.082038	3.082038		

Table 5.34 Cubic volumes of foods [cm^3/kg].

	food 1	food 2	food 3	food 4	food 5	food 6
cubic volume v_i	5000	3200	5000	1080	4000	2400
	food 7	food 8	food 9	food 10	food 11	food 12
cubic volume v_i	1000	24000	2000	16000	10000	4000
	food 13	food 14	food 15	food 16		
cubic volume v_i	1000	2000	1500	1500		

Table 5.35 Lower and upper limits of the foods [kg].

	food 1	food 2	food 3	food 4	food 5	food 6
lower limit D_i^L	4000	4000	2000	5000	10000	2000
upper limit D_i^U	5000	5000	2400	6000	14000	2500
	food 7	food 8	food 9	food 10	food 11	food 12
lower limit D_i^L	800	1500	3000	3000	1200	6000
upper limit D_i^U	1000	2000	4000	3600	1500	6600
	food 13	food 14	food 15	food 16		
lower limit D_i^L	12500	6000	4000	1000		
upper limit D_i^U	14500	7500	4800	1300		

Table 5.36 Caps on purchase for eight cities [yen].

	city 1	city 2	city 3	city 4	city 5	city 6	city 7	city 8
cap o_j	2000000	1500000	1500000	1500000	1500000	1500000	1500000	2000000

Table 5.37 Result of two-level purchase problem for food retailing.

	food 1	food 2	food 3	food 4	food 5	food 6
order quantity [kg]: x_i	4550	4000	2400	5000	10000	2000
purchase volume at city 1 [kg]: y_{1i}	4550	4000	0	0	0	2000
purchase volume at city 2 [kg]: y_{2i}	0	0	0	0	0	0
purchase volume at city 3 [kg]: y_{3i}	0	0	0	0	0	0
purchase volume at city 4 [kg]: y_{4i}	0	0	0	0	0	0
purchase volume at city 5 [kg]: y_{5i}	0	0	0	0	0	0
purchase volume at city 6 [kg]: y_{6i}	0	0	0	0	9031	0
purchase volume at city 7 [kg]: y_{7i}	0	0	0	0	0	0
purchase volume at city 8 [kg]: y_{8i}	0	0	2400	5000	969	0
lower limit [kg]: D_i^L	4000	4000	2000	5000	10000	2000
sum of purchase volumes [kg]: $\sum_{j=1}^{8} y_{ji}$	4550	4000	2400	5000	10000	2000
upper limit [kg]: D_i^U	5000	5000	2400	6000	14000	2500
	food 7	**food 8**	**food 9**	**food 10**	**food 11**	**food 12**
order quantity: x_i	800	1500	3000	3000	1200	6000
purchase volume at city 1: y_{1i}	0	0	0	0	0	5474
purchase volume at city 2: y_{2i}	0	0	0	0	0	0
purchase volume at city 3: y_{3i}	0	0	0	0	0	0
purchase volume at city 4: y_{4i}	0	0	0	0	0	0
purchase volume at city 5: y_{5i}	0	0	0	3000	0	526
purchase volume at city 6: y_{6i}	0	0	0	178	0	0
purchase volume at city 7: y_{7i}	0	0	0	0	0	0
purchase volume at city 8: y_{8i}	800	1500	3000	0	1200	0
lower limit: D_i^L	800	1500	3000	3000	1200	6000
sum of purchase volumes: $\sum_{j=1}^{8} y_{ji}$	800	1500	3000	3000	1200	6000
upper limit: D_i^U	1000	2000	4000	3600	1500	6600
	food 13	**food 14**	**food 15**	**food 16**	**amount**	**cap**
order quantity: x_i	14500	6000	4000	1000	—	—
purchase volume at city 1: y_{1i}	0	0	0	0	2000000	2000000
purchase volume at city 2: y_{2i}	0	1730	0	0	1500000	1500000
purchase volume at city 3: y_{3i}	0	2019	0	0	1500000	1500000
purchase volume at city 4: y_{4i}	13043	0	0	0	1500000	1500000
purchase volume at city 5: y_{5i}	1457	87	0	698	1500000	1500000
purchase volume at city 6: y_{6i}	0	0	4000	0	1500000	1500000
purchase volume at city 7: y_{7i}	0	2165	0	0	1500000	1500000
purchase volume at city 8: y_{8i}	0	0	0	302	2000000	2000000
lower limit: D_i^L	12500	6000	4000	1000	—	—
sum of purchase volumes: $\sum_{j=1}^{8} y_{ji}$	14500	6000	4000	1000	—	—
upper limit: D_i^U	14500	7500	4800	1300	—	—

usage of storehouse [cm³]: $\sum_{i=1}^{16} v_i x_i = 258,549,565$	capacity [cm³]: $W = 300,000,000$

	aggregate gain in sales [yen]	transportation cost [yen]	profit [yen]
Food retailer	$\sum_{i=1}^{n} a_i x_i = 8,717,310$	$\sum_{j=1}^{s}\sum_{i=1}^{n} b_{ji} y_{ji} = 372,835$	$z_1(\boldsymbol{x},\boldsymbol{y}) = 8,344,475$
	revenue from retailer [yen]	purchase cost [yen]	profit [yen]
Distributer	$\sum_{i=1}^{n} c_i x_i = 15,486,084$	$\sum_{j=1}^{s}\sum_{i=1}^{n} d_{ji} y_{ji} = 13,000,000$	$z_2(\boldsymbol{x},\boldsymbol{y}) = 2,486,084$

To examine the characteristics of the Stackelberg solution shown in Table 5.37, we give the profitability of each food for the food retailer, $(a_i - b_{ji})/c_i$, and the profit of each food per unit for the distributer, $c_i - d_{ji}$, in Tables 5.38 and 5.39,

Table 5.38 Profitability of each food for the food retailer.

	food 1	food 2	food 3	food 4	food 5	food 6
city 1	0.8671	0.6956	0.8632	0.5076	0.7336	0.4549
city 2	0.7350	0.5267	0.8492	0.5669	0.7305	0.4042
city 3	0.7853	0.5091	0.9792	0.6336	0.7369	0.4473
city 4	0.6781	0.4300	0.9215	0.6461	0.7360	0.4218
city 5	0.6312	0.4060	0.9481	0.6328	0.7758	0.4066
city 6	0.5010	0.4066	1.0710	0.7533	0.8519	0.3956
city 7	0.5863	0.5389	1.0641	0.7378	0.7817	0.4132
city 8	0.5870	0.3874	1.2295	0.8708	0.8870	0.4769
	food 7	food 8	food 9	food 10	food 11	food 12
city 1	0.3773	0.2087	0.5207	0.2293	0.6368	0.5520
city 2	0.4096	0.5599	0.4710	0.3824	0.8552	0.5188
city 3	0.4036	0.5056	0.5948	0.3241	0.8402	0.5615
city 4	0.3999	0.4736	0.4316	0.3555	0.7648	0.5702
city 5	0.4070	0.6276	0.5113	0.4049	0.7762	0.5109
city 6	0.4240	0.4960	0.5832	0.3592	0.7715	0.4684
city 7	0.4432	0.5636	0.6141	0.3803	0.8907	0.4792
city 8	0.5027	0.6275	0.7333	0.3504	1.0480	0.4970
	food 13	food 14	food 15	food 16		
city 1	0.8736	0.3140	0.4871	0.7540		
city 2	0.8429	0.3398	0.4943	0.6361		
city 3	0.9334	0.3965	0.5826	0.7935		
city 4	1.2076	0.3374	0.6136	0.7733		
city 5	0.9497	0.3166	0.5112	0.7771		
city 6	0.9438	0.3130	0.6241	0.7227		
city 7	0.7864	0.4232	0.5765	0.7427		
city 8	0.7398	0.3730	0.6376	0.9625		

Table 5.39 Profit of each food per unit of the distributer [yen/kg].

	food 1	food 2	food 3	food 4	food 5	food 6
city 1	35.25	54.15	−1.20	−20.30	1.45	23.55
city 2	12.25	24.15	−14.20	−13.30	−9.55	−7.45
city 3	17.25	21.15	0.80	−3.30	−8.55	10.55
city 4	7.25	6.15	−4.20	−1.30	−7.55	1.55
city 5	−4.75	−5.85	−5.20	−4.30	−5.55	−9.45
city 6	−20.75	1.15	10.80	10.70	8.45	−9.45
city 7	−1.75	30.15	11.80	9.70	1.45	0.55
city 8	5.25	5.15	26.80	21.70	17.45	28.55
	food 7	food 8	food 9	food 10	food 11	food 12
city 1	−28.65	−67.50	−2.25	−64.45	−62.85	24.10
city 2	−10.65	−6.50	−47.25	−9.45	22.15	−4.90
city 3	−14.65	−24.50	34.75	−60.45	16.15	14.10
city 4	−16.65	−26.50	−82.25	−21.45	−14.85	19.10
city 5	−13.65	−8.50	−18.25	−14.45	−18.85	−12.90
city 6	−0.65	−11.50	29.75	−12.45	−9.85	−28.90
city 7	11.35	23.50	46.75	17.55	43.15	−19.90
city 8	42.35	68.50	97.75	32.55	98.15	−3.90
	food 13	food 14	food 15	food 16		
city 1	−17.35	−38.70	−11.65	−18.50		
city 2	−25.35	20.30	−14.65	−77.50		
city 3	−9.35	144.30	15.35	−7.50		
city 4	24.65	15.30	24.35	−14.50		
city 5	−7.35	−46.70	−9.65	−14.50		
city 6	−7.42	−51.70	27.35	−34.50		
city 7	−36.35	193.80	15.35	−25.50		
city 8	−46.35	105.30	33.35	44.16		

respectively. As seen in Table 5.37, the profits of the food retailer and the distributer are $z_1(\boldsymbol{x},\boldsymbol{y}) = ¥8,344,475$ and $z_2(\boldsymbol{x},\boldsymbol{y}) = ¥2,486,084$. The order quantities of foods 3 and 13 reach the upper limit D_i^U, that of food 1 is between the upper limit D_i^U and the lower limit D_i^L, and those of the rest of the foods are at the lower limit D_i^L. The purchase costs in all the cities reach the budget caps. Although the wholesale prices d_{ji} of foods are greater than the selling price c_i to the food retailer in city 5, Tokyo, the distributer buys foods 10, 12, 13, 14, and 16 in order to fill the order from the food retailer. Basically, as seen in Tables 5.37, 5.38, and 5.39, the food retailer orders highly profitable foods at the upper limit, and the distributer buys high-margin foods in the corresponding cities within the budget caps. For example, food 3, cabbage, is most profitable, and then the food retailer orders food 3 up to the upper limit, 2400 units, and the distributer buys food 3 in city 8, Miyazaki, as expected.

5.3.3 Sensitivity analysis

First, from the viewpoint of the food retailer, we examine variations of the solutions when some parameters are changed. Changes in the cost of fuel for truck transportation is an issue of considerable concern for the management of the food retailer. Although we assume that the cost of fuel is ¥116 per liter in the previous subsection, we compute the Stackelberg solution for problem (5.50) again on the assumption that the cost of fuel is ¥150 per liter because the highest fuel price in 2008 in Japan was ¥148 per liter. In this case, the solution is the same as before, but the profit of the food retailer decreases from $z_1 = ¥8,344,475$ to $z_1 = ¥8,318,173$ by ¥26,302 because of the increase in the transportation costs.

Moreover, assume that the food retailer selects the most profitable food i, and increases its upper limit D_i^U of the order quantity by 100 units. Because the most profitable food for the food retailer is food 3, cabbage, as seen in Table 5.38, the upper limit D_3^U of food 3 is changed from 2400 to 2500 units. The Stackelberg solution to the slightly changed problem is shown in Table 5.40. The upper limit of D_3^U of food 3 is shown in a gray box, and the numbers changed from the original solution are marked with asterisks. The profit of the food retailer becomes $z_1(\boldsymbol{x},\boldsymbol{y}) = ¥8,346,364$, and it increases by about ¥2,000. In contrast, the profit of the distributer is $z_2(\boldsymbol{x},\boldsymbol{y}) = ¥2,483,259$, and it decreases by about ¥3,000. Because the whole order quantity increases, but the distributer must buy foods in cities in which the prices are relatively higher, the profit of the distributer decreases. Specifically, the expansion of the upper limit of food 3 increases the purchase volume of food 3 in city 8, decreases that of food 16 in city 8, increases that of food 16 in city 5, decreases that of food 12 in city 5, increases that of food 12 in city 1, and finally decreases that of food 1 in city 1.

Next, we conduct a sensitivity analysis from the viewpoint of the distributer. Assume that the distributer increases the budget cap of city 8, Miyazaki, where the prices of most foods are lower compared to the other districts, from ¥2,000,000 to ¥2,100,000. The Stackelberg solution to the problem with the larger budget cap is

Table 5.40 Sensitivity analysis for the food retailer.

	food 1	food 2	food 3	food 4	food 5	food 6
order quantity: x_i	4409*	4000	2500*	5000	10000	2000
purchase volume at city 1: y_{1i}	4409*	4000	0	0	0	2000
purchase volume at city 2: y_{2i}	0	0	0	0	0	0
purchase volume at city 3: y_{3i}	0	0	0	0	0	0
purchase volume at city 4: y_{4i}	0	0	0	0	0	0
purchase volume at city 5: y_{5i}	0	0	0	0	0	0
purchase volume at city 6: y_{6i}	0	0	0	0	9031	0
purchase volume at city 7: y_{7i}	0	0	0	0	0	0
purchase volume at city 8: y_{8i}	0	0	2500*	5000	969	0
lower limit: D_i^L	4000	4000	2000	5000	10000	2000
sum of purchase volumes: $\sum_{j=1}^{8} y_{ji}$	4409*	4000	2500*	5000	10000	2000
upper limit: D_i^U	5000	5000	2500	6000	14000	2500
	food 7	**food 8**	**food 9**	**food 10**	**food 11**	**food 12**
order quantity: x_i	800	1500	3000	3000	1200	6000
purchase volume at city 1: y_{1i}	0	0	0	0	0	5509*
purchase volume at city 2: y_{2i}	0	0	0	0	0	0
purchase volume at city 3: y_{3i}	0	0	0	0	0	0
purchase volume at city 4: y_{4i}	0	0	0	0	0	0
purchase volume at city 5: y_{5i}	0	0	0	3000	0	491
purchase volume at city 6: y_{6i}	0	0	0	178	0	0
purchase volume at city 7: y_{7i}	0	0	0	0	0	0
purchase volume at city 8: y_{8i}	800	1500	3000	0	1200	0
lower limit: D_i^L	800	1500	3000	3000	1200	6000
sum of purchase volumes: $\sum_{j=1}^{8} y_{ji}$	800	1500	3000	3000	1200	6000
upper limit: D_i^U	1000	2000	4000	3600	1500	6600
	food 13	**food 14**	**food 15**	**food 16**	**amount**	**cap**
order quantity: x_i	14500	6000	4000	1000	—	—
purchase volume at city 1: y_{1i}	0	0	0	0	2000000	2000000
purchase volume at city 2: y_{2i}	0	1730	0	0	1500000	1500000
purchase volume at city 3: y_{3i}	0	2019	0	0	1500000	1500000
purchase volume at city 4: y_{4i}	13043	0	0	0	1500000	1500000
purchase volume at city 5: y_{5i}	1457	87	0	729*	1500000	1500000
purchase volume at city 6: y_{6i}	0	0	4000	0	1500000	1500000
purchase volume at city 7: y_{7i}	0	2165	0	0	1500000	1500000
purchase volume at city 8: y_{8i}	0	0	0	271*	2000000	2000000
lower limit: D_i^L	12500	6000	4000	1000	—	—
sum of purchase volumes: $\sum_{j=1}^{8} y_{ji}$	14500	6000	4000	1000	—	—
upper limit: D_i^U	14500	7500	4800	1300	—	—

usage of storehouse: $\sum_{i=1}^{16} v_i x_i = 258,345,686$*	capacity of storehouse: $W = 300,000,000$

	aggregate gain in sales	transportation cost	profit
Food retailer	$\sum_{i=1}^{n} a_i x_i = 8,718,720$*	$\sum_{j=1}^{s}\sum_{i=1}^{n} b_{ji} y_{ji} = 372,356$*	$z_1(\boldsymbol{x},\boldsymbol{y}) = 8,346,364$*
	revenue from retailer	purchase cost	profit
Distributer	$\sum_{i=1}^{n} c_i x_i = 15,483,259$*	$\sum_{j=1}^{s}\sum_{i=1}^{n} d_{ji} y_{ji} = 13,000,000$	$z_2(\boldsymbol{x},\boldsymbol{y}) = 2,483,259$*

given in Table 5.41. The enlarged budget cap is shown in a gray box, and the numbers changed from the original solution are marked with asterisks. The profit of the food retailer becomes $z_1(\boldsymbol{x},\boldsymbol{y}) = ¥8,427,859$, and it increases by about ¥83,000.

Table 5.41 Sensitivity analysis for the distributer.

	food 1	food 2	food 3	food 4	food 5	food 6
order quantity: x_i	5000*	4000	2400	5000	10000	2000
purchase volume at city 1: y_{1i}	5000*	4000	0	0	0	2000
purchase volume at city 2: y_{2i}	0	0	0	0	0	0
purchase volume at city 3: y_{3i}	0	0	0	0	0	0
purchase volume at city 4: y_{4i}	0	0	0	0	0	0
purchase volume at city 5: y_{5i}	0	0	0	0	0	0
purchase volume at city 6: y_{6i}	0	0	0	0	9031	0
purchase volume at city 7: y_{7i}	0	0	0	0	0	0
purchase volume at city 8: y_{8i}	0	0	2400	5000	969	0
lower limit: D_i^L	4000	4000	2000	5000	10000	2000
sum of purchase volumes: $\sum_{j=1}^{8} y_{ji}$	5000*	4000	2400	5000	10000	2000
upper limit: D_i^U	5000	5000	2400	6000	14000	2500
	food 7	food 8	food 9	food 10	food 11	food 12
order quantity: x_i	800	2000*	3000	3000	1317*	6000
purchase volume at city 1: y_{1i}	0	0	0	0	0	5362*
purchase volume at city 2: y_{2i}	0	0	0	0	0	0
purchase volume at city 3: y_{3i}	0	0	0	0	0	0
purchase volume at city 4: y_{4i}	0	0	0	0	0	0
purchase volume at city 5: y_{5i}	0	0	0	3000	0	638*
purchase volume at city 6: y_{6i}	0	0	0	178	0	0
purchase volume at city 7: y_{7i}	0	0	0	0	0	0
purchase volume at city 8: y_{8i}	800	2000*	3000	0	1317*	0
lower limit: D_i^L	800	1500	3000	3000	1200	6000
sum of purchase volumes: $\sum_{j=1}^{8} y_{ji}$	800	2000*	3000	3000	1317*	6000
upper limit: D_i^U	1000	2000	4000	3600	1500	6600
	food 13	food 14	food 15	food 16	amount	cap
order quantity: x_i	14500	6000	4000	1000	—	—
purchase volume at city 1: y_{1i}	0	0	0	0	2000000	2000000
purchase volume at city 2: y_{2i}	0	1730	0	0	1500000	1500000
purchase volume at city 3: y_{3i}	0	2019	0	0	1500000	1500000
purchase volume at city 4: y_{4i}	13043	0	0	0	1500000	1500000
purchase volume at city 5: y_{5i}	1457	87	0	598*	1500000	1500000
purchase volume at city 6: y_{6i}	0	0	4000	0	1500000	1500000
purchase volume at city 7: y_{7i}	0	2165	0	0	1500000	1500000
purchase volume at city 8: y_{8i}	0	0	0	402*	2100000*	2100000
lower limit: D_i^L	12500	6000	4000	1000	—	—
sum of purchase volumes: $\sum_{j=1}^{8} y_{ji}$	14500	6000	4000	1000	—	—
upper limit: D_i^U	14500	7500	4800	1300	—	—

usage of storehouse: $\sum_{i=1}^{16} v_i x_i = 273{,}972{,}318$*		capacity of storehouse: $W = 300{,}000{,}000$

	aggregate gain in sales	transportation cost	profit
Food retailer	$\sum_{i=1}^{n} a_i x_i = 8{,}832{,}557$*	$\sum_{j=1}^{s}\sum_{i=1}^{n} b_{ji} y_{ji} = 404{,}717$*	$z_1(\boldsymbol{x},\boldsymbol{y}) = 8{,}427{,}859$*
	revenue from retailer	purchase cost	profit
Distributer	$\sum_{i=1}^{n} c_i x_i = 15{,}649{,}441$*	$\sum_{j=1}^{s}\sum_{i=1}^{n} d_{ji} y_{ji} = 13{,}100{,}000$*	$z_2(\boldsymbol{x},\boldsymbol{y}) = 2{,}549{,}441$*

The profit of the distributer is $z_2(\boldsymbol{x},\boldsymbol{y}) = ¥2{,}549{,}441$, and it also increases by about ¥63,000. The enlarged budget cap increases the purchase volumes of a couple of foods in city 8, and by these changes the purchase volumes of some foods in cities 1 and 5 are changed. Moreover, the order quantities of foods 1, 8, and 11 specified by the food retailer also increase, and therefore the profit of the food retailer increases.

5.3.4 Multi-store operation problem

We extend the two-level linear programming problem for food retailing purchases to cope with a multi-store operation in multiple regions in this subsection.

As in the single store problem, the food retailer deals with n kinds of vegetables and fruits, and it has r stores in different cities in Japan. Therefore, after buying vegetables and fruits ordered from the food retailer at the central wholesale markets in s cities, the distributer transports them by truck from each of the central wholesaler markets to the food retailer's storehouses in r cities.

Let x_{ki}, $i = 1,\ldots,n$, $k = 1,\ldots,r$ denote an order quantity of food i at store k, and the decision variables of the order quantities are also expressed by vectors $\boldsymbol{x}^T = (\boldsymbol{x}_1^T,\ldots,\boldsymbol{x}_r)$, $\boldsymbol{x}_k^T = (x_{k1},\ldots,x_{kn})$, $k = 1,\ldots,r$. The decision variables $\boldsymbol{y}^T = (\boldsymbol{y}_1^T,\ldots,\boldsymbol{y}_s)$, $\boldsymbol{y}_j^T = (y_{j1},\ldots,y_{jn})$, $j = 1,\ldots,s$ of purchase volumes are the same as those of the single store problem. In the extended problem, new decision variables on transportation are introduced, and let t_{jki} denote transportation volumes of food i from the central wholesaler market in city j to the storehouse for store k.

Let W_k denote the capacity of the storehouse of the food retailer for store k, $k = 1,\ldots,r$. The constraints for the storehouses are represented by

$$\sum_{i=1}^{n} v_i x_{ki} \leqq W_k,\ k = 1,\ldots,r. \tag{5.51}$$

With multiple stores, the lower limits and the upper limits of order quantities of foods are also specified for all the stores, and the constraints for the upper and lower limits are represented by

$$D_{ki}^L \leqq x_{ki} \leqq D_{ki}^U,\ i = 1,\ldots,n,\ k = 1,\ldots,r. \tag{5.52}$$

The distributer must purchase food i such that its volume is larger than or equal to the quantity ordered from the food retailer for all the stores at the central wholesaler markets in one or more cities, and then, the constraints for order quantities are represented by

$$\sum_{j=1}^{s} y_{ji} \geqq \sum_{k=1}^{r} x_{ki},\ i = 1,\ldots,n. \tag{5.53}$$

The constraints on financial resources of the distributer are the same as those (5.49) of the single store problem.

For the extended problem with multi-store operation, the profit of the food retailer is represented by

$$z_1(\boldsymbol{x},\boldsymbol{y}) = \sum_{i=1}^{n} a_i \sum_{k=1}^{r} x_{ki} - f(\boldsymbol{x},\boldsymbol{y};\boldsymbol{b}_{jk}), \tag{5.54}$$

where and a_i is the profit per unit of food i; $\boldsymbol{b}_{jk} = (b_{jk1},\ldots,b_{jkn})$ and b_{jki} is the transportation cost per unit of food i from city j to store k. The second term $f(\boldsymbol{x},\boldsymbol{y};\boldsymbol{b}_{jk})$ of the objective function (5.54) is the optimal value of the following linear programming problem:

$$\text{minimize} \;\; f(\boldsymbol{x},\boldsymbol{y};\boldsymbol{b}_{jk}) = \sum_{j=1}^{s}\sum_{k=1}^{r}\sum_{i=1}^{n} b_{jki} t_{jki} \tag{5.55a}$$

$$\text{subject to} \;\; \sum_{j=1}^{s} t_{jki} \geqq x_{ki}, \; k = 1,\ldots,r, \; i = 1,\ldots,n \tag{5.55b}$$

$$\sum_{k=1}^{r} t_{jki} \leqq y_{ji}, \; j = 1,\ldots,s, \; i = 1,\ldots,n \tag{5.55c}$$

$$t_{jki} \geqq 0, \; j = 1,\ldots,s, \; k = 1,\ldots,r, \; i = 1,\ldots,n. \tag{5.55d}$$

It follows that problem (5.55) is separable into the following subproblems for food i, $i = 1,\ldots,n$:

$$\text{minimize} \;\; f_i(\boldsymbol{x},\boldsymbol{y};\boldsymbol{b}_{jk}) = \sum_{j=1}^{s}\sum_{k=1}^{r} b_{jki} t_{jki} \tag{5.56a}$$

$$\text{subject to} \;\; \sum_{j=1}^{s} t_{jki} \geqq x_{ki}, \; k = 1,\ldots,r \tag{5.56b}$$

$$\sum_{k=1}^{r} t_{jki} \leqq y_{ji}, \; j = 1,\ldots,s \tag{5.56c}$$

$$t_{jki} \geqq 0, \; j = 1,\ldots,s, \; k = 1,\ldots,r. \tag{5.56d}$$

The profit of the distributer is represented by

$$z_2(\boldsymbol{x},\boldsymbol{y}) = \sum_{i=1}^{n} c_i \sum_{k=1}^{r} x_{ki} - \sum_{j=1}^{s}\sum_{i=1}^{n} d_{ji} y_{ji}, \tag{5.57}$$

where c_i is the selling price of food i to the food retailer, and d_{ji} is the buying price of food i at the central wholesale market in city j.

The extended problem with a multi-store operation for purchase in food retailing is formulated as follows:

$$\text{maximize} \quad z_1(\boldsymbol{x},\boldsymbol{y}) = \sum_{i=1}^{n} a_i \sum_{k=1}^{r} x_{ki} - f(\boldsymbol{x},\boldsymbol{y};\boldsymbol{b}_{jk}) \tag{5.58a}$$

where $\boldsymbol{y}$ solves

$$\text{maximize} \quad z_2(\boldsymbol{x},\boldsymbol{y}) = \sum_{i=1}^{n} c_i \sum_{k=1}^{r} x_{ki} - \sum_{j=1}^{s}\sum_{i=1}^{n} d_{ji} y_{ji} \tag{5.58b}$$

$$\text{subject to} \quad \sum_{i=1}^{n} v_i x_{ki} \leqq W_k,\ k = 1,\dots,r \tag{5.58c}$$

$$D_{ki}^L \leqq x_{ki} \leqq D_{ki}^U,\ i = 1,\dots,n,\ k = 1,\dots,r \tag{5.58d}$$

$$\sum_{j=1}^{s} y_{ji} \geqq \sum_{k=1}^{r} x_{ki},\ i = 1,\dots,n \tag{5.58e}$$

$$\sum_{i=1}^{n} d_{j2i} y_{ji} \leqq o_j,\ j = 1,\dots,s \tag{5.58f}$$

$$\boldsymbol{x} \geqq \boldsymbol{0},\ \boldsymbol{y} \geqq \boldsymbol{0}. \tag{5.58g}$$

Because the objective function (5.54) includes the minimization problem (5.55), problem (5.58) becomes a three-level linear programming problem, and it can be transformed into the following single level mathematical programming problem where the Kuhn-Tucker conditions for optimality of the linear programming problems at the second and the third levels are involved in its constraints:

$$\text{maximize} \quad z_1(\boldsymbol{x},\boldsymbol{y}) = \sum_{i=1}^{n} a_i \sum_{k=1}^{r} x_{ki} \tag{5.59a}$$

$$\text{subject to} \quad \sum_{i=1}^{n} v_i x_{ki} \leqq W_k,\ k = 1,\dots,r \tag{5.59b}$$

$$D_{ki}^L \leqq x_{ki} \leqq D_{ki}^U,\ i = 1,\dots,n,\ k = 1,\dots,r \tag{5.59c}$$

$$\sum_{j=1}^{s} y_{ji} \geqq \sum_{k=1}^{r} x_{ki},\ i = 1,\dots,n \tag{5.59d}$$

$$\sum_{i=1}^{n} d_{j2i} y_{ji} \leqq o_j,\ j = 1,\dots,s \tag{5.59e}$$

$$\boldsymbol{x} \geqq \boldsymbol{0},\ \boldsymbol{y} \geqq \boldsymbol{0} \tag{5.59f}$$

$$\boldsymbol{y} \in KT_2 \tag{5.59g}$$

$$\boldsymbol{t} \in KT_3, \tag{5.59h}$$

where KT_2 is a set of $\boldsymbol{y}$ satisfying the Kuhn-Tucker optimality condition for the second level problem (5.58b)–(5.58g); $\boldsymbol{t}$ is a vector of variables t_{jki}, $j = 1,\dots,s$, $k = 1,\dots,r$, $i = 1,\dots,n$, and KT_3 is a set of $\boldsymbol{t}$ satisfying the Kuhn-Tucker optimality condition for the third level problem (5.55). Although problem (5.58) can be solved by directly applying the Bard method (1984) for three-level linear programming problems if the size of the problem is not large, as might be expected, it becomes difficult to solve it when the numbers of foods and stores are large. Because problem (5.59) can be transformed into a mixed zero-one programming problem, a computational method based on genetic algorithms seems to be promising as we have given the computational results on performance of the solution method for obtaining Stackelberg solutions to two-level linear programming problems.

References

Anandalingam, G. (1988) "A mathematical programming model of decentralized multi-level systems," *Journal of Operational Research Society* 39, 1021–1033.

Anandalingam, G., Mathieu, R., Pittard, C.L., and Sinha, N. (1989) "Artificial intelligence based approaches for solving hierarchical optimization problems," *Impacts of Recent Computer Advances on Operations Research,* Sharda, Golden, Wasil, Balci and Stewart (eds.), North-Holland, 289–301.

Arrow, K.J. (1963) *Social Choice and Individual Values,* Second edition, Wiley, New York.

Azad, M.A.K., Sakawa, M., Kato, K., and Katagiri, H. (2005) "Interactive fuzzy programming for two-level nonlinear integer programming problems through genetic algorithms," *Asia Pacific Management Review: An International Journal* 10, 70–77.

Azad, M.A.K., Sakawa, M., Kato, K., and Katagiri, H. (2006) "Interactive fuzzy programming for multi-level nonlinear integer programming problems through genetic algorithms," *Scientiae Mathematicae Japonicae* 63, 305–318.

Azad, M.A.K., Kato, K., Katagiri, H., Sakawa, M., and Uno, T. (forthcoming) "Interactive fuzzy programming for multiobjective two-level nonlinear integer programming problems through genetic algorithms," *Scientiae Mathematicae Japonicae.*

Bard, J.F. (1983) "An efficient point algorithm for a linear two-stage optimization problem," *Operations Research* 38, 556–560.

Bard, J.F. (1983b) "Coordination of a multidivisional firm through two levels of management," *Omega* 11, 457-465.

Bard, J.F. (1984) "An investigation of the linear three-level programming problem," *IEEE Transactions on Systems, Man, and Cybernetics* SMC-14, 711–717.

Bard, J.F. (1998) *Practical Bilevel Optimization: Algorithms and Applications*, Kluwer Academic Publisher, Dordrecht.

Bard, J.F. (1991) "Some properties of the bilevel programming problem," *Journal of Optimization Theory and Applications* 68, 371–378.

Bard, J.F., and Falk, J.E. (1982) "An explicit solution to the multi-level programming problem," *Computers and Operations Research* 9, 77–100.

Bard, J.F., and Moore, J.T. (1990a) "A branch and bound algorithm for the bilevel programming problem," *SIAM Journal on Scientific and Statistical Computing* 11, 281–292.

Bard, J.F., and Moore, J.T. (1990b) "The mixed integer linear bilevel programming problem," *Operations Research* 38, 911–921.

Bard, J.F., and Moore, J.T. (1990c) "Production planning with variable demand," *Omega* 18, 35-42.

Bard, J.F., and Moore, J.T. (1992) "An algorithm for the discrete bilevel programming problem," *Naval Research Logistics* 39, 419–435.

Bard, J.F., Plummer, J., and Sourie, J.C. (1997) "Determining tax credits for converting nonfood crops to biofuels: an application of bilevel programming," *Multilevel Optimization: Algorithms and Applications* A. Migdalas, P. Pardalos and P. Värbrand (Eds.), Kluwer Academic Publishers, Dordrecht, 23–50.

Baucells, M. and Shapley, L.S. (2006) "Multiperson utility," Social Science Electronic Publishing.

Bellman, R.E. and Zadeh, L.A. (1970) "Decision making in a fuzzy environment," *Management Science*, 17, 141–164.

Ben-Ayed, O., Boyce, D.E., and Blair, C.E. (1988) "A general bilevel linear programming formulation of the network design problem," *Transportation Research* 22B, 311–318.

Berkelaar, M. "lp_solve 2.0," ftp://ftp.es.ele.tue.nl/pub/lp_solve.

Bialas, W.F., and Karwan, M.H. (1984) "Two-level linear programming," *Management Science* 30, 1004–1020.

Bitran, G.R. and Novaes, A.G. (1973) "Linear programming with a fractional objective function," *Operations Research* 21, 22–29.

Chankong, V. and Haimes, Y.Y. (1983) *Multiobjective Decision Making: Theory and Methodology,* North-Holland, Amsterdam.

Charnes, A. and Cooper, W.W. (1962) "Programming with linear fractional functions," *Naval Research Logistic Quarterly* 9, 181–186.

Charnes, A. and Cooper, W.W. (1963) "Deterministic equivalent for optimizing and satisficing under chance constraints," *Operations Research* 11, 18–39.

Delgado, M., Kacprzyk, J., Verdegay, J.-L., and Vila, M.A. (eds.), (1994) *Fuzzy Optimization: Recent Advances*, Physica-Verlag, Heidelberg.

Dempe, S. (2002) *Foundations of Bilevel Programming*, Kluwer Academic Publishers, Dordrecht.

Dubois, D. and Prade, H. (1978) "Operations on fuzzy numbers," *International Journal of Systems Science* 9, 613–626.

Dubois, D. and Prade, H. (1980) *Fuzzy Sets and Systems: Theory and Applications,* Academic Press, New York.

Dyer, J.S. (2005) "MAUT—Multiattribute utility theory," *Multiple Criteria Decision Analysis:State of the Art Surveys*, Figueira, J., Greco, S. and Ehrgott, M. (eds.), Springer, New York, 265–296.

Dyer, J.S. and Sarin, R.K. (1979) "Group preference aggregation rules based on strength of preference," *Management Science* 25, 822–832.

Fishburn, P.C. (1969) "Preferences, summation, and social welfare functions," *Management Science* 16, 179–186.

Fortuny-Amat, J., and McCarl, B. (1981) "A representation and economic interpretation of a two-level programming problem," *Journal of the Operational Research Society* 32, 783–792.

Gendreau, M., Marcotte, P. and Savard, G. (1996) "A hybrid Tabu-ascent algorithm for the linear bilevel programming problem," *Journal of Global Optimization* 8, 217–233.

Geoffrion, A.M. (1967) "Stochastic programming with aspiration or fractile criteria," *Management Science* 13, 672–679.

Goldberg, D.E. (1989) *Genetic Algorithms in Search, Optimization, and Machine Learning*, Addison-Wesley, Reading.

Goldberg, D.E. and Lingle, R. (1985) "Alleles, loci, and the traveling salesman problem," *Proceedings of the 1st International Conference on Genetic Algorithms and Their Applications,* Lawrence Erlbaum Associates Publishers, New Jersey, 154–159.

Haimes, Y.Y. and Chankong, V. (1979) "Kuhn-Tucker multipliers as trade-offs in multiobjective decision-making analysis," *Automatica* 15, 59–72.

Hansen, P., Jaumard, B., and Savard, G. (1992) "New branch-and-bound rules for liner bilevel programming," *SIAM Journal of Scientific and Statistical Computing* 13, 1194–1217.

Harsanyi, J.C. (1955) "Cardinal welfare, individualistic ethics, and interpersonal comparison of utility," *Journal of Political Economy* 63, 309–321.

Holland, J.H. (1975) *Adaptation in Natural and Artificial Systems*, University of Michigan Press, Ann Arbor; MIT Press, Cambridge, 1992.

Jeroslow, R.G. (1985) "The polynomial hierarchy and s simple model for competitive analysis," *Mathematical Programming* 32, 146–164.

Júdice, J.J., and Faustino, A.M. (1992) "A sequential LCP method for bilevel linear programming," *Annals of Operations Research* 34, 89–106.

Katagiri, H., Nishizaki, I., Sakawa, M., and Kato, K. (2007) "Stackelberg solutions to stochastic two-level linear programming problems," *First IEEE Symposium on Computational Intelligence in Multi-Criteria Decision-Making*, Hawaii, CD-ROM.

Kataoka, S. (1963) "A stochastic programming model," *Econometrica* 31, 181–196.

Kato, K., Sakawa, M., and Nishizaki, I. (2002) "Interactive fuzzy decentralized two-level linear fractional programming through decomposition algorithms," *Knowledge-based Intelligent Information Engineering Systems & Allied Technologies KES 2002*, 1227–1231.

Kato, K., Wang, J., Katagiri, H., and Sakawa, M. (2004) "Interactive fuzzy programming for two-level linear programming problems with random variable coefficients based on fractile criterion optimization model," *The 47th IEEE International Midwest Symposium on Circuits and Systems*, Hiroshima, 65–68.

Kaufmann, A., and Gupta, M. (1988) *Fuzzy Mathematical Models in Engineering and Management Science*, North-Holland, Amsterdam.

Kaufmann, A., and Gupta, M. (1991) *Introduction to Fuzzy Arithmetic*, Van Nostrand Reinhold, New York.

Keeney, R.L. and Kirkwood, C. (1975) "Group decision making using cardinal social welfare functions," *Management Science* 22, 430–437.

Keeney, R.L. and Raiffa, H. (1976) *Decisions with Multiple Objectives, Preferences and Value Tradeoffs*, John Wiley and Sons, New York.

Kidachi, M. (2006) "Evolution and development of retailing-oriented distribution systems," *Theory, History and Analysis on Distributive Trades,* Kidate and Tatsuma (eds.), Chuo University Press, 133–174 (in Japanese).

Kornbluth, J.S.H. and Steuer, R.E. (1981b) "Goal programming with linear fractional criteria," *European Journal of Operational Research* 8, 58–65.

Kornbluth, J.S.H. and Steuer, R.E. (1981b) "Multiple objective linear fractional programming," *Management Science* 27, 1024–1039.

Kuhn, H.W. and Tucker, A.W. (1951) "Nonlinear programming," *Proceedings of the 2nd Berkeley Symposium on Mathematical Statistics and Probability*, Berkeley, California, 481–492.

Lai, Y.J. (1996) "Hierarchical optimization: a satisfactory solution," *Fuzzy Sets and Systems* 77, 321–335.

Lai, Y.J. and Hwang, C.L. (1994) *Fuzzy Multiple Objective Decision Making: Methods and Applications*, Springer-Verlag, Berlin.

Lemke, C.E. (1965) "Bimatrix equilibrium points and mathematical programming," *Management Science* 11, 681–689.

Luhandjula, M.K. (1984) "Fuzzy approaches for multiple objective linear fractional optimization," *Fuzzy Sets and Systems* 13, 11–23.

Malakooti, B. (1989) "Identifying nondominated alternatives with partial information for multi-objective discrete and linear programming problems," *IEEE Transactions on Systems, Man, and Cybernetics* 19, 95–107.

Marccote, P. (1986) "Network design problem with congestion effects: a case of bilevel programming," *Mathematical Programming* 34, 142–162.

Marmol, A.M., Puerto, J., and Fernandez, F.R. (1998) "The use of partial information on weights in multicriteria decision problems," *Journal of Multi-Criteria Decision Analysis* 7, 322–329.

Michalewicz, Z. (1995) "Genetic algorithms, numerical optimization and constraints," *Proceedings of the Sixth International Conference on Genetic Algorithms*, 151–158.

Michalewicz, Z. (1996) *Genetic Algorithms + Data Structures = Evolution Programs*, 3rd revised and extended edition, Springer-Verlag, Berlin (the first edition, 1992; 2nd extended edition, 1994).

Michalewicz, Z. and Janikow, C.Z. (1991) "Handling constraints in genetic algorithms," *Proceedings of the Fourth International Conference on Genetic Algorithms*, 151–157.

Michalewicz, Z., Logan, T., and Swaminathan, S. (1994) "Evolutionary operators for continuous convex parameter spaces," *Proceedings of the Third Annual Conference on Evolutionary Programming*, World Scientific Publishers, River Edge, 84–97.

Michalewicz, Z. and Nazhiyath, G. (1995) "Genocop III: a co-evolutionary algorithm for numerical optimization problems with nonlinear constraints," *Proceedings of the Second IEEE International Conference Evolutionary Computation*, IEEE Press, 647–651.

Michalewicz, Z. and Schoenauer, M. (1996) "Evolutionary algorithms for constrained parameter optimization problems," *Evolutionary Computation* 4, 1–32.

Ministry of Agriculture, Forestry and Fisheries of Japan (2008) "Summary of report on price formation in each stage of food distribution 2008," Statistics of agriculture, forestry and fisheries, Ministry of Agriculture, Forestry and Fisheries of Japan.

Moore, J.T., and Bard, J.F. (1990) "The mixed integer liner bilevel programming problem," *Operations Research* 38, 911–921.

Nishizaki, I., and Sakawa, M. (1999) "Stackelberg solutions to multiobjective two-level linear programming problems," *Journal of Optimization Theory and Applications* 103, 161–182.

Nishizaki, I., and Sakawa, M. (2000) "Computational methods through genetic algorithms for obtaining Stackelberg solutions to two-level mixed zero-one programming problems," *Cybernetics and Systems: An International Journal* 31, 203–221.

Nishizaki, I., and Sakawa, M. (2005) "Computational methods through genetic algorithms for obtaining Stackelberg solutions to two-level integer programming problems," *Cybernetics and Systems: An International Journal* 36, 565–579.

Nishizaki, I., Sakawa, M., and Katagiri, H. (2003) "Stackelberg solutions to multiobjective two-level linear programming problems with random variable coefficients," *Central European Journal of Operations Research* 11, 281–296.

Niwa, K., Nishizaki, I., and Sakawa, M. (2001) "Computational methods for two-level linear programming problems through genetic algorithms," *Proceedings of FUZZ-IEEE '01*, Melbourne, 1211–1214.

Perkgoz, C., Sakawa, M., Kato, K., and Katagiri, H. (2003) "Interactive fuzzy programming for two-level linear programming problems involving random variable coefficients through a probability maximization model," *Proceedings of the 10th IFSA World Congress*, Istanbul, 547–550.

Sakawa, M. (1993) *Fuzzy Sets and Interactive Multiobjective Optimization,* Plenum Press, New York.

Sakawa, M. (2000) *Large Scale Interactive Fuzzy Multiobjective Programming: Decomposition Approaches*, Physica-Verlag, Heidelberg.

Sakawa, M. (2001) *Genetic Algorithms and Fuzzy Multiobjective Optimization,* Kluwer Academic Publishers, Boston.

Sakawa, M. (2002) "Fuzzy multiobjective and multilevel optimization," *Multiple Criteria Optimization: State of the Art Annotated Bibliographic Surveys*, M.

Ehrgott and X. Gandibleux (eds.), Kluwer Academic Publishers, Boston, Chapter 4, 171–226.

Sakawa, M., and Kato, K. (2000) "Integer programming through genetic algorithms with double strings based on reference solution updating," *Proceedings of 2000 IEEE International Conference on Industrial Electronics, Control and Instrumentation,* Nagoya, 2744–2749.

Sakawa, M., and Kato, K. (2003) "Genetic algorithms with double strings for 0-1 programming problems," *European Journal of Operational Research* 144, 581–597.

Sakawa, M., Kato, K., Katagiri, H., and Wang, J. (2003) "Interactive fuzzy programming for two-level linear programming problems involving random variable coefficients through a probability maximization model," *Proceedings of the 10th IFSA World Congress*, Istanbul, 555–558.

Sakawa, M., Kato, K., and Nishizaki, I. (2001) "Interactive fuzzy decentralized two-level linear programming through Dantzig-Wolfe decomposition," *Knowledge-Based Intelligent Information Engineering Systems & Allied Technologies*, Osaka-Nara, 1365–1369.

Sakawa, M., Kato, K., Sunada, H. and Shibano, T. (1997) "Fuzzy programming for multiobjective 0-1 programming problems through revised genetic algorithms," *European Journal of Operational Research* 97, 149–158.

Sakawa, M., and Nishizaki, I. (1999) "Interactive fuzzy programming for cooperative two-level linear fractional programming problems with multiple decision makers," *International Journal of Fuzzy Systems* 1, 48–59.

Sakawa, M., and Nishizaki, I. (2001a) "Interactive fuzzy programming for two-level linear fractional programming problems," *Fuzzy Sets and Systems* 119, 31–40.

Sakawa, M., and Nishizaki, I. (2001b) "Interactive fuzzy programming for multi-level nonconvex nonlinear programming problems through genetic algorithms," *Dynamical Aspects in Fuzzy Decision Making*, Y. Yoshida (ed.), Physica-Verlag, Heidelberg, Chapter 5, 99–116.

Sakawa, M., and Nishizaki, I. (2002a) "Interactive fuzzy programming for decentralized two-level linear programming problems," *Fuzzy Sets and Systems* 125, 301–315.

Sakawa, M., and Nishizaki, I. (2002b) "Interactive fuzzy programming for two-level nonconvex programming problems with fuzzy parameters through genetic algorithms," *Fuzzy Sets and Systems* 127, 185–197.

Sakawa, M., Nishizaki, I., and Hitaka, M. (1999) "Interactive fuzzy programming for multi-level 0-1 programming problems through genetic algorithms," *European Journal of Operational Research* 114, 580–588.

Sakawa, M., Nishizaki, I., and Hitaka, M. (2001) "Interactive fuzzy programming for multi-level 0-1 programming problems with fuzzy parameters through genetic algorithms," *Fuzzy Sets and Systems* 117, 95–112.

Sakawa, M., Nishizaki, I., and Oka, Y. (2000) "Interactive fuzzy programming for multiobjective two-level linear programming problems with partial information of preference," *International Journal of Fuzzy Systems* 2, 79–86.

Sakawa, M., Nishizaki, I., and Shen, Y. (2001) "Interactive fuzzy programming for multiobjective two-level linear fractional programming problems with partial information of preference," *International Journal of Fuzzy Systems* 3, 452–461.

Sakawa, M., Nishizaki, I., and Uemura, Y. (1998) "Interactive fuzzy programming for multi-level linear programming problems," *Computers & Mathematics with Applications* 36, 71–86.

Sakawa, M., Nishizaki, I., and Uemura, Y. (2000a) "Interactive fuzzy programming for multi-level linear programming problems with fuzzy parameters," *Fuzzy Sets and Systems* 109, 3–19.

Sakawa, M., Nishizaki, I., and Uemura, Y. (2000b) "Interactive fuzzy programming for two-level linear fractional programming problems with fuzzy parameters," *Fuzzy Sets and Systems* 115, 93–103.

Sakawa, M., Nishizaki, I., and Uemura, Y. (2001) "Interactive fuzzy programming for two-level linear and linear fractional production and assignment problems: a case study," *European Journal of Operational Research* 135, 142–157.

Sakawa, M., Nishizaki, I., and Uemura, Y. (2002) "A decentralized two-level transportation problem in a housing material manufacturer —Interactive fuzzy programming approach—," *European Journal of Operational Research* 141, 167–185.

Sakawa, M. and Shibano, T. (1996) "Interactive fuzzy programming for multiobjective 0-1 programming problems through genetic algorithms with double strings," *Fuzzy Logic Foundations and Industrial Applications,* Da Ruan (Ed.), Kluwer Academic Publishers, Boston, 111–128.

Sakawa, M. and Shibano, T. (1997) "Multiobjective fuzzy satisficing methods for 0-1 knapsack problems through genetic algorithms," *Fuzzy Evolutionary Computation,* W. Pedrycs (ed.), Kluwer Academic Publishers, Boston, 155–177.

Sakawa, M. and Yano, H. (1988) "An interactive fuzzy satisficing method for multiobjective linear fractional programming problems," *Fuzzy Sets and Systems* 28, 129–144.

Sakawa, M. and Yano, H. (1990) "An interactive fuzzy satisficing method for generalized multiobjective linear programming problems with fuzzy parameters," *Fuzzy Sets and Systems* 35, 125–142.

Sakawa, M., Yano, H. and Yumine, T. (1987) "An interactive fuzzy satisficing method for multiobjective linear programming problems and its application," *IEEE Transactions on Systems, Man, and Cybernetics* SMC-17, 654–661.

Sakawa, M. and Yauchi, K. (1998) "Coevolutionary genetic algorithms for nonconvex nonlinear programming problems: Revised GENOCOP III," *Cybernetics and Systems: An International Journal* 29, 885–899.

Sakawa, M. and Yauchi, K. (1999) "An interactive fuzzy satisficing method for multiobjective nonconvex programming through floating point genetic algorithms," *European Journal of Operational Research* 117, 113–124.

Sakawa, M. and Yauchi, K. (2000) "Interactive decision making for multiobjective nonconvex programming Problems with fuzzy parameters through coevolutionary genetic algorithms," *Fuzzy Sets and Systems* 114, 151–165.

Sakawa, M. and Yumine, T. (1983) "Interactive fuzzy decision-making for multiobjective linear fractional programming problems," *Large Scale Systems* 5, 105–113.

Seo, F., and Sakawa, M. (1988) *Multiple Criteria Decision Analysis in Regional Planning: Concepts, Methods and Applications*, D. Reidel Publishing, Dordrecht.

Shih, H.S., Lai, Y.J., and Lee, E.S. (1996) "Fuzzy approach for multi-level programming problems," *Computers and Operations Research* 23, 73–91.

Shimizu, K., Ishizuka, Y., and Bard, J.F. (1997) *Nondifferentiable and Two-Level Mathematical Programming*, Kluwer Academic Publishers, Boston.

Simaan, M., and Cruz, J.B. (1973a) "On the Stackelberg strategy in nonzero-sum games," *Journal of Optimization Theory and Applications* 11, 533–555.

Simaan, M., and Cruz, J.B. (1973b) "A Stackelberg solution for games with many players," *IEEE Transactions on Automatic Control* AC-18, 322–324.

Slowinski, R. and Teghem, J. (eds.) (1990) *Stochastic versus Fuzzy Approaches to Multiobjective Mathematical Programming Problems under Uncertainty*, Kluwer Academic Publishers, Norwell.

Steuer, R.E. (1986) *Multiple Criteria Optimization: Theory, Computation, and Application* Wiley, New York.

Ünlü, G. (1987) "A linear bilevel programming algorithm based on bicriteria programming," *Computers and Operations Research* 14, 173–179.

Vicente, L., Savard, G., and Judice, J. (1996) "Discrete linear bilevel programming problem," *Journal of Optimization Theory and Applications* 89, 597–614.

Wen, U.-P., and Bialas, W.F. (1986) "The hybrid algorithm for solving the three-level linear programming problem," *Computers and Operations Research* 13, 367–377.

Wen, U.-P., and Hsu, S.-T. (1991) "Efficient solutions for the linear bilevel programming problem," *European Journal of Operational Research* 62, 354–362.

Wen, U.-P., and Hsu, S.-T. (1991) "Linear bi-level programming problems—a review," *Journal of Operational Research Society* 42, 125–133.

Wen, U.-P., and Lin, S.-F. (1996) "Finding an efficient solution to linear bilevel programming problem: an effective approach," *Journal of Global Optimization* 8, 295–306.

Wen, U.-P., and Yang, Y.-H. (1990) "Algorithms for solving the mixed integer two-level linear programming problem," *Computers and Operations Research* 17, 133–142.

Wierzbicki, A.P. (1979) "The use of reference objectives in multiobjective optimization: Theoretical implication and practical experiences," WP-79-66, International Institute for Applied Systems Analysis, Laxenburg, Austria.

White, D.J., and Anandalingam, G. (1993) "A penalty function approach for solving bi-level linear programs," *Journal of Global Optimization* 3, 397–419.

Wolfe, P. (1959) "The simplex method for quadratic programming," *Econometrica* 27, 382–398.

Yu, P.-L. (1985) *Multiple-Criteria Decision Making,* Plenum Press, New York.

Zeleny, M. (1982) *Multiple Criteria Decision Making*, McGraw-Hill, New York.

Zimmermann, H.-J. (1976) "Description and optimization of fuzzy systems," *International Journal of General Systems*, 2, 209–215.

Zimmermann, H.-J. (1978) "Fuzzy programming and linear programming with several objective functions," *Fuzzy Sets and Systems*, 1, 45–55.

Zimmermann, H.-J. (1987) *Fuzzy Sets, Decision-Making and Expert Systems*, Kluwer Academic Publishers, Norwell.

Zimmermann, H.-J. (1996) *Fuzzy Set Theory and Its Applications*, Kluwer Academic Publishers, Norwell, 1985; 2nd edition, 1991, 3rd edition, 1996.

Index

GPSR Compliance

The European Union's (EU) General Product Safety Regulation (GPSR) is a set of rules that requires consumer products to be safe and our obligations to ensure this.

If you have any concerns about our products, you can contact us on ProductSafety@springernature.com

In case Publisher is established outside the EU, the EU authorized representative is:

Springer Nature Customer Service Center GmbH
Europaplatz 3
69115 Heidelberg, Germany

Batch number: 10682213

Printed by Printforce, the Netherlands